Lecture Notes in Chemistry

Volume 79

For further volumes:
http://www.springer.com/series/632

The Lecture Notes in Chemistry

The series Lecture Notes in Chemistry (LNC) reports new developments in chemistry and molecular science—quickly and informally, but with a high quality and the explicit aim to summarize and communicate current knowledge for teaching and training purposes. Books published in this series are conceived as bridging material between advanced graduate textbooks and the forefront of research. They will serve the following purposes:

- provide an accessible introduction to the field to postgraduate students and nonspecialist researchers from related areas,
- provide a source of advanced teaching material for specialized seminars, courses and schools, and
- be readily accessible in print and online.

The series covers all established fields of chemistry such as analytical chemistry, organic chemistry, inorganic chemistry, physical chemistry including electrochemistry, theoretical and computational chemistry, industrial chemistry, and catalysis. It is also a particularly suitable forum for volumes addressing the interfaces of chemistry with other disciplines, such as biology, medicine, physics, engineering, materials science including polymer and nanoscience, or earth and environmental science.

Both authored and edited volumes will be considered for publication. Edited volumes should however consist of a very limited number of contributions only. Proceedings will not be considered for LNC.

The year 2010 marks the relaunch of LNC.

Evangelia A. Varella
Editor

Conservation Science for the Cultural Heritage

Applications of Instrumental Analysis

Springer

Editor
Evangelia A. Varella
Department of Chemistry
Aristotle University of Thessaloniki
Thessaloniki
Greece

ISSN 0342-4901 ISSN 2192-6603 (electronic)
ISBN 978-3-642-30984-7 ISBN 978-3-642-30985-4 (eBook)
DOI 10.1007/978-3-642-30985-4
Springer Heidelberg New York Dordrecht London

Library of Congress Control Number: 2012949084

Printed on acid-free paper

Springer is part of Springer Science+Business Media (www.springer.com)

Contents

Chapter 1
General Considerations

Evangelia A. Varella

1.1 On Materials and Artefacts

Abstract This introductory chapter deals with the materials and techniques used through the centuries for creating works of art and artefacts. The implementation of new natural or synthetic substances is concisely discussed, and the development of technical innovations elucidated. The gradual establishment of novel materials, as well as their multifaceted and complex applications in arts and crafts, is systematically though briefly exposed, in order to smooth identification studies, and permit chronological benchmarking of objects and monuments. The analysis encompasses constitutive core elements of objects, materials used for surface artistic interventions and all types of binding media.

1.2 Constitutive Core Elements of Objects

1.2.1 Fibrous Materials

Fibrous materials have been utilised for creating indispensable artefacts and elaborate works of art since the dawn of civilisation. While wood is used as such, and leather is maintaining its distinct appearance even as parchment, textiles and paper are the result of sophisticated techniques respectively based on spinning and weaving yarns into fabrics or forming a pulp. Fibres—natural or synthetic—are typically organic,asbestos, the silicate mineral, being the sole notable exception. Whereas plant

E. A. Varella (✉)
Department of Chemistry, Aristotle University of Thessaloniki, 54124, Thessaloniki, Greece
e-mail: varella@chem.auth.gr

E. A. Varella (ed.), *Conservation Science for the Cultural Heritage*,
Lecture Notes in Chemistry 79, DOI: 10.1007/978-3-642-30985-4_1,

fibres are composed of cellulose, those of animal origin are built of proteins, ranging from wool keratin and silk sericin and fibroin to leather collagen, elastin and keratins.

1.2.1.1 Plant Fibres

They are mostly derived from the industrial crops cotton, flax and hemp. The cellulose content reaches 80–90 % in cotton and flax, compared to 70 % for hemp. Protein, pectin and wax traces are removed by scouring and bleaching. The monomer unit is in all cases the disaccharide cellobiose, and the degree of polymerisation varies from 5,000 units for cotton to 18,000 for flax. Consequently, the crystalline structure, achieved by means of hydrogen bonds, is highest in flax, hence the fibre is extremely durable with a high tensile strength.

Known in archaic Mediterranean as an Oriental or occasionally African product, cotton has been for centuries largely imported from India, to be cultivated in Europe only from late antiquity on. The main plant yarn of pre Babylonian Mesopotamia and Pharaonic Egypt, flax, was later used up to the northern regions along with hemp, actually known to Black Sea Scythes since remote times. In both cases, the sprouts undergo lengthy hydrolytic decay to gain the threads, which are dried and vehemently separated from wooden remnants.

Paper is produced by extracting plant cellulose fibres, allowing them to swell in water, and collecting the suspended pulp on a screen. Upon drying by pressing, the fibres are randomly interwoven, in a way that micro fibrils form physical and chemical links with each other. Fillers, such as chalk or the silico–aluminium mineral kaolinite, may ameliorate the quality of paper.

Invented in China by the beginning of the Christian era, paper reached the Arab world in the late eighth century, to gradually fully replace parchment, papyrus and other locally used writing materials.

1.2.1.2 Animal Fibres

They all have protein structure, and differ according to the material. Thus, wool fibres are constituted of keratin, which has helical configuration, and is further composed of 20 amino acids of variable sequence, with a prevalence of arginine, cystine and glutamic acid. Rearranging the protein tertiary structure by boiling wool under acidic conditions creates felt; the technique is probably chronologically preceding proper spinning and weaving. Silk is produced in the posterior silk glands of the larva of the silkworm. A cocoon is made of a single silk thread usually up to one kilometre long. A silk strand consists of two silk filaments encased by sericin, which contains a high ratio of hydroxyl amino acids, and composed by fibroin, representing 80 % of the fibre, and built by 16 amino acids, with a high occurrence of alanine, glycine and serine. The system is up to 70 % crystalline and bears a substantial tensile strength.

Although the Greco-Roman world was familiar with both Chinese silk textiles and the relevant thread-acquiring techniques, only emperor Justinian in the sixth century AD succeeded in importing the worm and establishing a long-lasting monopoly of purple dyed silk fabrics.

Leather proteins, conferring to the multi-layer material—consisting of epidermis, dermis and subcutis—its peculiar properties, are fibrous. Most abundant is collagen, containing substantial ratios of glycine, proline, hydroxyproline and hydroxylysine, and having high tensile strength. It is accompanied by elastic elastin. Rich in cysteine, keratins form inter-molecular disulphide bonds, conferring exceptional strength and rigidity. Leather may be used solely if subjected to tanning, a time-honoured chemical process traditionally involving the presence of plant tannins or later chromates.

1.2.1.3 Man-Made Fibres

The first ever not naturally occurring fibre, regenerated cellulose, was introduced by Hilaire de Chardonnet in the 1860s, under the name of artificial silk, later *rayon*. The process, consisting in forcing a viscous fluid through a special type of needle, is still used in its principles for producing synthetic textile fibres. Viscose and acetate *rayon* were followed in 1939 by *nylon*, and subsequently by a constantly expanding series of low-cost polymeric derivatives, often with interesting stipulated properties.

1.2.2 Stone, Ceramics and Glass

Stone manufacturing, the earliest technique providing man with fundamental weapons and utensils, was in rather early days complemented by pottery—the first ever procedure durably transforming inorganic materials by means of high temperatures, and smoothing thus the way to somehow more demanding glass manufacture and even to metal treatment. Nowadays, the elemental composition of these non-metallic solids permits drawing provenance schemes and understanding trade routes of the remote and more recent past.

1.2.2.1 Stone

From ingenious flint stone instruments to glorious marble sculptures, and from gigantic town walls to sophisticated dwellings, stone is omnipresent and vital in every era and site. Nevertheless, it hardly undergoes any significant alterations, and actually maintains the overall physicochemical characteristics of the unprocessed matter.

1.2.2.2 Ceramics

The difficulty in dealing with stone, a least versatile material, led from the fifth millennium BC to the production of ceramic objects, later recognised as a major factor in the establishment of an agricultural way of living. In the broadest sense, ceramics are inorganic, non-metallic solids prepared by the action of heat and subsequent cooling. Nevertheless, in the area of cultural heritage the definition is restricted to crystalline artefacts and works of art made of clay—composed usually of the silico-aluminium mineral kaolinite or iron-containing illite accompanied by various admixtures—and fired under various conditions, conferring the requested properties; while amorphous ceramics are classified as glasses. Wet clays are characterised by plasticity due to water caged among the plate crystals, and depending on granule size. Ceramic materials are inert, brittle, hard and strong in compression, withstanding high temperatures and chemical erosion.

Ceramic pastes are traditionally produced in the dry or—for more sophisticated results—the humid way, and are moulded by hand or already from the fourth millennium BC using a wheel. Firing the dry object may proceed in four different ways, namely open or closed reductive, oxidising and semi-oxidising. The products are roughly classified as non-glazed terracotta and earthenware, the latter having lower porosity and thus better quality; and as glazed earthenware, stoneware and practically non-porous Chinese, soft French or hard German porcelain.

Known for more than 3,000 years, ceramic glazes are vitreous coatings containing silica, eutectic metal oxides and colour-bearing metal oxides, and are usually pre-fabricated and then applied to the surface of the raw object. Sole genuine porcelain is glazed upon firing without addition of any external material. Salt glaze developed in fourteenth century Germany, and lead glaze was perfected in sixteenth century France; while tin glaze, born as Iranian lustreware, arrived in Europe though Arab Spain in the form of painted *majolica* or *faience*. Decoration encompasses *sgraffito*, largely used in Byzantine ceramics; and painting, which may proceed under or over the glaze. The limited palette of *grand feu* colours, fired at 1,000 °C, are cobalt blue, manganese purple, copper green, antimony yellow and iron red/brown, finally chrome salts after 1840. Over-glaze *petit feu* painting, fired at 750–850 °C, permits operating with a sophisticated variety of pigments.

1.2.2.3 Glass

Glass an amorphous ceramic, is obtained via a super-cooled liquid frozen as rapidly as to permit the disordered atomic configuration to freeze into the solid state. It is composed of silica and a eutectic alkali carbonate; lime, magnesium oxide and aluminium oxide improve chemical resistance.

Glass beads and seals—colourless transparent or coloured with metal oxides—were documented in Mesopotamia from around the mid-third millennium BC. Cast glass seems to be an Egyptian innovation; while classical Greece developed glass blowing, and both Romans and Arabs proceeded to highly sophisticated products. By the turn of the first millennium BC, Northern Europe replaced plant or naturally encountered sodium carbonate with wood ash potassium carbonate, creating a standard geographical differentiation among glass objects.

1.2.3 Metals

Aesthetically appealing and socially treasured, metals have simultaneously been for millennia almost awful to man, since the long process from the subterranean ore to the sophisticated artefact always required outstanding expertise and constantly developing skills, actually elevating the artisan to the status of a *demiurge*—a god-like creator, able to extort the precious substance from the cavernous profundities, to purify it literally by force using his arcane knowledge, and to transform it into a durable yet versatile brilliant material. This metaphysical aspect easily elucidates the mythical dimension of metals, resulting in Hellenistic and medieval alchemy. Nonetheless, even in the most pragmatic approach metals define the level of technical and societal evolution, thus civilisation periods are pursuing their appearance—the age of copper, bronze or iron. Their importance has hardly diminished with the recent dawn of the synthetic material era.

1.2.3.1 Non-Ferrous Metals

Gold, silver, copper, tin and lead have been used in Egypt, the Middle East and Eastern Mediterranean for the creation of both utensils and decorative objects at least since the third millennium BC, while precious metals and copper were locally processed in Europe from the second millennium BC. Copper alloys—bronze, containing ca. 5–10 % tin, and brass—were documented shortly afterwards.

Even in pre-classical times bronze statuettes were cast in clay matrices. Elaboration of the method led to the renowned hollow casting process, in which a coarse clay replica of a statue is covered with wax, which is then finely modelled, and subsequently coated with clay. The two clay layers are securely mounted to each other, and bronze is poured between them. Further ameliorations of the technique permit preliminary casting of statue elements that are then assembled into the final artwork.

In Medieval and Renaissance times metallic artefacts—chiefly manufactured from precious metals or copper/bronze—were the products of highly developed techniques. Thus, in the early sixteenth century simple washing as a standard procedure for acquiring pure gold was replaced all over Europe by the amalgamation process, and before long also applied overseas in the gold and silver mines of Peru, Mexico and Bolivia. In the same period, copper/silver minerals were melted with lead to efficiently isolate the constituents.

1.2.3.2 Iron

Actually a benchmark for technical progress, iron was largely introduced in post Mycenaean Greece, without however replacing bronze, which remained for centuries the predominant material for manufacturing precision instruments or works of art. In classical times, iron oxides underwent reduction in ovens at approximately 1,300 °C. Following a technique diverse from modern steel production, this raw iron was hardened through addition of <2 % carbon, in a procedure yielding from the fourth century the celebrated Damascus swords.

Already in the high Middle Ages, iron smelting and forging was widely known, mechanical bellows and sophisticated ovens easily yielding temperatures over 1,100 °C. On the whole, iron is forged into sheets, usually under heating; is pulled to wire; or is casted, using the time-honoured technique already present in sixth century BC China. Finally, iron artefacts are protected from rust by means of a tin coating.

1.3 Materials Used for Surface Artistic Interventions

1.3.1 Dyes

The use of natural organic dyes, extracted from plant and eventually animal sources, has been intrinsic to the cultural identity of all civilizations for millennia and up to the mid-nineteenth century, when synthetic substances gradually expelled traditional materials from the market.

1.3.1.1 Natural Dyes

The reconstruction of time-honoured manufacturing processes and dyeing procedures, founded on the meticulous trial of ancient and traditional recipes, permits implementing a detailed sequence of reproducible optimised operations, essential for the preservation of the polychrome tangible heritage. A comprehensive codification of materials and dyeing techniques, as they were applied through the ages, leads to the subsequent synoptic notions for a series of commonly encountered colouring principles.

Saffron. Although ancient dyeing recipes were mainly based on hydro solubility, allowing direct application on the yarn under moderate warming, interesting results were further generated by elaborate precipitation over alum or tin under boiling. Tints fluctuate from bright sulphur yellow to yellowish orange.

Turmeric. At all times the procedure is based on direct dyeing, or eventually alum mordanting, in the lukewarm bath, carefully equilibrated carbonates and organic acids account for neutral conditions, and create vivid yellows. Tin salts lead to orange hues, chrome derivatives to olive greens, alkalis and iron to reddish browns.

Weld. The long-established practise involves moderate warming, eventually in presence of chalk or lime, short boiling and addition of mordant and yarn—mostly wool or silk—to produce pale flax to vivid lemon tones. The weight relation may reach 1:1, whilst in the case of alum complexation is facilitated by means of tartar. Copper sulphate or tin dichloride mordants result in greenish yellows.

Persian berries. Even ancient recipes took into account differentiations brought about by hydrolysing the glycosides. Thus, when unripe dry buckthorn fruits are brought slowly to boil and then employed on fibres treated with tin chloride and tartar, an orange hue is obtained. Should preliminary abrupt heating destroy the enzymes, the intact glycosides will result in a clear yellow hue. Addition of bluish pigments or copper sulphate yields olive greens, while well-matured berries usually lead to earth browns. Recent studies ascribe hue variations to alkalinity changes.

Henna. Since ancient times textile dyeing has commonly addressed wool or silk, and consists in boiling the powdered dry leaves even without mordant, except for the inherent tannins, in a slightly acidic water bath. Depending on tannin concentration, the tints are reported to range from reddish brown to orange yellowish.

Safflower. In recipe collections, the procedure mainly consists in removing any water-soluble yellow ingredients. Hence, the flowers—in an almost 1:1 weight relation to the fibres—are thoroughly washed, and the residual reddish colouring agent extracted from a dilute alkali bicarbonate solution, acting further as a mordant. The yarn is added in presence of vinegar or lemon juice, essentially accounting for the adsorption rate.

Dyer's bugloss. The ancient practise consists in moderately warming the bath under eventual neutralisation by means of urine or lime. Alum, copper salts and nitrates serve as standard mordants, and the solution is buffered with vinegar or sour fruit juices. Dark reddish violet in presence of alum, the hue will turn to dark brown when mordanted with iron salts.

Madder. Successfully practised since prehistoric periods, madder dyeing techniques require initial treatment with an organic fatty compound in presence of chalk, and rely on the use of mordants and the formation of an alum lake on the yarn. According to the standard procedure, an aqueous dispersion of the slightly acidic powdered colourant and alum is alkalized by means of chalk or lime. After introducing the fibres, the bath is slowly heated and then stirred at temperatures close to boiling for 2 h. Subsequent vinegar addition decelerates alum hydrolysis, and soap washing removes yellowish components; while gallnuts support complexation with cellulose fibres, and animal blood albumin—or simply hot water—may adsorb immanent plant resins. The brilliant colour turns darker in the presence of metal salts to become violet by addition of iron sulphate. Elaborated extraction procedures may even yield warm pink or brownish hues. In recent times, tin chloride has replaced alum, and tartar vinegar.

Scale insects. Ancient and traditional recipes allow reproducing the commonest hues—scarlet or purple violet with an eventual tint of orange or brown—and offer a variety of further alternatives. Ranging from pink to bluish red, these are due to slight differentiations in the dyeing procedures. For millennia, recipes are based on the same empirical techniques. The insects are dried in the sun, eventually after treatment with vinegar, and dyeing proceeds in boiling water with alum as the usual mordant. Already, Hellenistic collections were documented on ingenious variations, while the Middle Ages preferred an almost neutral environment, and the support of vinegar or tartar. Later options experimented with copper or iron salts.

Sappan and Brazil wood. Used since time immemorial in cotton dyeing and textile printing, and equally convincing for both Old and New World barks, the procedure involves lengthy fermentation and treatment with boiling water. The commonest mordants have been alum, occasionally with an addition of tartar or tin salts, and in later periods copper nitrate or lead acetate; slightly alkaline or eventually acidic conditions produce subtle variations in dark scarlet tone.

Woad and indigo. Indigenous in the Mediterranean, woad is encountered in Neolithic paints, and provides antique and medieval Europe with a standard organic violet blue, bearing its precious Oriental counterpart in indigo. In Modern Times woad contributes in the establishment of flourishing textile manufactures, principally in Italian Genoa and French Nimes. Hellenistic and Roman technical manuals meticulously describe dyeing procedures, almost identical to those perceived later by Marco Polo on the Western Indian coast. Maceration of the leaves for 12 h at about 40 °C results in enzymatic hydrolysis of the glycosides to water soluble monomers; whereas bacteria are converting glucose into lactic acid. Under anaerobic conditions, the latter yields butyric acid and releases nascent hydrogen hindering oxidation, while ley neutralises free lactic acid. The product is commercialised as a paste, maintaining the requested reductive conditions by enzymatic activity and the additional support of urine. For dyeing purposes, the fibres are kept in the bath for 12 h, and then exposed to air to allow oxidation of the monomers. Mordants may lead to variations ranging from pink to cerulean, and from azure to lilac and black.

Logwood. Indigene in Mesoamerica and originally monitored by a Spanish monopoly, logwood initiated quarrels among the colonial powers, resulting in 1871 in the establishment of British Honduras. The dyeing procedure involves lengthy fermentation and treatment with boiling water, eventually in presence of tannins and a yellow colourant. The commonest mordants leading to the celebrated black are alum, occasionally with an addition of tartar or tin salts, and in later periods copper or lead acetate, tin salts and chromates.

1.3.1.2 Synthetic Dyes

Low cost standardised synthetic dyes became of increasing importance in artistic creation since their implementation in the last decades of the nineteenth century.

Among the first to be broadly accepted were laboratory substitutes of natural colourants, *e.g.* alizarin and indigo, as well as purely new compounds, such as the triphenylmethane compounds fuchsine and rhodamine, or the azo-derivatives chrysoidine and Hansa yellow.

Thus around 1870, C. Graebe and C. Liebermann, and independently W. Perkins, provided markets with synthetic alizarin; while A. von Bayer succeeded in the year 1882 in preparing indigo, which became commercially available by the turn of the century. As early as 1858 the Renard brothers and Franc, established in Lyon, patented the treatment of impure aniline with stannic chloride yielding brilliant red fuchsine. The first member of the rhodamine group was invented by M. Ceresole in 1887. In the same line, the bright orange mono-azo compound chrysoidine was synthesised in 1875 by H. Caro and O.N. Witt. Crowning a number of attempts, arylide Hansa yellows were developed by Hoechst AG from 1909 on.

1.3.2 Pigments and Lakes

Pigments are since time immemorial the main materials man utilised for painting purposes. Since remote antiquity, however, natural lakes have widely been used as exquisite alternatives to mineral colourants in artistic procedures. Creating the most subtle and rich tunes known in visual arts, lakes are organic dyes adsorbed on an inert substrate in order to gain the requested physical properties.

1.3.2.1 Pigments

Inorganic pigments may be roughly divided into natural and artificial ones, the former deriving from surface or underground deposits, and easily recuperated by crushing, washing and powdering. Calcination may at times lead to complete colour change.

Artificial pigments are obtained by a process of fabrication, the dry way of implying chemical transformations due to high temperature, and the humid one presuming precipitation of products formed by chemical reaction within a solution. In a broad sense, the production date (or period) of a synthetic pigment may serve as a *terminus post quem* for the chronological classification of an artefact.

Among most ancient artificial pigments may be classified Egyptian blue, manufactured in Pharaonic times by heating a mixture of quartz, malachite, calcium carbonate and sodium carbonate, as well as verdigris, a greenish copper acetate hydrate well known to Greco-Roman antiquity. Recipes for acquiring Mars yellow, a hydrated iron/aluminium oxide, are traced to the late seventeenth century; while Prussian blue was prepared around 1704/1706 by J.J. Diesbach and was available for artistic purposes by 1724. Later, antimony white was introduced in

1920 by the British Cookson Lead & Antimony Co; and Norwegian/U.S.A. titanium white did not enter the oil-painting palette until 1921. Interesting furthermore is the synthetic production of long-established natural minerals. Thus, the rarely encountered and extravagantly precious ultramarine blue has been utilised since Greco-Roman times, to be replaced after 1828 by a related synthetic compound, processed by J.B. Guimet in France and independently by C.G. Gmelin in Germany.

1.3.2.2 Colouring Lakes

Permitting the routine use of widely used organic colourants in pigment form, lakes are traditionally prepared by precipitation of the dye, usually after adsorption on chalk and alum, or at times on a colloid base; treatment with metallic salts may on occasion yield insoluble salts as well. While a palette deriving from synthetic dyes—and launched essentially with fuchsine, rhodamine and alizarine—appeared solely in the late nineteenth century. They are prepared according to elaborate recipes, and are mainly encountered in the chromatic regions of blue/violet, red and yellow; green pigments being almost entirely of mineral origin.

Saffron. Since antiquity and up to the nineteenth century, saffron was utilised to imitate the colour of gold. The dye is adsorbed on alum or tartar in hot water, eventually in presence of ley and vinegar or bile. They may also undergo complexation with zinc or tin.

Persian berries. For millennia, Persian berries lakes were used for creating gold-like surfaces of any type. As a general rule the powered dye is precipitated over alum or copper salts in presence of ley or sodium sub carbonate. Copper sulphate yields olive greens, while sodium sub carbonate creates a brownish tint; and alum yields both yellow and green products, according to pH variations. Plant gums standardise the viscosity.

Madder. Well known in the Orient and Mediterranean since antiquity, madder painting lakes conquered Renaissance Central Europe. In the eighteenth century the powdered dye was adsorbed on chalk, argyle and later alum. The choice of the alkalising agent—urine, ley or alkali sub carbonates— affects the hue and brilliance, the latter being improved by treatment with soap water. Eventual cleansing of the root may proceed by washing with the adequate albumin emulsion; while impurities are preferably treated with a weak chalk or soap solution. The tints extend from pink to violet, and from reddish brown to purple.

Scale insects. For more than two millennia, the preparation of scale insect lakes kept increasing in sophistication, without modifying the core of the technique. Thus, antique recipes propose adsorption on chalk and alum, while medieval manuals occasionally replace alum with tartar or orpiment, and introduced a series of ingredients affecting hue and stability—vinegar and lemon juice; urine or lime; varnish-forming gum Arabic and wild fennel resin; viscous plantain water, sugar syrup or corn flour; astringent essential oils. The ingredients are dissolved in water and eventually heated to boil. The colour extends from dark purple to vivid scarlet.

Woad and indigo. The bluish lakes are constantly present in the Old World cultural history, the recipes being for millennia interchangeable, or the materials appearing simultaneously. The dyes are precipitated over alum or chalk, whereas orpiment and copper sulphate plays a hardly less significant role as mordants. Highly celebrated is further Maya blue, encountered in Mesoamerica as an extremely stable greenish blue lake, obtained by treating West Indian indigo leaves extract at ca. 190° with the 10-fold amount of attapulgite argyle, and adding gum Nopal. The pigment owes its durability and brilliance to the peculiar structure of the mineral, permitting the formation of cages enclosing zeolithic water, which is replaced under heating by indigo molecules and iron nanoparticles.

1.3.3 Varnishes

Used for protecting and embellishing paintings or non-polychrome wooden and metal surfaces, varnishes are translucent and generally colourless films, offering a consistent and uniform coverage. By smoothing heterogeneities and improving optical characteristics, they are thus critically intervening in the final visual aspect of artefacts. They are applied as colloid viscous solutions of resins into siccative oils or—in later times—common organic solvents. Their gradual drying permits the coating to develop.

Ancient varnishes were usually prepared by dissolving natural resins under heating in burnt linseed oil. Drying involves oxidation and polymerisation of the oil, creating a permanent film.

On the contrary, the use of volatile solvents—essentially ethanol—creates a thin elastic reversible coating, and is therefore straightforwardly implemented upon diffusion of distillation processes.

Natural resins are water-insoluble amorphous substances, bearing colloidal solubility in organic solvents and oils, and characterised upon heating by a softening range. Constituted by terpene mixtures and inert polymers, they may contain low molecular weight olephinic substances open to oxidation and polymerisation, or carboxylic groups undergoing esterification and saponification.

They may be classified as vegetal resins—mastic, dammar, colophony, sandarac, dragon's blood—oil resins and balsams—Venetian turpentine, gum elemi, Copaiba balsam, Canada balsam—and animal resins—shellac. Numerous types of synthetic poly addition or poly condensation resins are continually being introduced after the mid-twentieth century; at any rate they are adopted for painting varnishes only with considerable hesitation, although they actually play a major role in coatings for monochrome artistic surfaces.

1.4 Binding Media

1.4.1 Painting Binding Media

Binding media are responsible for the mutual joining of pigment particles, and at the same time for their firm adhesion on the support in the form of a thin layer. The powered colourant is therefore typically applied as a homogeneous suspension into the liquid binder, suitable viscosity being the main guarantee for durable flexibility and controlled stability.

Binders should be chemically and photo chemically unchangeable, and compatible to the colouring materials, while remaining entirely transparent and colourless. These factors are adequately addressed by traditional materials, but are not always taken into account when synthetic compounds are utilised. In any case, all optical prerequisites can hardly be fulfilled; hence binding media necessarily contribute to the final aesthetic effect, actually forming the crucial factor in discerning painting techniques. In the following paragraphs they are codified according to this pattern.

1.4.1.1 Fresco

The technique *a fresco* is slightly diverging from the general methodology, since the binding medium—a sand and lime mortar—is applied in the form of a thin layer on the wall acting as a support before the actual artistic intervention. Pigments suspended in water are then promptly put on the wet plaster. Upon drying, air carbon dioxide reacts with lime to form calcium carbonate, which permanently fixes the pigments.

1.4.1.2 Encaustic

In the Greco Roman world, encaustic was first documented around 400 BC, and was continuously operated and highly estimated up to the eighth century AD. The sophisticated technique is commonly applied on wood, metal and stone preliminarily treated with animal glue. It consists in adding whitened natural wax to the powdered colour on the metal palette, heating and inserting a natural resin—such as mastic—to the melt, and finally painting while the mixture is still liquid.

1.4.1.3 Tempera

Ancient and medieval tempera techniques are basically applied on wood preliminarily treated with animal glue and gypsum. Egg yolk in water—and rarely milk or casein—serves as the binder, since the fatty compounds contained guarantee formation of a coherent and flexible film. The *al secco* method proposes a similar procedure for wall paintings. On the contrary, parchment and paper supports request plant gums and/or egg white, with an eventual addition of glue. Finally, to aquarelle, gouache and pastel colourants are added plant gums, e.g. amorphous water-soluble polysaccharides such as gum Arabic or tragacanth.

1.4.1.4 Oil Painting

Introduced in the European North from the thirteenth century, and soon implemented in Renaissance Italy, the technique uses siccative oils as binding media. These aliphatic compounds, almost exclusively composed of oleic, linoleic and linolenic acid triglycerides, are initially auto-oxidised and subsequently polymerised to a grid enclosing any still liquid non-oxidised triglyceride parts. Thus, linseed oil—or eventually poppy and walnut oil—is transformed upon drying to a film bearing exceptional mechanical and optical properties. Since the procedure is very time-consuming, both stages are accelerated, metal cations—mainly bivalent lead, cobalt or manganese—serve as auto-oxidation regulators, and preliminary boiling of the oil is used to speed up polymerisation. Temperatures should surpass 100 °C, and actually reach 250–300 °C in absence of air to manufacture the newly introduced *stand oil*.

The transition from egg tempera to oil painting was produced in the European Renaissance hybrid binding media, combining oils with egg, with the assistance of lecithin, the natural yolk emulsifier. Extremely flexible and stable, but very slow to dry, the resulting film actually bears oil painting properties.

1.4.1.5 Synthetic Materials

Acryl and vinyl polymers and co-polymers—such as poly (methyl methacrylate) or poly (vinyl acetate) were introduced as binding media in the mid-twentieth century, and have in fact generated the development of new painting techniques, based on water emulsions able to create a homogeneous grid, where pigments are dispersed. Vinyl resins stable towards organic solvents serve as varnishes protecting the film.

1.4.2 Mortars

Mortars result by thoroughly mixing sand and an appropriate binder with water. The adhesive properties are essentially due to hydration of the binder molecules, and are improved when adding a natural viscid substance, and lately industrial by-products or organic polymers.

Already in the third millennium BC binders displayed a notorious diversity, ranging from lime or pitch in Mesopotamia to clay, gypsum or lime in Egypt; and from gypsum or bitumen in India to lime or lime and rice extract paste in China. In spite of their remarkable mechanical characteristics, all these materials, however, harden solely in air. Constructions employing hydraulic mortars were first documented in Greece by the turn of the sixth century BC. Addition of volcanic ashes to lime permitted in fact underwater hardening, and inaugurated the era of *pozzolanic* mortars. Constantly ameliorating the properties and manufacturing methods in the Roman and Byzantine times, hydraulic mortars have been mostly forgotten in the Occident up to Modern Times. In the nineteenth century, and as a consequence of long trials, Portland cement has been revolutionising building concepts in both linking stones/bricks and covering wall surfaces.

On the whole, mortar binders encountered in cultural heritage preservation are lime, lime and pozzolanic materials, clay, gypsum, organic resinous compounds, and more recently lime and cement. Lime—actually quicklime—hardening is based on the colloidal properties of calcium hydroxide water emulsions, while the eventual presence of carbon dioxide reversibly yields the relevant carbonate. Gypsum for mortar manufacturing is heated and rehydrated to sharp crystals generating dense easy-hardening structures. Pozzolanic binders—unstable systems containing silica and silico-aluminium derivatives accompanied by iron oxides—react with lime in the presence of water to form hydrated calcium silicates bearing outstanding mechanical properties. On the other hand, cement hardening takes place through hydrolysis and hydration under development of fibrous gel micro formations creating cohesive structures.

Chapter 2
Separation Techniques

Catherine Vieillescazes, Isabel Sierra and Sonia Morante-Zarcero

2.1 High Performance Liquid Chromatography and Gas Chromatography

Vieillescazes Catherine

Abstract This chapter deals with the importance of analytical chemistry and more particularly the separation techniques in the field of conservation science. High performance liquid chromatography (HPLC) and gas chromatography (GC) are mainly developed. The principles of the methods, some information about the methodology and their evaluation are discussed.

2.1.1 In Which Cases is the Method Used in Conservation Science

Chemistry was first applied to the conservation field starting back in the eighteenth century, and, due to the increased number of collections exhibited in the museums of Europe, gradually assumed a fundamental role. Nowadays, analytical chemistry is central to resolve topics of interest in the cultural heritage field, as it offers

---

C. Vieillescazes (✉)
Laboratory of Chemistry Applied to Art and Archaeology, 33 rue Louis Pasteur,
F-84000 Avignon, France
e-mail: cathy.vieillescazes@univ-avignon.fr

I. Sierra · S. Morante-Zarcero
Inorganic and Analytical Chemistry Department, ESCET, Rey Juan
Carlos University, Móstoles, Spain
e-mail: isabel.sierra@urjc.es

E. A. Varella (ed.), *Conservation Science for the Cultural Heritage*,
Lecture Notes in Chemistry 79, DOI: 10.1007/978-3-642-30985-4_2,

precious elements about the nature and composition of the original materials, as well as their degradation processes having occurred throughout the years. The chemical characterisation of the constituting materials allows discovering the abundant information enclosed in a work of art, thus providing a better knowledge.

Chromatographic techniques [1, 2] are widely used in analytical chemistry [3–6]. Employed in the study of organic materials, they allow the identification of a large number of them. Among the various processes, gas chromatography (GC) [7] and high performance liquid chromatography (HPLC) [8, 9] are two methods frequently used in the study of cultural heritage objects [10, 11]. Indeed, chemistry, in its analytical part, brings information, both qualitative and quantitative, essential for a better understanding of these materials, mainly constituted of organic compounds [12, 13]. That is the reason why GC can successfully be applied to the study of archaeological adhesives, balms of mummification [14], inks [15] and asphalts present in the composition of museum objects [16], binding media and wax sculptures [17]. GC can apply only to volatile samples or easily vaporised without decomposition. Therefore, the analysis of complex organic compounds with high molecular weight or high polarity cannot be realised. HPLC is applied to pigments [18], colouring agents in textiles [19, 20], tannins [21, 22], resins [23, 24] and binding media [25, 26].

In fact, GC and HPLC are two complementary techniques. Each one has its own specificity, the choice of the method depending on the analysed sample, the chosen protocol and the relevant targeted information.

However GC offers a further advantage as it can be coupled with a large number of detectors and more particularly, with spectrometric techniques [27].

2.1.2 Principles and Concepts

Analytical chemistry is a branch of pure chemistry. Its main objective is the study of the separation, identification and quantification of the chemical components of natural and artificial materials. Qualitative analysis gives an indication of the identity of the chemical species in the sample and quantitative analysis determines the concentration of one or more of these components. In fact, the separation of compounds is often performed prior to analysis. Finally, analytical chemistry refers to the study of the methods and their different techniques to solve analytical problems (Fig. 2.1).

Concerning the separation of complex mixtures, chromatography and/or electrophoresis methods are the most generally used. The creation of new measurement tools provides better chemical information. Since Tswett's work (a Russian botanist 1906), preparative chromatography has been used for the separation and purification of the different components of a mixture. Later, analytical chromatography has been employed for micro separation.

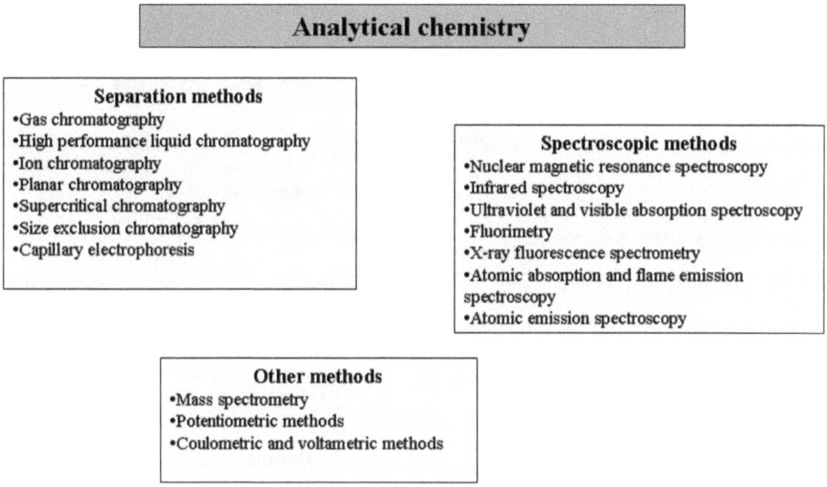

Fig. 2.1 Analytical chemistry classification

Fig. 2.2 General principle of chromatography

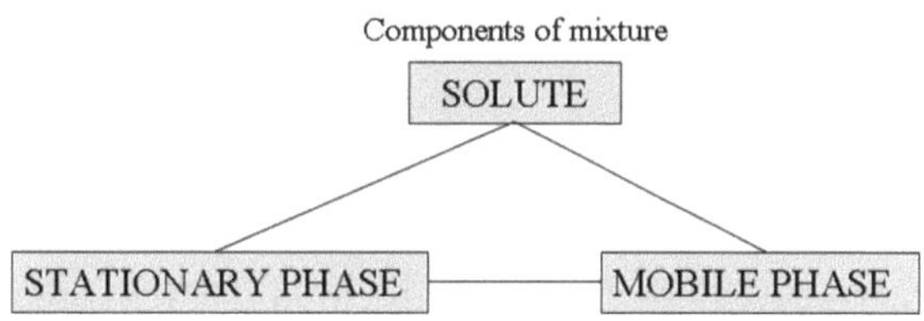

In chromatography, the principle is based on the concentration equilibrium of the compounds between two phases: a stationary one which is immobilised in a column, and the other one, mobile, going through the system (Fig. 2.2).

The differential separation of the several constituents or analytes all along the column leads to their separation. Sensitive detectors permit the analysis of very small quantities (a few nanograms). The signal is registered and leads to a chromatogram indicating the variation of the composition of the mobile phase with time at the end of the column.

HPLC [6, 8] also called High Pressure (*versus* high performance) Liquid Chromatography some years ago, is an instrumental system composed of several modules making up an integrated unit (Fig. 2.3).

According to the general principle, HPLC is based on the selective partitioning of the molecules of interest between two different phases. The partition mechanisms between analytes, mobile phase and stationary phase are based on coefficients of adsorption or partition. The isocratic mode means that the mobile phase (solvent or combination of solvents) has a fixed composition. A gradient elution has the composition of the mobile phase varying with time. While travelling through the column under high pressure (50×10^5–150×10^5 Pa), analytes are separated selectively between the mobile phase and stationary phase. If the

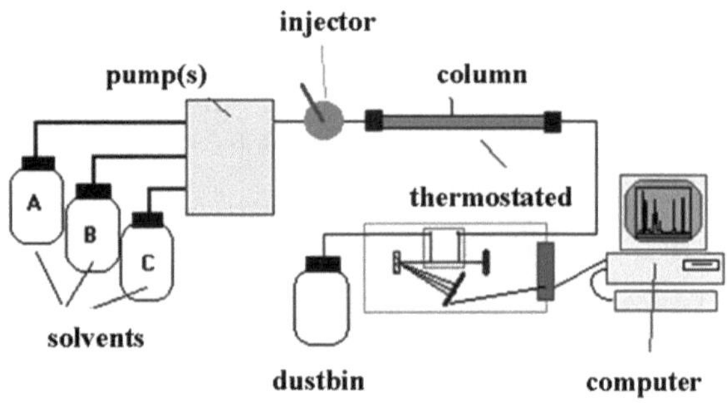

Fig. 2.3 HPLC model

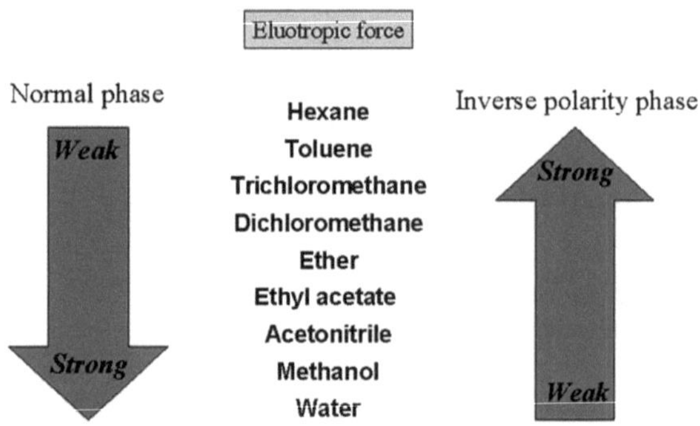

Fig. 2.4 Polarity and eluotropic properties of main solvents in normal and reverse phase

stationary phase is polar (silica gel), then a less polar mobile phase is used and the technique is called normal phase chromatography. If the stationary phase is non polar (bonded silica), a polar mobile phase is then used, generally water with co-solvent (methanol or acetonitrile) and the technique is referred to as reversed phase chromatography (Fig. 2.4).

A detector is placed at the end of the column to identify the eluted components. The most widely used detectors in HPLC are based on the optical properties of the analytes: refractive index, absorption and fluorescence. The refractive index detector is a universal detector, responding to any eluted compound. Detectors based on the absorption of light in the ultraviolet and visible ranges (UV/VIS detectors) are the most commonly used, responding to a wide variety of compounds with good to excellent sensitivity. More particularly, the photodiode array detector (PDA) not only permits to obtain the retention time but also the UV/VIS spectrum of each of the components being analysed. It is essential of course that the mobile phase be transparent in such conditions. Fluorescence detectors are very sensitive (about 1000 × more sensitive than UV detection). Finally, the mass spectrometer detector provides both quantitative information and in most cases a definitive identification of each component (qualitative information) (Fig. 2.5).

GC [6, 7] is an instrumental technique used mainly in the analyses of organic compounds. In fact, GC exploits the fundamental properties common to all types

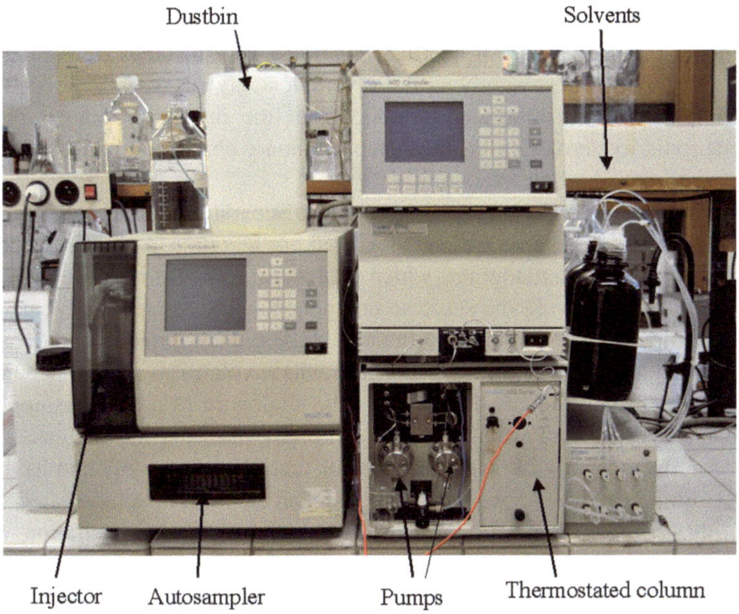

Fig. 2.5 HPLC-PDA apparatus. High-pressure ternary pump waters 600, vacuum degasser, waters 717 autosampler and photodiode array detection system waters 2996

Fig. 2.6 GC system

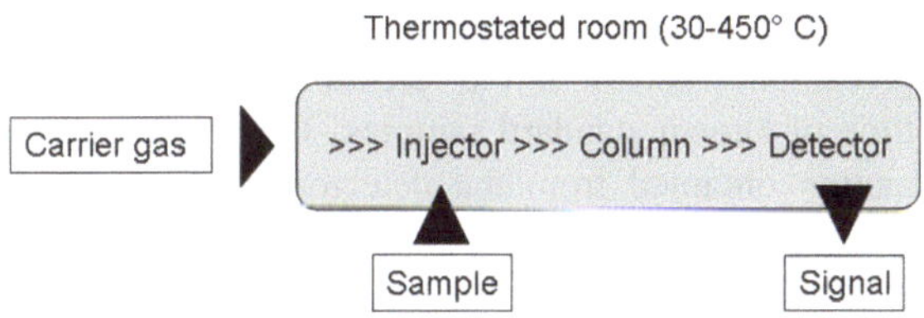

of chromatography. The separation is based on the selective partitioning of compounds between two relevant phases. The first one is an inert gas like helium (He), dihydrogen (H_2), dinitrogen (N_2) or argon (Ar) constituting the mobile phase. In this particular technique, it is important to note that there is no interaction solute/mobile phase. That is why the mobile phase is also termed the carrier gas. The stationary phase is coated on a solid porous support material (Fig. 2.6).

Globally, three types of columns can be used in GC. Packed columns are the oldest. Around the mid-1980s, column chromatography began to give way to capillary columns, in which the liquid phase is coated onto the inner walls of a thin capillary tube that can be anywhere from 15 to 100 m long, also wound into a coil. Capillary column chromatography represented a significant advance in the field and greatly improved the ability of columns to separate the multiple components found in complex samples. Wide bore or '530' columns are the third category.

If more than 100 stationary phases of different types have been possible for packed columns, it is not the same for capillary columns. Polysiloxanes (low polarity) or silicone oils are the most widely used. Polyethylene glycols (carbowax family) are polar polymers. It is important to notice that the only well-defined phase is squalene (polarity zero). On this reference stationary phase, compounds are eluted according to their boiling temperature.

The purpose of the gas chromatograph is to separate mixtures into individual components that can be detected and measured one at a time. A plot of the detector output is called a chromatogram, which charts the response of the detector as a function of time, showing the separate components. The separation is carried out based on differences in affinities for the two phases. The sample is introduced into the GC column by way of a heated injector, which volatilises all components and introduces them into the gas flowing over the stationary phase. For example, if a compound has the least affinity for the stationary phase, it moves ahead of the other components and will reach the detector first. And the compound having the greatest affinity for the stationary phase will spend the most time associated with it. Hence, this compound will be the last to reach the detector. Separation will be achieved based on the different affinities of the different types of molecules found in the analysed sample.

In most GC applications, a sample is prepared by dissolving it in a solvent, and the solution is injected into the heated injector port of the instrument using a syringe. The mobile phase gas also enters the injector port, picking up the volatilised sample and introducing it into the column where the separation process is performed. Each component of the mixture will emerge from the column at a given time (known as the retention time) that can be compared to the retention time of known standard samples. The retention times in combination with information obtained from the detector are used to identify the compounds of the mixture. Another method of sample introduction for GC is called pyrolysis, in which a solid sample such as a fibre or paint chip is heated in a special sample holder to extreme temperatures, causing the sample to decompose into gaseous components that can then be introduced into the GC. Pyrolysis is used when the sample is not readily soluble in common GC solvents. The relative abundance of the individual components can be useful to obtain a chemical signature or fingerprint of that particular sample.

A number of different detectors are available for use in GC. The most commonly used are mass spectrometry detector, flame ionisation detector (FID) and nitrogen-phosphorus detector (NPD). The MS detector permits definitive identification of compounds along with quantitative information. The FID has a great sensitivity to hydrocarbons. The NPD is mainly used in drug analysis and toxicology.

One of the advantages of chromatographic systems such as GC is the ability to provide both qualitative information (identification of individual components) and quantitative information (concentrations of individual components) (Fig. 2.7).

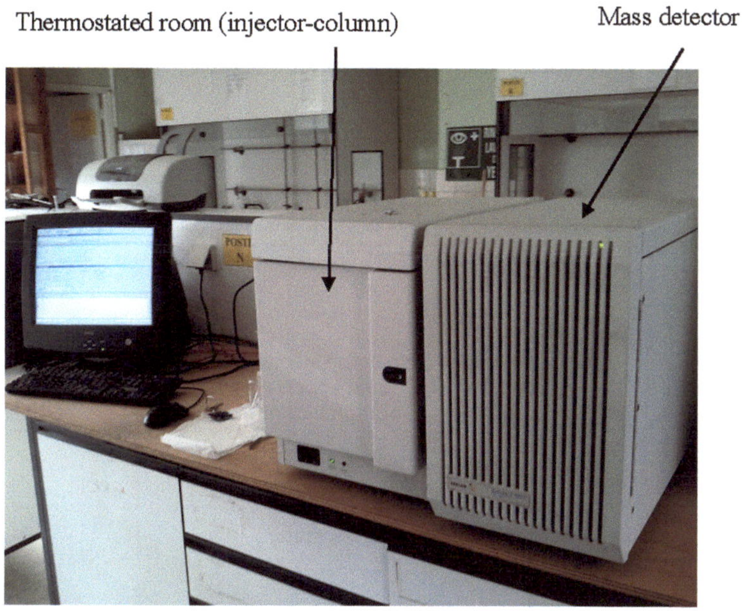

Fig. 2.7 GC apparatus Varian saturn 3900, Varian 1177 injector, coupled with a Varian 2100 T ion trap mass spectrometer

Fig. 2.8 Optimization law

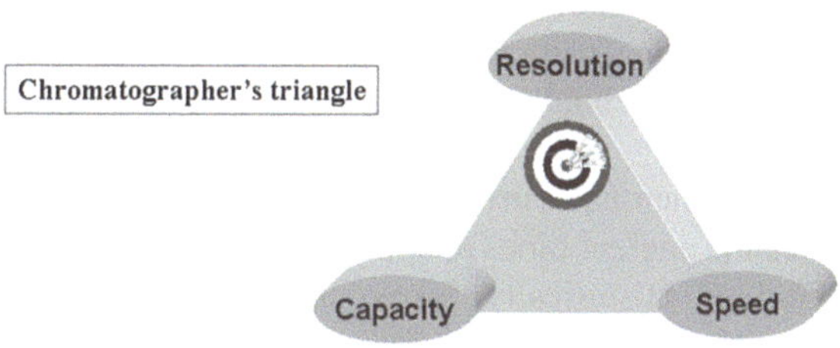

2.1.3 Description of Methodology

In liquid chromatography, the separation method is chosen as a function of the molar mass, solubility and the polarity of the compounds to be separated. In gas chromatography, all parameters have an impact on the separation. Hence, the analysis has to be optimised. In this view, resolution and elution time are the most important things to check out: the best resolution in a minimum time.

To perform this goal, the chromatographer's triangle is based on resolution, speed and capacity (Fig. 2.8).

Talking about the constitutive materials of painting matters, such matrices are not easy to be studied, given the difficulties of physically separating the different thin layers. Another issue is that in this field it is of outmost importance to reduce

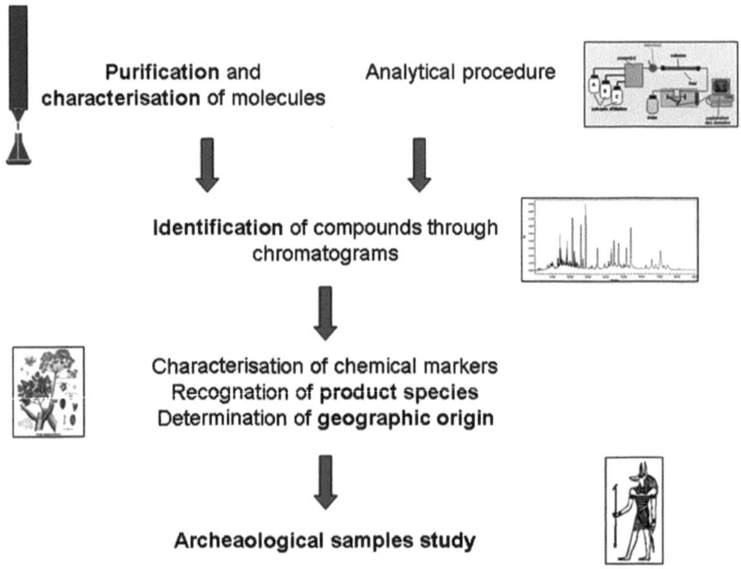

Fig. 2.9 Experimental process

the number and amount of samples taken from the object. For this reason, it is recommended to follow a methodological approach in which different techniques can combine in an integrated way.

Concerning natural compounds involved in archaeological materials, organic dyes and terpenoids are usually studied following a specific experimental process (Fig. 2.9).

Craftsmen have used organic dyes extracted from plants, insects and shellfish since the ancient times and, painters have used plant lacquer as a medium for their paintings. Historically, lakes have been prepared by complexing dyes with an inorganic substrate (alum for example). In this case the complex dye/substrate is insoluble and requires extraction of the colourant before analysis. Moreover, these natural substances were also used for dyeing textiles in many parts of the world. They generally have high tinting strength, directly related to high extinction coefficients in the visible range, and they could be used at low concentrations. That is why the identification of dyes in cultural heritage objects implies so many difficulties. High performance liquid chromatography is a good tool for the analysis of such materials. Highly selective, sensitive and specific this technique has demonstrated the most consistent results for dye analysis. In particular, reverse phase liquid chromatography–photodiode array detection method is reported for the characterisation of the biomarkers in ancient samples of dyes. After extraction of dyes, different compounds (anthraquinonic, flavonoidic) can be identified through the chemical constituents present in the samples. These compounds are identified through retention times and UV spectra in comparison with pure standards.

Several extraction techniques have been experimented (maceration, reflux and ultrasounds) to obtain the best dye yields. Maceration is the most ancient extraction manner to extract dye from plants. Currently, the most common extraction method of dyes corresponds to a heating reflux with a water–alcohol mixture for 1 h in general and extractions using an ultrasonic bath are also performed. Ultrasound processes are developed to determine the best extraction yield of such molecules in preserving the native chemical population of the plant, reducing considerably the experimentation time and the energy cost [28].

The study of archaeological sample extractives is often a challenge for a chemist because of the wide range of compound classes involved, the frequent occurrence of degradation products and the small quantities of ancient matter available. In this context, in order to characterise resinous materials without losing information about other hypothetical original ingredients, a simple procedure using trimethylsilylation followed by GC–MS analysis, has been developed.

Concerning terpenoids, since the amount of archaeological material available was more often than not very low, a GC–MS method of chemical analysis was chosen. As triterpenes contain oxygenated function(s) and because of their relatively high molecular weight, these compounds should be derivatised to increase their volatility when using such a technique. Moreover, trimethylsilylation performed at room temperature is preferred to methylation since the former method should allow determination of naturally occurring methyl esters in archaeological samples. Injector parameters and GC operating conditions are adjusted in order to obtain the best separation for triterpenes as well as for other types of chemical compounds (sesqui- and diterpenes, hydrocarbons, fatty acids, …) which could be of highest interest in an archaeological context.

Current application of any analytical method is always easier when a pre-treatment process takes into account the analyte, the small amount of sample and the measurement technique chosen. Therefore, the quality of the result depends on the preparation of the sample.

Solid-phase extraction (SPE) consists in having a liquid sample pass over a solid sorbent. In such conditions, the analytes are retained and the other components are eliminated by rinsing the column. Finally, the analytes of interest are desorbed using an appropriated solvent (Fig. 2.10).

Headspace is a very useful sampling preparation using gas chromatography. It concerns the analysis of volatile compounds involved in a matrix not directly analysed by GC. Static or dynamic modes are employed, depending on the basic principle.

In the same way, solid-phase microextraction (SPME) has given good results [29].

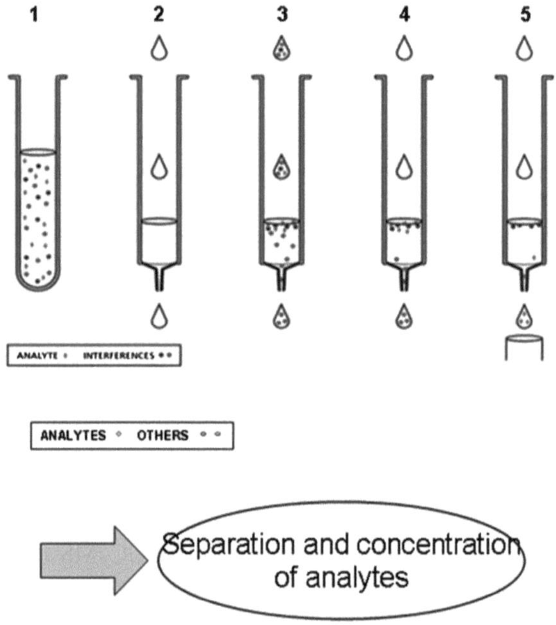

Fig. 2.10 Solid-phase extraction using columns or discs *1*. Sample preparation *2*. Conditioning *3*. Introduction of the sample *4*. Rinsing *5*. Elution

2.1.4 Evaluation of the Method as Used in Conservation Science

The link between art and science is very strong since artists possess great knowledge of the properties of the materials used. Through experimentation with recipes, they have mixed natural ingredients and obtained new materials or coloured matters.

Moreover, chemistry has achieved a very important role in artwork restoration and conservation in the area of cultural heritage. In collaboration with restorers, art historians, techniques historians, archaeologists and museum curators, chemists have investigated art and archaeological materials. They have studied the basic processes ruling their different properties. The topics of interest within the chemistry of cultural heritage are related to a better knowledge of both the nature and composition of original materials, as well as a better comprehension of the causes and mechanisms of their alteration by studying the different chemical and physical factors responsible for the oxidation processes, which lead to the formation of degradation products.

It appears that a detailed knowledge of the composition and structure of the materials used by artists must prevail as a fundamental component of successful and durable conservation and restoration methods. Current scientific research is focused on analytical techniques for diagnostic purpose, and on the identification of the causes and effects of deterioration processes. Such a research is especially developing micro sampling techniques while the chemical characterisation of the organic natural materials is still very important. The significant advances in the procedures used in coupled methods like GC–MS have greatly improved the possibility to distinguish complex mixtures of biomarkers and alteration markers in micro samples [30, 31].

In the same way, it is generally admitted that vegetal colour is very difficult to study because of its different parameters (species, quality, age…). The identification of historical dyes has required small amounts of samples in order to be analysed by high performance liquid chromatography. Specific protocols have been successfully reported [20, 28].

The contribution of conservation sciences to the characterisation of the constitutive materials proves very precious through geographically and historically well-defined museum objects. The results and information are useful for restoration and conservation purposes. In this view, the development of separation methods and more particularly chromatographic techniques have turned conservation sciences into a keystone for cultural heritage.

2.2 Capillary Electrophoresis

Isabel Sierra and Sonia Morante-Zarcero

2.2.1 Capillary Electrophoresis in Conservation Science

Capillary electrophoresis (CE) has proven to be a very powerful tool for the separation of inorganic and organic compounds since its introduction in the early 1980s by Jorgeson and Lukas. Many types of materials have been used to make cultural objects like books, textiles, ceramics, paintings, drawings, maps, etc. An important aspect is the knowledge of the composition of these objects, in order to identify and prevent degradation processes, to decide on conservation procedures, to identify its provenance and to verify its authenticity. CE is a modern and interesting instrumental analytical technique to analyse cultural objects, because it is non-destructive and requires a very small amount of sample. For these reasons, continuous improvements in the development of CE methods are being carried out.

CE can be used to identify different types of binding media, in paintings and other objects of cultural interest, such as animal glues, plant gums, drying oils and natural resins. Gröβl et al. [32] developed a method based on the use of a chromophoric background electrolyte to determinate sugar monomers and uranic acids in different plant gums. The characterisation of different types of proteinaceous media (egg, collagen, casein) by CE has been carried out by Kaml et al. [33] without derivatisation of amino acids. More recently, Harrison et al. [34] studied the potential interferences in CE of the amino acids obtained from the hydrolysis of animal glues due to drying oil and plant gums. The characterisation by CE of some resins from pinaceae families, based on the pattern of their diterpenoid constituents, was carried out by Findeisen et al. [35]. Dying oils and their degradation products formed on ageing have been characterised by Surowiec et al. [36] and Harrison et al. [37] in mixtures with other polysaccharide or proteinaceous media by CE using conductivity and indirect UV detection.

In addition, CE can be used to identify different types of pigments and dyes in textiles and paintings and, in combination with MS, has proved to be an effective technique for investigating historic cellulose textiles. Maguregui et al. [38] used SDS as surfactant to determine a broad range of red organic pigments in paint samples. Puchalska et al. [39] used CE with UV–Vis diode-array detection and electrospray mass spectrometry detection for the identification of anthraquinone colour components of cochineal, lac-dye and madder, and natural red dyestuffs. Kouznetsov et al. [40] used CE to carry out a comparative study on the possible chemical modification of cellulose from eight archaeological textiles dated between 1200 BC and 1500 AD. López-Montes et al. [41] identified five natural dyes (carmine, indigo, saffron, gamboges and *Rubia tinctoria* root) present in historical maps and drawings using a simple and rapid CE method with UV detection. More recently, the same authors [42] developed a CE–UV method for sepia ink identification in ancient graphic documents.

2.2.2 Principles and Concepts of Capillary Electrophoresis

CE has emerged in recent decades as a very efficient technique (in 1987 the first instrument was commercialised). CE uses a capillary that allows much higher electric fields to be applied, resulting in better resolution and a shorter analytical time. The main advantages of CE over GC and HPLC are: the extremely high peak efficiency (the most important), the lower cost associated with the analysis, the few volume of sample required and that it is an environmentally friendlier technique than HPLC. In addition, the basic CE instrument is not more expensive than comparable CG and HPLC systems and its maintenance is in most cases easier [43]. The different separation mechanism of each CE mode allows this technique be used for a wide range of substances with the same equipment, or only with slight modifications, so CE is a very flexible and cost-effective technique (capillary zone electrophoresis, CZE, capillary electro kinetic chromatography, CEKC, capillary gel electrophoresis, CGE, capillary isotachophoresis, CITP, capillary isoelectric focusing, CIEF, capillary electrochromatography, CEC).

In a separation by Capillary Zone Electrophoresis (CZE), the simplest and most popular mode, the components of a solution are separated by applying a voltage from end to end of a capillary tube filled with an electrolyte (Fig. 2.11). Different solutes have different mobilities and therefore migrate through the capillary at different speeds. Migration of analytes in the background electrolyte to the end of the capillary takes place by two driving forces: electrophoresis (electrophoretic mobility, μep) and electroosmosis (electroosmotic mobility, μeo).

Fig. 2.11 Separation by CZE

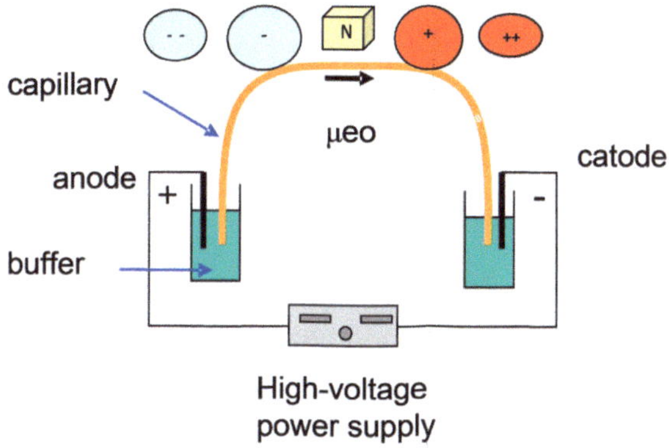

CZE routinely produce 50,000–500,000 theoretical plates, which is an order-of-magnitude better performance than chromatography. For analytes with low diffusion coefficient as proteins it can be expected more than 3 million plates. The high plate count means that bands are very narrow and resolution between adjacent bands is excellent. Figure 2.12 shows a basic CE instrument. The main components in this instrument are: the high-voltage power supply (HVPS), the capillary, the background electrolyte, the thermostating/cooling system, the injection system and the detector.

In most CE systems, the HVPS allows voltages up to 30 kV to be applied and the current can reach 150–300 μA. The HVPS must allow performing analyses in constant-voltage or constant-current mode and to programme changes to the running voltage, current or field strength during an analysis. The capillary is made of fused-silica (SiO_2) because it is transparent across the UV–Vis regions, has low fluorescence, good thermal conductivity and is flexible when coated with polyimide. In addition, the inside capillary wall may be modified with neutral hydrophilic substituents, for example, to reduce the wall charge near to zero. The selection of the BGE is very important to optimise separations. The sensitivity of both electrophoretic mobility of the charged analytes and electroosmotic flow to the pH requires the use of buffers that maintain a constant pH. The general requirements to a CE buffer are: high buffer capacity, low response to the detector used, low mobility to minimise current generation, availability in highly pure reagent grade and low temperature coefficient. It is desirable to use a CE instrument equipped with an effective cooling system. Most commercially available CE instruments use air or liquid cooling systems because the lack of temperature control can lead to problems with reproducible sample injection, sample stability, bubble formation and band broadening (Fig. 2.13).

CE is a microanalytical separation technique, hence volumes between 1 and 50 nl are required that correspond to an injection length of a few mm. This is an advantage when the sample volume is limited but on the other hand it can cause problems for the detection sensitivity and reproducibility of the volume injected. There are two injection modes: hydrodynamic injection and electrokinetic injection (Fig. 2.14).

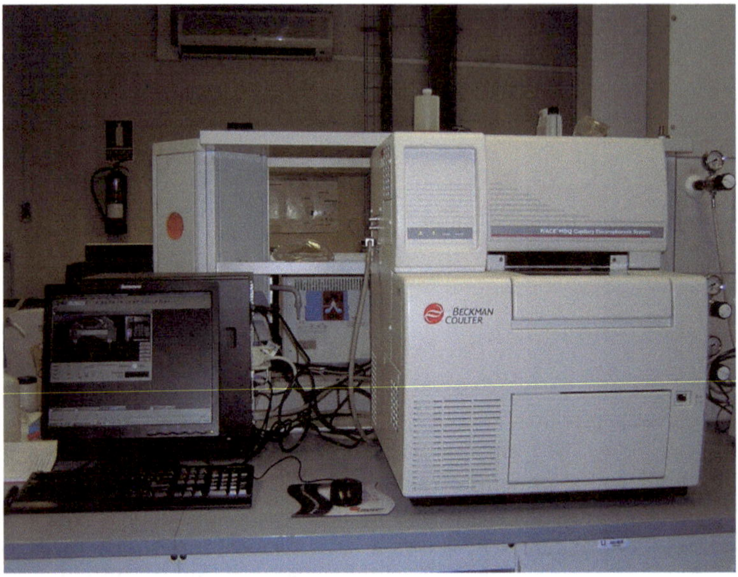

Fig. 2.12 Basic CE instrument

Fig. 2.13 Capillary, thermostating/cooling system and detection window

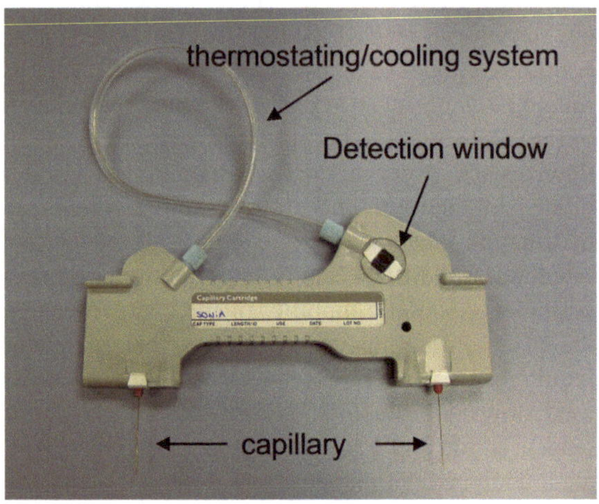

Most important detection systems are optical, electrochemical and mass spectrometry (CE/MS). UV–Vis absorption detector is the most commonly used because it is applicable to organic compounds and inorganic ions in the indirect detection mode. With optical detectors detection is carried out "on capillary" so optical path-length is very low (equal to the internal diameter of the capillary) and this reduces the sensitivity in concentration terms.

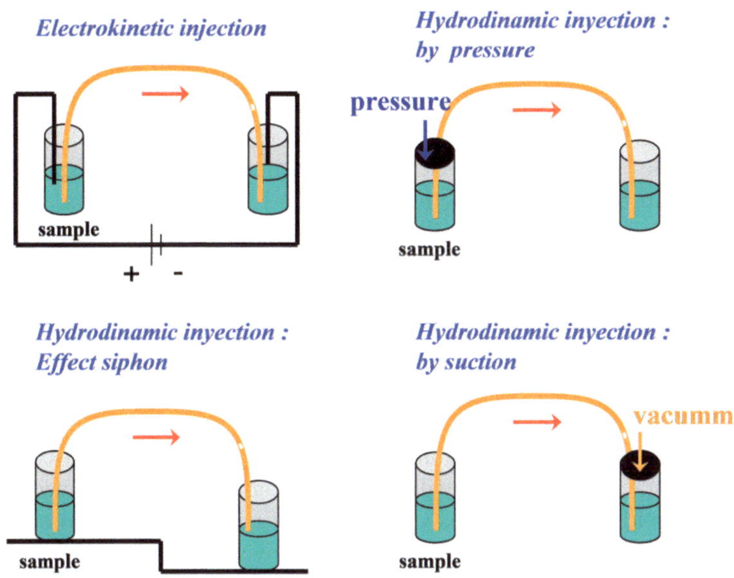

Fig. 2.14 Injection modes

2.2.3 Capillary Electrophoresis Methods in Conservation Science

Plant gums consist of polysaccharides, which are exuded by many trees and plants when their barks are wounded. Gum arabic and gum tragacanth are the most important and representative of this family. They have been used since the second century BC as binding media in Egyptian ointments for mummifications, in mural paintings in Christian catacombs, in paintings on silk and in manuscripts in the Middle Ages. Gums are hydrolysed to monosaccharides prior to chromatographic analysis (TLC, anion-exchange chromatography and GS/MS). CE was introduced as an alternative technique for the analysis of plant gums, because monosaccharides have alcoholic (−OH) and acidic (−COOH) groups that can be ionised at the appropriate pH. The monosaccharide and uranic acid constituents of different plant gums relevant for museum objects were separated by CE and detected with indirect UV absorbance by Grö*β*l et al. [32] using a fused-silica capillary (total length 36 cm, effective length 28 cm, 50 μm ID) and a BGE at pH 12.1 consisting of 20 nm 2,6-pyridinedicarboxylic acid as chromophore for detection and 0.5 mM CTAB as additive to reverse the electroosmotic flow (20 °C, −20 kV, hydrodynamic injection 150 mbar s).

Animal glues are proteins, mainly stemming from three sources: egg, collagen and casein and are an important class of binding media found in objects of cultural and historical interest. For its identification two approaches are used: thermal analysis and FT-IR (without hydrolysis) and TLC, HPLC (UV), and GC/MS (with hydrolysis that delivers free amino acid). From the chemical point of view, it is evident that capillary electrophoresis is appropriate for amino acid analysis due to the ionisable functional groups that the amino acid contain. Animal glues can be identified using CZE by determining its characteristic amino acids (hydroxiproline, proline, glycine, glutamic acid, serine and valine). The main advantage of these methods over HPLC and GC is that no derivatization is required when indirect UV absorption or conductivity detection is used. In addition, other constituents of plant gums or drying oils never interfered with the identification of the animal glues. For example, Kaml et al. [33] developed a method for the characterisation of the three types of animal glues based on CZE with conductivity detection using a fused-silica capillary (total length 60.2 cm, 25 μm ID, 375 μm OD), a BGE composed of 51.9 mM chloroacetic acid and 3.05 mM lithium hydroxide as BGE at pH 2.26 and 25 °C, +20 kV, hydrodynamic injection 90 mbar. The different binder classes are characterised and identified with the aid of the relative peak areas of the amino acids in the electropherograms obtained under appropriate pH conditions. Under the same experimental conditions, Harrison et al. [34] obtained electropherograms to identify collagen as a constituent of a filling material from an eighteenth century polychrome painted sculpture "Revived Christ".

Natural resins have been used in artistic works as glues, varnishes or plasticizers. There are products of certain trees from which they exude, the higher molecular constituents being dissolved in the mono- and sesquiterpenoid essential oils. From a chemical point of view it is possible to differentiate the natural resins according to the number of C atoms from which their molecules are built up (diterpenoids = 20 C, triterpenoids = 30 C). The characterisation by CZE of some resins, based on the pattern of their diterpenoid constituents, was carried out by Findeisen et al. [35]. Separation conditions were a fused-silica capillary (total length: 48.5 cm, effective length 40 cm, 50 μm ID), 20 °C, +25 kV, hydrodynamic injection 100 mbar, and a BGE composed of borate buffer at pH 9.25 with 10 mM sulphobutylether-β-CD and 6.6 mM methyl-β-CD as additives to increase selectivity and enhance the solubility of the analytes. The UV-absorbance was measured at different wavelengths, 200, 250 and 270 nm. The electrophoretic patterns determined served for the characterisation of the relevant diterpenoids resins, balsams and copals.

Drying oils are being used since centuries for coating; especially in Western art they are applied as binding media for pigments in oil paintings. These oils have in common a high content of unsaturated fatty acids: linoleic, linolenic, oleic, etc. The drying process itself results in a polymerisation upon uptake of oxygen. As a result in aged samples dicarboxylic acids are formed as decomposition products. Long-chain fatty acid composition of vegetable oils and their degradation products can be analysed by CE. Early attempts to analyse drying oils were made with TLC, HPLC and especially GC (after derivatization) after triglycerides hydrolysis. Other techniques used for direct analysis are MS, FTIR, NMR and thermal analysis. CE has some potential for the analysis of fatty acids, because they form anions at intermediate or high pH (pKa $\approx$ 4.8). In addition, dicarboxylic acids can be identified as well. Surowiec et al. [36] developed a method to analyse the long-chain fatty acids and the dicarboxylic acids from a sample taken from the painting "still life with flowers" using contactless conductivity detection. These authors used a fused-silica capillary (total length 48.5 cm, 75 μm ID, 375 μm OD) and 10 mM salicylic acid, 20 mM histidine buffer, pH 5.85, as BGE (25 °C, −10 kV, and hydrodynamic injection 15 mbar s). Separation of the dicarboxylic acids was further improved by dynamically coating the capillary wall with 0.2 mM cetyltrimetilammonium bromide. With this method the long-chain fatty acids and their degradation products could be determined in recent and aged samples of drying oils and in samples taken from two paintings of the nineteenth century. Harrison et al. [34] developed a CE method for the determination of the constituents of drying oils in mixtures with other binders (animal and plant glues). The practical applicability of the method was demonstrated in the identification of drying oils as binding medium in a very complex real sample, namely in a paint layer of a mask of Papua New Guinea. The relevant fatty acids found and the dicarboxylic acids formed by oxidative degradation indicate that the most probable source for the drying oil used was from the kernels of the candlenut tree (*Aleurites moluccana*).

Colour has always been an important element in the cultures of people all over the world. Natural colouring agents are of inorganic or organic origin. Pigments and dyes may be characterised for purposes of restoration, to determine what is original, to repair damaged areas and to compensate for missing portions of a painted surface. Natural red pigments (madder, cochineal, sandalwood, etc.) have been extensively used since the sixteenth century. They are all substituted derivatives of naphthoquinones and anthraquinones. Maguregui et al. [38] used SDS as surfactant in MECK to determine a broad range of red organic pigments in paint samples. The main advantages of this method were less analysis time than in HPLC methods and the reduced amount of sample needed to carry out an analysis. Electropherograms showing the presence of madder in an oil painting were obtained using a fused-silica capillary (total length: 58 cm, 75 μm ID, 20 mM borax (pH 9), 20 mM SDS, 10 % ACN as BGE (25 °C, +20 kV, hydrodynamic injection 20 mbar s) and UV–vis detection at 214/254 nm. All colourants were separated in less than 13 min with a good baseline resolution and the method was applied to the analysis of paint samples obtained from the Diocesan Museum of Holy Art of Bilbao. Puchalska et al. [39] used CE with UV–Vis diode-array detection and electrospray mass spectrometry detection for the identification of anthraquinone colour components of cochineal, lac-dye and madder, and natural red dyestuffs. Separation conditions were a fused-silica capillary (total length: 55 cm for UV detection and 85 cm for MS detection, 75 μm ID), a BGE composed of 5 mM sodium phosphate at pH = 8.5 for UV detection and 20 mM ammonium carbonate at pH = 9.0 for MS detection and hydrodynamic injection of 50 mbar for 4 s. The method developed allows the rapid, direct and straightforward identification and quantification of components of natural products used in art and could be very helpful in restoration and conservation procedures. López-Montes et al. [41] developed a simple and rapid CE method with UV detection for the identification of natural dyes. The separation was performed in a fused-silica capillary of 64.5 cm length and 50 μm id. The running buffer was

40 mM sodium tetraborate (pH 9.25). The method was applied to the identification of carminic acid, gambogic acid, crocetin, indigotin, alizarin and purpurin in the collection of drawings and maps at the Royal Chancellery Archives in Granada (Spain). Sepia ink characterisation from *Sepia officinalis* (cuttlefish) was successful with a simple and rapid CE-DAD method [42]. Separation was performed in a fused-silica capillary and the running buffer was 20 mM sodium tetraborate (pH 9.2). In the collection of maps under study, sepia ink was detected in the writing, decorative elements and drawings, mainly applied with watercolour technique. Sepia ink was found on 9 maps dated 1570–1817.

Detailed chemical studies on archaeological textiles could provide important information about the lost, archaic technological processes practiced long ago in different geographical areas and should be important for development of methods for the conservation of relics. Textile samples differ from each other by alkyl groups (acetyl, carboxyl or methyl) as well as by the structure and abundance of the alkylated glucose residues. Kouznetsov et al. [40] used CE to carry out a comparative study on the possible chemical modification of cellulose from eight archaeological textiles dated between 1200 BC and 1500 AD which were kindly provided by several European museums. Electropherograms were obtained using a fused-silica capillary (total length 70 cm, effective length 55 cm, 50 μm ID) with a BGE composed of 100 mM tetraborate buffer at pH = 9.0, 27 °C, +14 kV and mass spectrometry detection.

The evaluation of the state of conservation of aged paper artefacts in relation to paper lifespan prediction is currently a primer issue in the cultural heritage field. In this sense, Dupont et al. [44] developed different CZE methods for a comprehensive characterisation of several classes of degradation products from old books. Three categories of organic compounds were investigated (low molar mass organic acids, aromatic lignin derivatives and simple carbohydrates) because the occurrence of these compounds in paper depended on the state of degradation and on the type of pulp. All the organic compounds identified were found in a panel of papers from various origins and diverse states of deterioration. The study proved the usefulness of CZE in the comprehensive analysis of organic compounds found in aqueous extracts of papers and originating during their degradation.

2.2.4 Evaluation of Capillary Electrophoresis Methods in Conservation Science

Since the first reported studies in conservation science carried out in the late eighteenth century, analytical methods applied to the study of cultural objects have constantly grown. Characterisation of these objects has a double use. On one side, it allows us to put them in their historical context, and on the other it could be useful for a possible restoration or conservation process. The most recent advances in the identification and determination of organic compounds were reviewed by Doménech-Carbó [45]. A variety of instrumental analytical methods have been proposed in an attempt to improve precision, accuracy, sensitivity, selectivity, etc., of the results. In this sense, the use of novel analytical techniques as CE in conservation science has notably expanded in recent years. CE is a fast technique allowing the combination of short analysis times and high separation efficiency for the analysis and has proved to be an excellent alternative to chromatographic (HPLC) techniques in conservation science. The resolution of CE procedures is greater than that of HPLC and the precision is of the same order. CE is less costly than HPLC and requires only modest quantities of samples. In addition, as CE separations are based on totally or partially uncorrelated physical–chemical principles compared to chromatographic ones, they represent an additional dimension in the analytical process. For this reason, in recent years some CE methods have been developed to obtain analytical information as to the kind and composition of the binding media, pigments and dyes, etc., used in cultural objects. This article reviews some developments concerning the analysis of cultural objects by CE. To give the most relevant information on this topic, the experimental conditions employed to achieve the analysis of these compounds in cultural objects by CE are provided.

References

1. Burgot G, Burgot JL (2002) Méthodes Instrumentales d'Analyse Chimique. Tec et Doc, Paris
2. Poole CF (2003) The essence of chromatography. Elsevier, Bristol
3. Chavanne M, Beaudoin GJ, Jullien A, Flamand E (1999) Chimie organique expérimentale. Belin, Paris
4. Defranceschi M (1990) 144 manipulations de chimie générale et minérale. Ellipses, Paris
5. Douglas A, Skoog F, Holler J, Timothy A, Nieman T (2003) Principes d'Analyse Instrumentales De Boeck. Paris et, Bruxelles
6. Rouessac F (2004) Analyse chimique (méthodes et techniques instrumentales modernes). Masson, Paris
7. Arpino P, Prevôt A, Serpinet J, Tranchant J, Vergnol A, Wittier P (1995) Manuel pratique de chromatographie en phase gazeuse. Masson, Paris
8. Rosset R, Caude M, Jardy A (1991) Chromatographies en phase liquide et supercritique. Masson, Paris
9. Martin AJP, Synge RLM (1941) A new form of chromatogram employing two liquid phases. 1. A theory of chromatography. 2. Application of the microdetermination of the higher monoaminoacids in proteins. Biochem J 35:1358–1368
10. Mills JS, White R (1994) The organic chemistry of museum objects. Butterworth Heinemann, London
11. Shedrinsky AM, Wampler TP, Indictor N, Baer NS (1989) Application of analytical pyrolysis to problems in art and archaeology: a review. J Anal Appl Pyrol 15:393–412
12. Hovaneissian M, Archier P, Mathe C, Vieillescazes C (2006) Contribution de la chimie analytique à l'étude des exsudats végétaux styrax, storax et benjoin. C R Chim 9(9):1192–1202
13. Regert M, Rolando C (1996) Archéologie des résidus organiques: De la chimie analytique à l'archéologie : un état de la question. Techné 3:118–128

14. Mejanelle P, Bleton J, Goursaud S, Tchapla A (1996) Analyse de baumes de momification d'Egypte ancienne par chromatographie gazeuse capillaire: spectrométrie de masse. Analusis Magazine 24(7):16–20
15. Bleton J, Coupry C, Sansoulet J (1996) Approche d'étude des encres anciennes. Stud Conserv 41:95–108
16. Connan J, Deschene O (1996) Le bitume à Suse: Collection du Musée du Louvre Réunion des Musées Nationaux et Elf Aquitaine Production. Paris et Pau, Paris
17. Colinart S (1987) Chapter IV: Matériaux constitutifs. In: Sculptures en cire de l'ancienne Egypte à l'art abstrait, Ministère de la Culture et de la Communication. Editions de la réunion des musées nationaux, Paris PP 29–57
18. Ploeger R, Scalaron O, Chiantore O (2008) The characterization of commercial artists' alkyd paints. J Cult Herit 9(4):412–419
19. Quye A, Wouters J, Boon JJ (1996) A preliminary study of light-ageing effect on the analysis of natural flavanoid-dye wools by photodiode array HPLC and by direct temperature mass spectrometry. In: Commitee for Conservation. ICOM, Edimbourg 2:704–713
20. Szostek B, Orska-Gawrys J, Surowiec I (1012) Trojanowicz M (2003) Investigation of natural dyes occurring in historical Coptic textiles by high-performance liquid chromatography with UV–Vis and mass spectrometric detection. J Chromatogr 2:179–192
21. Wallert A (1996) Tannins of the parchment of the Dead Sea scrolls. In: committee for conservation ICOM. Edimbourg, 2:560–564
22. Richardin P, Copy S, Chahine C, Saltron F, Bonnassies TS (1996) GC and GC/MS characterization of degradation amino acids in naturally and artificially aged leathers. J Am Leather Chem Assoc 91(1):2–17
23. Pastore P, Magno F, Volpin S, Biscontin G (1991) Chromatographic determination of oxalate ion in patinas covering ancient materials. Ann Chim (Rome) 81(5–6):233–241
24. Vieillescazes C, Coen S (1993) Charactérisation de quelques résines utilisées en Egypte ancienne. Stud Conserv 38:255–264
25. Casoli A, Mirti P, Palla G (1995) Characterization of medieval proteinaceous painting media using gas chromatography and gas chromatography- mass spectrometry. Fresenius' J Anal Chem 352(3–4):372–379
26. Halpine SM (1992) Amino acid analysis of proteinaceous media from Cosimo Tura's The Annunciation with Saint Francis and Saint Louis of Toulouse. Stud Conserv 37(1):22–38
27. Bouchonnet S (2009) La spectrométrie de masse en couplage avec chromatographie en phase gazeuse. Tec and Doc Lavoisier, Paris
28. Cuoco G, Mathe C, Archier P, Vieillescazes C (2009) A multivariate study of the performance of an ultrasound-assisted madder dyes extraction and characterization by liquid chromatography-photodiode array detection. Ultrason sonochem 16:75–82
29. Bleton J, Tchapla A (2009) SPME/GC-MS in the caracterisation of terpenic resins. In: Organic mass spectrometry in art and archeology. J Wiley 10:261–302
30. Colombini MP, Modugno F, Giannarelli S, Fuoco R, Matteini M (2000) GC-MS characterization of paint varnishes. Microchem J 67:385–396
31. Colombini MP, Andreotti A, Bonaduce I, Modugno F, Ribechini E (2009) Analytical strategies for characterizing organic paint media using gas chromatography/mass spectrometry. Acc Chem Res 43(6):715–727
32. Gröβl M, Harrison S, Kaml I, Kenndler E (2005) Characterisation of natural polysaccharides (plant gums) used as binding media for artistic and historic works by capillary zone electrophoresis. J Chromatogr 1077:80–89
33. Kaml I, Vcelakova K, Kenndler E (2004) Characterisation and identification of proteinaceous binding media (animal glues) from their amino acid profile by capillary zone electrophoresis. J Sep Sci 27:161–166
34. Harrison S, Kaml I, Prokoratova V, Mazanek M, Kenndler E (2005) Animal glues in mixtures of natural binding media used in artistic and historic objects: identification by capillary zone electrophoresis. Anal Bional Chem 382:1520–1526

35. Findeisen A, Kolivoska V, Kaml I, Baatz W, Kenndler E (2007) Analysis of diterpenoic compounds in natural resins applied as binders in museum objects by capillary electrophoresis. J Chromatogr 1157:454–461
36. Surowiec I, Kaml I, Kenndler E (2004) Analysis of drying oils used as binding media for objects of art by capillary electrophoresis with indirect UV and conductivity detection. J Chromatogr 1024:245–254
37. Harrison S, Kaml I, Rainer F, Kenndler E (2005) Identification of drying oils in mixtures of natural binding media used for artistic and historic works by capillary electrophoresis. J Sep Sci 28:1587–1594
38. Maguregui M, Alonso R, Barandiaran M, Jimenez R, García N (2007) Micellar electrokinetic chromatography method for the determination of several natural red dyestuff and lake pigments used in art work. J Chromatogr 1154:429–436
39. Puchalska M, Orlinska M, Ackacha M (2003) Identification of anthraquinone coloring matters in natural red dyes by electrspray mass spectrometry coupled to capillary electrophoresis. J Mass Spectrom 38:1252–1258
40. Kouznetsov DA, Ivanov A, Veletsky P (1994) Detection of alkylated cellulose derivatives in several archaeological linen textile samples by capillary electrophoresis/mass spectrometry. Anal Chem 66:4359–4365
41. López-Montes A, Blanc R, Espejo T, Huertas-Pérez J, Navalón A, Vilchez JL (2007) Simultaneous identification of natural dyes in the collection of drawings and maps from the Royal Chancellery Archives in Granada (Spain) by CE. Electrophoresis 28:1243–1251
42. López-Montes A, Blanc R, Espejo T, Navalón A, Vilchez JL (2009) Characterization of sepia ink in ancient graphic documents by capillary electrophoresis. Microchim J 93:121–126
43. Chankvetadze B (1997) Capillary electrophoresis in chiral analysis. Willey, New York
44. Dupont A-L, Egasse C, Morin A, Vasseur F (2007) Comprehensive characterisation of cellulose- and lignocellulose- degradation products in agent papers: Capillary zone electrophoresis of low-molar mass organic acids, carbohydrates, and aromatic lignin derivatives. Carbohydr Polym 68:1–16
45. Doménech-Carbó M (2008) Novel analytical methods for characterising binding media and protective coatings in artworks. Anal Chim Acta 621:109–139

Chapter 3
Identification Techniques I

Ioannis A. Kozaris, Eleni Pavlidou, Reiner Salzer, D. Capitani, A. Spinella and E. Caponetti

I. A. Kozaris (✉)
Department of Chemistry, Aristotle University of Thessaloniki, 54124 Thessaloniki, Greece
e-mail: ikozaris@chem.auth.gr

E. Pavlidou (✉)
Department of Physics, School of Science, Aristotle University of Thessaloniki, 54124 Thessaloniki, Greece
e-mail: elpavlid@auth.gr

R. Salzer (✉)
Department of Chemistry, Dresden University of Technology, 01062 Dresden, Germany
e-mail: reiner.salzer@tu-dresden.de

D. Capitani (✉)
Magnetic Resonance Laboratory "Annalaura Segre", Institute of Chemical Methodologies (IMC), National Research Council (CNR), Via Salaria Km 29, 300 00015 Monterotondo, RM, Italy

A. Spinella · E. Caponetti
Centro Grandi Apparecchiature, Università di Palermo, Via Marini 14, 90128 Palermo, PA, Italy

E. Caponetti
Dipartimento di Chimica "S. Cannizzaro", Università di Palermo, Parco d' Orleans II, Viale delle Scienze pad.17, 90128 Palermo, PA, Italy

E. A. Varella (ed.), *Conservation Science for the Cultural Heritage*,
Lecture Notes in Chemistry 79, DOI: 10.1007/978-3-642-30985-4_3,

3.1 Imagine Techniques

Ioannis A. Kozaris

3.1.1 Introduction

Imaging is the representation or reproduction of an object's form and *imaging technology* is the application of materials and methods to create, preserve or duplicate images. Imaging spectroscopy is similar to colour photography, but each pixel acquires many bands of light intensity data from the spectrum, instead of just the three bands of the RGB color model. More precisely, it is the simultaneous acquisition of spatially co-registered images in many spectrally contiguous bands. We can define the *imaging spectroscopy* (also *spectral imaging* or *chemical imaging*) as the analytical capability to create a visual image of components distribution from simultaneous measurement of spectra and spatial, time information.

Imaging instrumentation is composed of three components:

- a radiation (energy) source to "illuminate" the sample,
- a spectrally selective element, which interacts with the radiation (wave-matter interaction),
- and usually a detector array to collect the images (interactions with the radiation) (Fig. 3.1).

3.1.2 Energy Sources

Radiation is a form of energy. There are two basic types of radiation. One kind is *particulate radiation*, which involves tiny fast-moving particles that have both energy and mass. Particles that can carry 'radiation' energy are:

- *Electrons*, sometimes called beta minus particles, are small mass, negatively charged particles.
- *Positrons* which are positively charged electrons.
- *Protons* which are larger mass positively charged particles.
- *Neutrons* have a similar mass to protons, but are uncharged.
- *Alpha particles* are particularly stable groups of two protons and two neutrons. All types of atomic particles can carry energy.

The second basic type of radiation is *electromagnetic radiation*. This kind of radiation is pure energy with no mass and is like vibrating or pulsating waves of electrical and magnetic energy. In addition to acting like waves, *quantum*

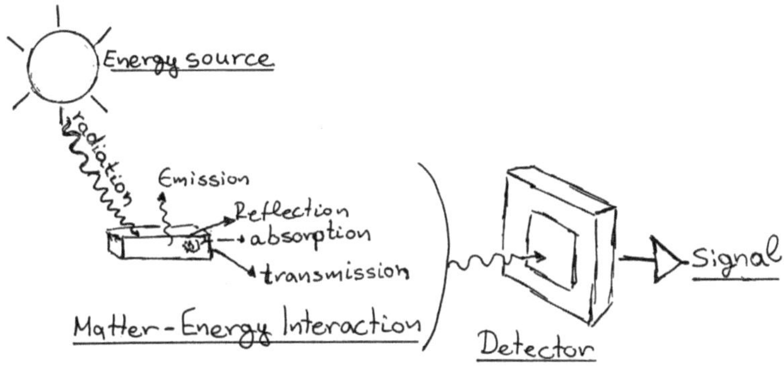

Fig. 3.1 Conceptual diagram of imaging instrumentation

mechanics predicts that electromagnetic radiation behave as uncharged energy particles, called *photons*. The mass-less photon particles travel in a wave-like pattern at the speed of light. Each photon contains a certain amount (or bundle) of energy, and all electromagnetic radiation consists of these photons. The only difference between the various types of electromagnetic radiation is the amount of energy found in the photons. The electromagnetic spectrum is the range of all possible frequencies of electromagnetic radiation (Fig. 3.2).

3.1.3 Wave-Matter Interaction

The ways in which radiation interacts with matter are reflected both in the *nature of the radiation* itself, and in the *structure of atoms*. Radiation is classified into two main categories, *non-ionising* and *ionising*, depending on its ability to ionise matter.

Non-ionising radiations have not sufficient energy to cause ionisation in matter that is, to completely remove an electron from an atom or molecule. Instead of producing charged ions when passing through matter, the energy is sufficient only for excitation, the movement of an electron to a higher energy state.

Ionising radiation is composed of particles with energy that individually can liberate an electron from an atom or molecule. The degree and nature of such ionisation depends on the energy of the individual particles (including photons), and not on their number (intensity). In general, particles with energies above about 10 eV are considered ionising.

To study the penetration of radiation into matter, particles classified according to their electrical behaviour (charge). Electrically neutral, uncharged, are particles with zero electrical charge, such particles are photons which they are used in imaging instrumentation as energy source. Photons from ultraviolet and above, belongs to the non-ionising radiations and from ultraviolet and below, belongs to

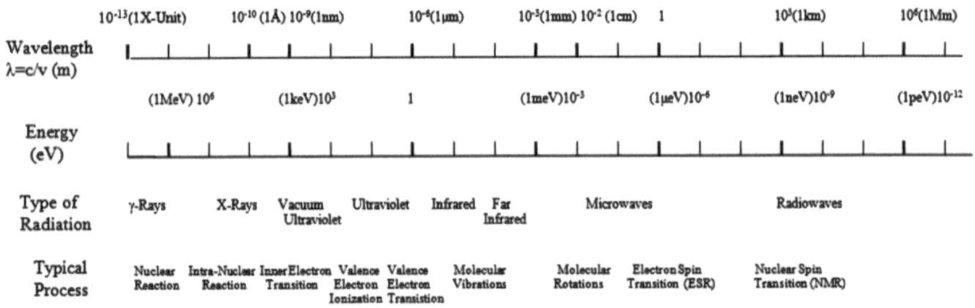

Fig. 3.2 The electromagnetic spectrum, in order of increasing frequency and decreasing wavelength, consists of radio waves, microwaves, infrared radiation, light, ultraviolet radiation, X-rays and gamma rays

indirectly ionising radiation. Particles with electrical charge which used in imaging instrumentation are mainly electrons. All charged particles used as energy source belong to the directly ionising radiations.

3.1.3.1 Photons Penetration

Being electrically neutral, a photon does not experience Coulomb forces at large distances that almost constantly reduce its energy as it traverses matter in the way that a charged particle does. What is observed, is that a photon of given energy in a uniform medium proceeds with a certain probability per unit distance of having an interaction or collision. The energy levels for all physical processes at the atomic and molecular levels are quantised, and if there are no available quantised energy levels with spacings which match the quantum energy of the incident radiation, then the material will be transparent to that radiation, and it will pass through.

Photons are thus electromagnetic radiation with zero mass, zero charge, and a velocity that is always the speed of light. Because they are electrically neutral, they do not steadily lose energy via coulombic interactions with atomic electrons as do charged particles. Instead, they travel some considerable distance before undergoing a more "catastrophic" interaction. How far a given photon will penetrate, depends on the specific medium traversed and on the photon energy. All of the photon interactions of interest to us lead to partial or total transfer of the photon energy to electron energy. Thus the history of a photon in material is characterised by the sudden disappearance of the photon or by scattering through significant angles with significant energy loss.

When the photon interacts:

- It can penetrate the section of matter without interacting
- It can interact with the matter and be completely absorbed by depositing its energy
- It can interact and be scattered or deflected from its original direction and deposit part of its energy (Fig. 3.3).

There are three main interaction mechanisms of photons with matter and three of them are the means by which photons are detected. The predominant mode of interaction depends on the energy of the incident photons and the atomic number of the material with which they are interacting. At low energies, in high atomic number materials, the photoelectric effect is the main interaction of photons with the material. At intermediate energies in low atomic number materials the dominant interaction is Compton scattering. At very high energies, the main mechanism by which photons are detected is pair production.

The different parts of the electromagnetic spectrum have very different effects upon interaction with matter

Microwave Interactions. The energy of microwave radiation (photons) is in the range 0.00001–0.001 eV which is in the range of energies separating the quantum states of molecular rotation and torsion. The interaction of microwaves with polar molecules, with positively and negatively charged ends, will be to rotate molecules and produce heat as result of that molecular motion (Fig. 3.4).

Infrared Interactions. The energy of infrared radiation (photons) is in the range 0.001–1.7 eV which is in the range of energies to excite vibrational and rotational states of a molecule. For example, simple diatomic molecules have only one bond allowing only stretching vibrations. More complex molecules may have many bonds, and vibrations can be conjugated. The atoms in a CH_2 group, commonly found in organic compounds, can vibrate in six different ways; symmetrical and antisymmetrical, scissoring, rocking, wagging, and twisting. Infrared is absorbed more strongly than microwaves, but less strongly than visible light (Fig. 3.5).

Visible Light Interactions. The primary mechanism for the absorption of visible light photons is the elevation of electrons to higher energy levels. There are many available states, so visible light is absorbed strongly. While exposure to visible light causes heating, it does not cause ionisation with its risks (Fig. 3.6).

Ultraviolet Interactions. The near ultraviolet is absorbed very strongly from matter by electron transitions. As you go to higher energies, the ionisation energies for many molecules are reached and the more photoionisation processes take place (Fig. 3.7).

X-ray Interactions. Since the energies of X-ray radiation (photons) are much too high to be absorbed in electron transitions between states for most atoms, they can interact with an electron only by knocking it completely out of the atom. That is, all X-rays are classified as ionising radiation. This can occur by giving all of the energy to an electron (photoionisation) or by giving part of the energy to the electron and the remainder to a lower energy photon (Compton scattering). At sufficiently high energies, the X-ray photon can create an electron positron pair.

The principal mechanisms of energy deposition by photons in matter are:

- Photoelectric absorption,
- Compton scattering,
- Pair production and photonuclear reactions (Fig. 3.8).

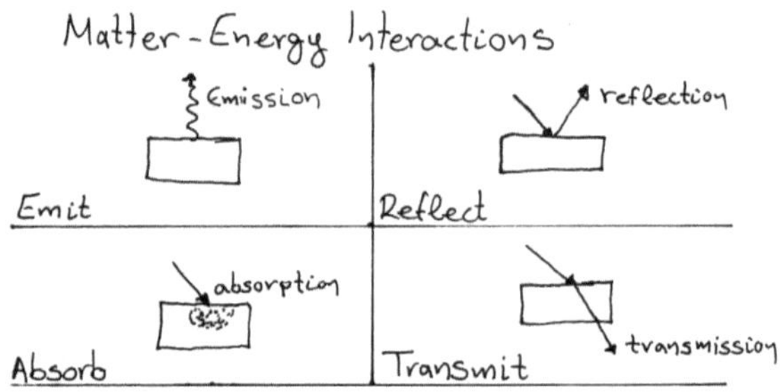

Fig. 3.3 Photons interaction with matter

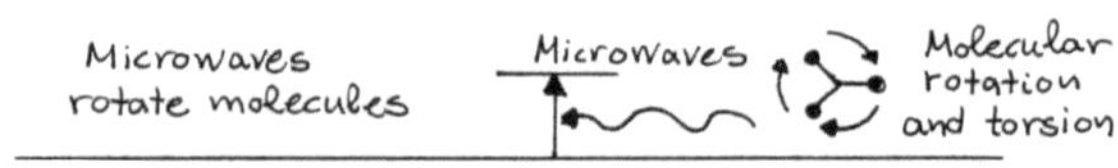

Fig. 3.4 Microwave interactions

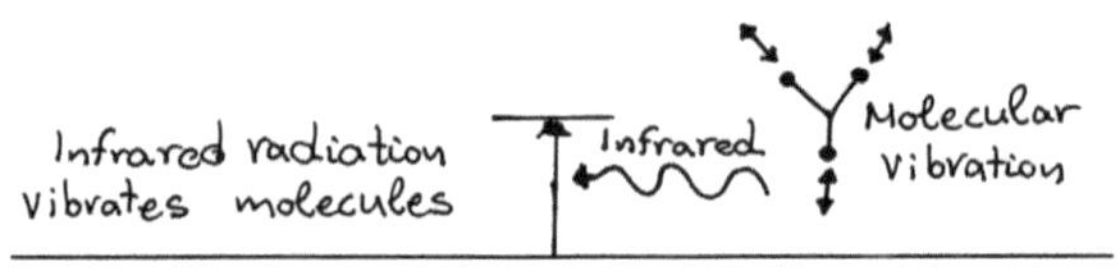

Fig. 3.5 Infrared interactions

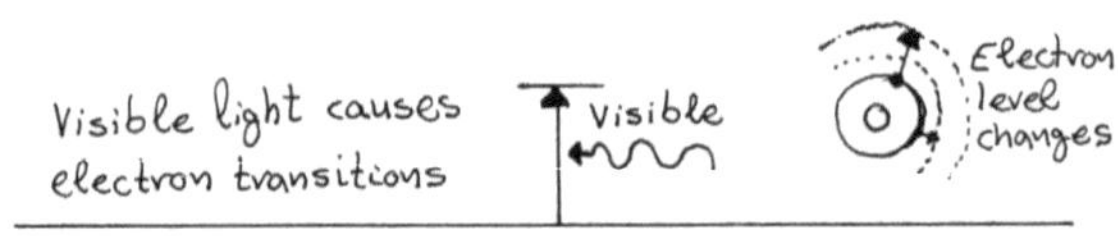

Fig. 3.6 Visible light interactions

3.1.3.2 Electrons Penetration

The interaction of the electrons with the matter in our sample produces different signals (electrons, photons from the infrared through visible till the X-ray range, etc.) These signals can be used for imaging different sample characteristics, like surface morphology or distribution of electrically active crystal defects or local composition, etc.

When an electron hits onto a material, different interactions can occur, the interactions can be classified into two different types, namely elastic and inelastic interactions.

(i) Elastic interactions, in this case, no energy is transferred from the electron to the sample. As a result, the electron leaving the sample still has its original energy. Of course, no energy is transferred if the electron passes the sample without any interaction at all. Such electrons contribute to the direct beam which contains the electrons that passes the sample in direction of the incident beam.

Furthermore, elastic scattering happens if the electron is deflected from its path by Coulomb interaction with the positive potential inside the electron cloud. By this, the primary electron loses no energy or—to be accurate—only a negligible amount of energy. These signals are mainly exploited in TEM and electron diffraction methods (Fig. 3.9).

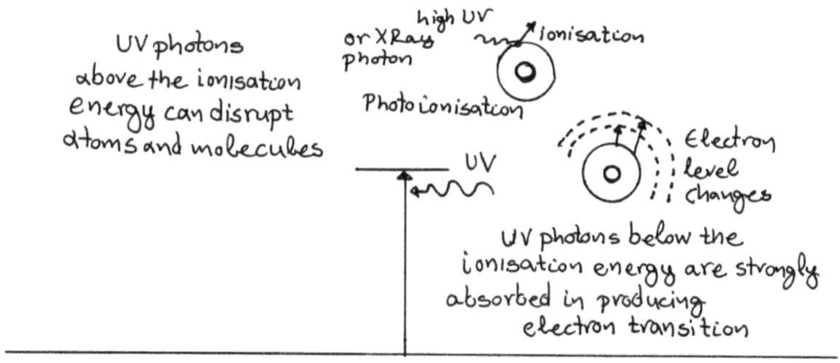

Fig. 3.7 Ultraviolet interactions

(ii) Inelastic Interactions. If energy is transferred from the incident electrons to the sample, then the electron energy of the electron after interaction with the sample is consequently reduced:

The energy transferred to the specimen can cause different signals, such as X-rays, Auger or secondary electrons, plasmons, phonons, UV quanta or cathode luminescence. Signals caused by inelastic electron-matter interactions are predominantly utilised in the methods of analytical electron microscopy (AEM) (Fig. 3.10).

3.1.4 Detectors

Devices used to detect and measure radiation characteristically emitted in a processes are known as radiation detectors. This can be achieved by observation of the electrical disturbance created by a particle or a photon as it passes through the radiation detector. For detailed observation of individual photons or particles, a pulse detector is used to convert each such event (that is, photon or particle) into an electrical signal. The detection sensitivity depends on:

- fluctuations in the detector
- fluctuations in the electronics.

Energy is converted into an electrical signal, either directly or indirectly. Each detected particle will appear as a pulse of electric charge.

Direct conversion: incident radiation ionises atoms/molecules in absorber, creating mobile charges that are detected.

Indirect conversion: incident radiation excites atomic/molecular states that decay by emission of light, which in a second step is converted into charge. The energy deposited by the primary radiation in a phosphor or scintillator is imaged through either fibre optics or a lens on to a detector. Incoming photons hit the photo-sensitive element of the system, which will generate photo-electrons. These photo-electrons are then guided by an electric field toward the amplification stage.

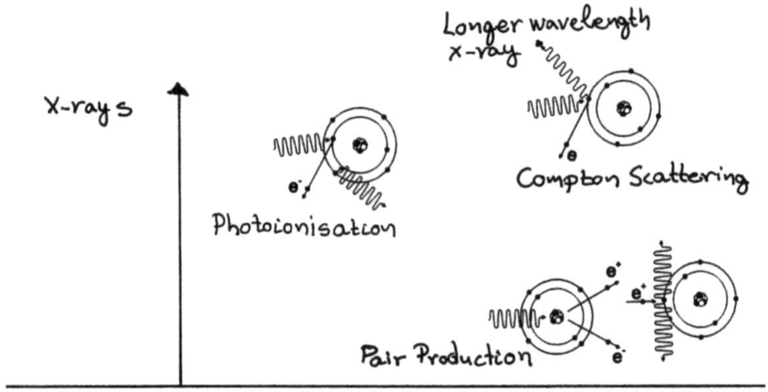

Fig. 3.8 X-ray interactions

Fig. 3.9 Elastic and bremsstrahlung collisions

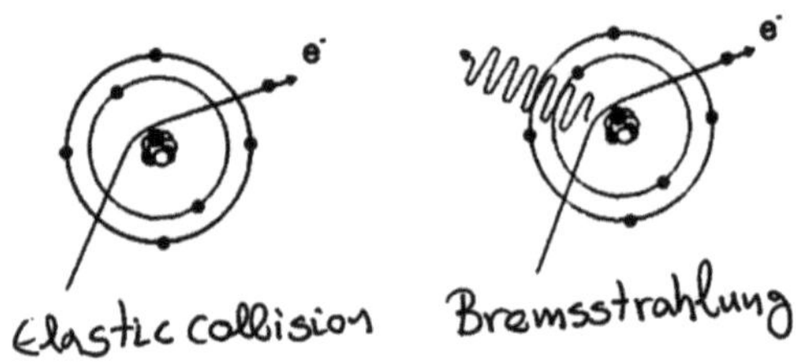

Fig. 3.10 Inelastic collisions

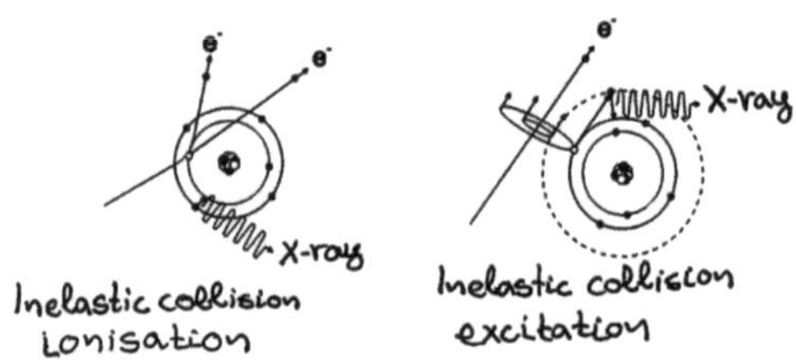

In general, detectors can be broken down into several classes and subclasses. Two typical subclasses are photo emissive and solid state.

In photoemissive detectors when a photon with sufficient energy strikes a surface, it can cause the ejection of an electron from the surface into a vacuum. A photoemissive diode consists of a surface (photocathode) appropriately treated to permit the ejection of electrons by low-energy photons and a separate electrode (the anode) on which electrons are collected, both sealed within an evacuated glass envelope. Photoemissive detectors use various photocathodes that are tailored to be sensitive to a specific, limited-wavelength interval, while solid state detectors, which absorb the photons into the semiconductor substrate, have sensitivities over a broad range of wavelengths.

Semiconductors (solid state detectors) have been used in particle or photon identification for many years. The first practical devices were small germanium surface barrier devices made in the 1950s. When first introduced in the 1960s, semiconductor detectors were used primarily for high-resolution X-ray and gamma spectroscopy. Semiconductor, or solid state detector, is a term which is usually meant to exclude scintillation counters. A semiconductor detector consists of a radiation-sensitive element whose valence and conduction energy bands are separated by a small energy difference (band gap). Radiation penetrating the semiconductor can provide the energy necessary to promote electrons from the valence band to the conduction band. The promoted electrons, and the holes created in the valence band, provide a quantifiable charge proportional to the radiation flux. This charge is transferred to a readout amplifier for amplification before digitisation by an analogue-to-digital converter (ADC).Today we find single element detectors and multi-element detector arrays, both linear and two dimensional.

The simplest type of semiconductor radiation detector involves observing the change in conductivity in a semiconductor due to the creation of additional charge carriers by an incident electromagnetic field. These devices are commonly known as photoconductors.

A semiconductor diode is a type of photodetector capable of converting radiation into either current or voltage, depending upon the mode of operation. Semiconductor junction photodiodes can exhibit dramatically improved performance over simple photoconductor detectors Photodiodes may be designed with improved response times, greater sensitivity, decreased thermal sensitivity, linearity over 9–10 orders of magnitude, and large internal amplification.

Photodiodes can be operated in two very different modes:

Photovoltaic operation, an external circuit is used to measure the small change in potential that develops across the diode. The photovoltaic effect occurs at a junction in a semiconductor. The junction is the boundary between a region where the conductivity is due to electrons and a region where the conductivity is due to holes (the absence of electrons). This is called a pn junction. At the junction, an electric field is present internally because there is a change in the level of the conduction and valence bands. This change leads to the familiar electrical rectification effect produced by such junctions. The photovoltaic effect is the generation of a voltage when light strikes a semiconductor pn junction. However, the dependence of this voltage on the radiation power is nonlinear, and the dynamic range is fairly small.

Photoconductive operation, the current generated by the electron/hole pairs is measured. A semiconductor in thermal equilibrium contains free electrons and holes. The concentration of electrons and holes is changed if light is absorbed by the semiconductor. The light must have photon energy large enough to produce free electrons within the material. The increased number of charge carriers leads to an increase in the electrical conductivity of the semiconductor. The change in electrical conductivity leads to an increase in the current flowing in the circuit, and hence to a measurable change in the voltage drop across the load resistor. The device is used in a circuit with a bias voltage and a load resistor in series with it. In both configurations, the measured signal is proportional to the instantaneous radiation flux. Additional circuitry may be used to integrate the signal (Fig. 3.11).

The detector's material determines the sensitivity and wavelength response. Semiconductor materials can generally be grouped into 2 categories. These are the

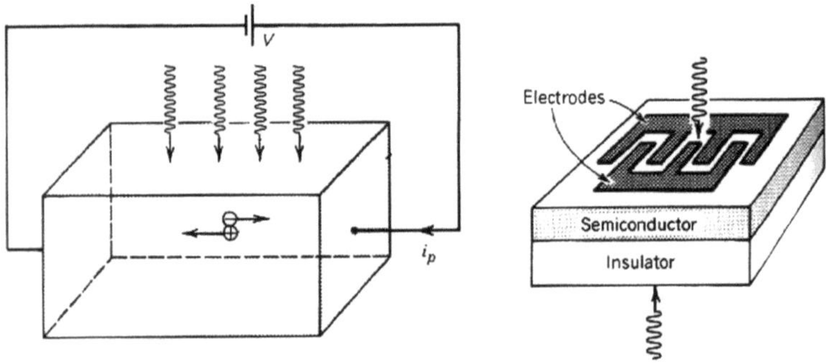

Fig. 3.11 Schematic diagram of semiconductor

elemental semiconductors, Si and Ge, which are derived from elements in Group IV of the periodic table and compound semiconductors—the most common of which are derived from groups III and V (e.g., GaAs, InP) and groups II and VI (e.g., CdTe) of the periodic table. In addition, to binary materials (such as GaAs or InP), most compounds are also soluble within each other, making it possible to synthesise ternary (e.g., AlGaAs, HgCdTe), quaternary(e.g., InGaAsP, InGaAlP) and higher order solutions, simply by alloying binary compounds together. Different semiconductor materials can be used to cover different spectral ranges. Silicon-based semiconductors, with an energy gap of approximately 1 eV, have been shown to have a useful response from the far ultraviolet (UV) to the near-infrared (near-IR) and have also been used in the X-ray region. Semiconductors with lower band gaps are effective in the near- and mid-IR. Compound semiconductors are so useful because of the shear range of compounds available.

Typical photodiode materials are:

silicon (Si): low dark current, high speed, good sensitivity between roughly 400 and 1,000 nm (best around 800–900 nm)

germanium (Ge): high dark current, slow speed due to large parasitic capacity, good sensitivity between roughly 900 and 1,600 nm (best around 1,400–1,500 nm)

indium gallium arsenide phosphide (InGaAsP): expensive, low dark current, high speed, good sensitivity roughly between 1,000 and 1,350 nm (best around 1,100–1,300 nm)

indium gallium arsenide (InGaAs): expensive, low dark current, high speed, good sensitivity roughly between 900 and 1,700 nm (best around 1,300–1,600 nm) (Fig. 3.12)

The performance of detectors is commonly characterised by a number of different parameters. There are two fundamental parameters that determine Sensitivity.

The first is the most commonly recognised one—*Quantum Efficiency*. The Quantum Efficiency measures the percentage probability of a photon to be converted into a detectable signal. The other photons are lost either by being reflected away, absorbed somewhere they do not produce a signal or even just pass through the sensor without being detected. Quantum efficiency only gives an indication on

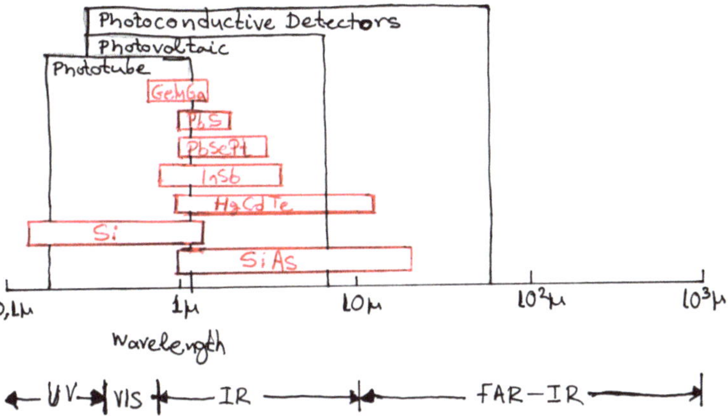

Fig. 3.12 Approximate spectra range and operation mode of various semiconductors

how good a detector is at capturing incoming radiation, not how good it is at detecting them. Therefore a complementary parameter to assess the Sensitivity of a detector is in fact the *Signal to Noise ratio.*

The second, however, is just as important as Quantum Efficiency but is often insufficiently understood. It is the *noise floor* determined by the detector itself. A sensitive device must be designed to optimise each of these parameters.

3.2 The Role of Microscopy Techniques in the Study of Cultural Heritage Materials

Eleni Pavlidou

3.2.1 In Which Ways Microscopy Techniques are Used in Conservation Science?

Conservation is a profession devoted to the preservation of cultural heritage for the future. Conservation activities include examination, documentation, treatment and preventive care. All of this work is supported by research and education. The traditional definition of the role of the conservator involves the examination, conservation and preservation of cultural heritage using any methods that have been proved to be effective in keeping artefacts' properties as close to its original condition as possible for as long as possible.

Microscopy includes a few techniques that allow samples to be viewed at a scale convenient for examination and analysis. Formed by various means, an image can be formed by direct imaging, electronic processing or a combination of these methods. The most familiar type of microscope is the *optical* or light microscope, in which lenses are used to form the image. Other major types of microscopes, such as *scanning* electron microscopes and *transmission* electron microscopes, use the wave nature of various physical processes. Optical and Electron microscopy can be considered amongst the most used non-destructive methods of analysis for the examination of paintings, wall paintings, ceramics, glasses, coins, bronzes and many other materials.

The assessment of the present status of works of art and archaeology is of great interest. Nowadays, scientists working in the field of artistic and historic objects conservation and preservation are able to choose the best combination of techniques and analytical methods from a wide range of procedures and technologies that are now available. However, at the same time they are constantly challenged by ageing and deterioration; preserved items are not only constantly exposed to the ambient atmosphere but also have to be regularly exhibited in museum collections and exhibitions or simply stored in museum depots. During their lifetime, artefacts constantly react with their environment through mechanisms having reaction times of several hundreds or even thousands of years, and probably through degradation mechanisms not yet known to modern materials science. This is the reason why there is still intensive ongoing research to investigate and observe the degradation mechanisms of artefacts in contact with soil or the ambient atmosphere.

3.2.2 Principles and Concepts of Microscopy

The image of a light absorbing specimen is formed due to diffraction. Some of the light will pass through the specimen undeviated and will only give rise to a uniformly bright image. The diffracted light carries the information about the structure of the specimen. Over the past decades, the number of applications of optical microscopy has grown enormously, and is now found in almost any field of science and industry. The microscope must accomplish three main tasks: produce a magnified image of the specimen, detect and enhance the details in the image, and render the details visible to the human eye or camera.

Microscopy is a very useful tool and is widely used by the scientific community to study paintings, in which the identification of the pigments and the painting technique is of great importance in order to determine the value and the appearance of a monument. Microscopy is also used to record the conservation status, and is also employed for the detailed description and placement of damages, weathering or interferences for the effective conservation of a painting or a wall painting. Pigments, as the basis of all paints which have been used for millennia, are ground coloured materials. Early pigments were simply ground earth or clay, and were made into paint using spit or fat. Modern pigments are often sophisticated masterpieces of chemical engineering. Paints used in paintings and other painted objects consist simply of pigments suspended in a substance that can easily be observed using a microscope. That substance can vary, from oil or egg yolk in paintings, to plaster in frescos, or sophisticated plastics in automobile finishes. When incident white light encounters a pigment, some of its wavelengths are absorbed by the chemical bonds and substituents of the pigment, while others are reflected.

In Fig. 3.13 the three types of microscopes that are discussed here are schematically presented [1]. The optical microscope (OM) relies on the following principle: the light passing through a specimen is magnified through glass lenses. When observing a specimen with a transmission electron microscope (TEM), it undergoes a proper preparation that is necessary in order for it to become thin enough to allow the transmission of an electron beam. The electron beam which has passed the specimen is then scattered and observed after being magnified by electromagnetic lenses. In a scanning electron microscope (SEM) observation, an electron beam finely focused by electromagnetic lenses is scanned over the specimen and the brightness on a monitor or computer screen is modulated by the signals obtained. So, in order to comprehend the operation of SEM and TEM microscopes, the concepts of electron-matter interactions should be briefly mentioned (Fig. 3.13).

When an electron beam is irradiated on a specimen surface, a number of different interactions between the electron beam and the atoms composing the specimen occur, producing various kinds of information, as can be seen in Fig. 3.14. The different types of information that can be obtained are of course closely related to the regions from which they are produced.

3.2.3 Description of the Methodology

3.2.3.1 Optical Microscope

Optical microscopes are used today in numerous applications and they are amongst the most commonly used tools in order to inspect the microstructure of a great range of materials. It is crucial to use the appropriate mode for the specimen, choosing from reflected-light or transmitted-light modes. Reflected-light microscopy is used for a range of materials, including metals, ceramics and composites. Different phases, grain orientations or boundary regions result in variations in surface topography and differences in reflectivity which leads to contrast between different regions when viewed in reflected light. These features are revealed by a series of specimen preparation techniques which, when carried out with care, can produce useful, high quality images. Transmission mode can be used when the specimen is transparent. The specimen is usually in the form of a thin slice, preferably a few tens of microns thick. Contrast arises from the differences in the absorption of light through the different regions of the sample. This method is used for the examination of minerals and rocks, as well as glasses, ceramics and polymers. In addition, the transmission mode can often be further enhanced with the use of polarised light. In polarised light microscopy, contrast is due to

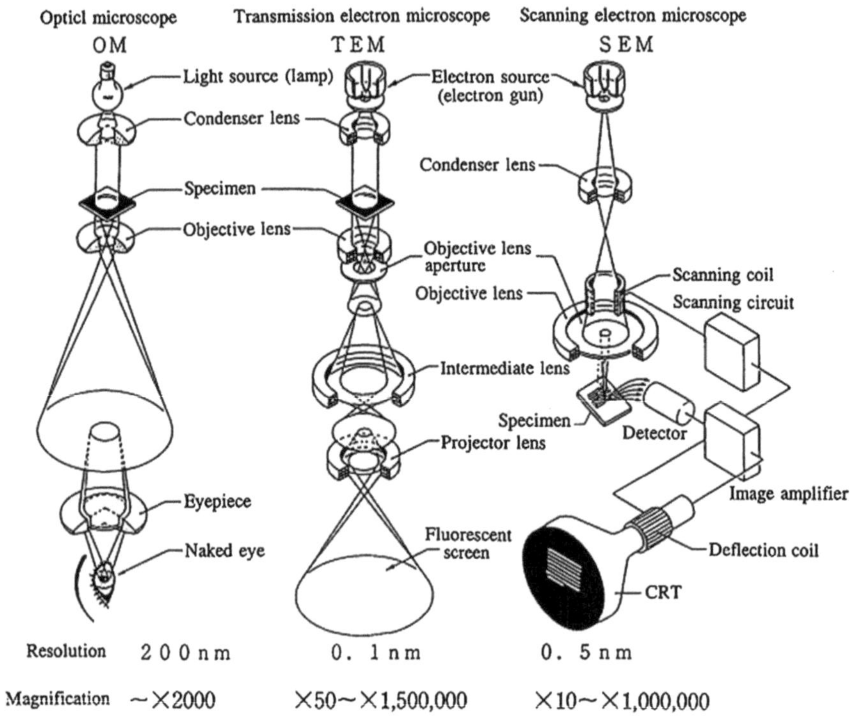

Fig. 3.13 Schematic diagram of an optical microscope (*left*), transmission electron microscope (*middle*) and scanning electron microscope (*right*). Image by permission of Jeol Ltd

differences in the birefringence and thickness of the specimen. This technique allows the observation of grains, grain orientation and thickness.

A basic light microscope uses a light source, usually an incandescent lamp, to illuminate the specimen, a nose piece to hold 4–5 objective lenses used in changing the viewing magnification, an aperture diaphragm to adjust the resolution and contrast, a field diaphragm to adjust the field of view, an eye piece to magnify the objective image and a x–y stage for manipulating the specimen position.

The light reflected off a sample forms the image we get looking down a reflection microscope. On the contrary, transmission light microscopes are used to look at very thin sections of a specimen that must be able to transmit the light. However, both types of microscope are used in very similar ways. The specimen is mounted and placed on the sample stage (Fig. 3.13) and the investigation usually begins by slowly increasing the power of the light source until there is a bright spot visible on the sample. Then, the magnification and the objective lenses are appropriately focused in order for the specimen to be studied. When a representative area is found and focused, a digital camera can be used to take a photo.

Optical microscopes are commonly classified as either low-power or high-power microscopes. Low-power microscopes are those which typically magnify the

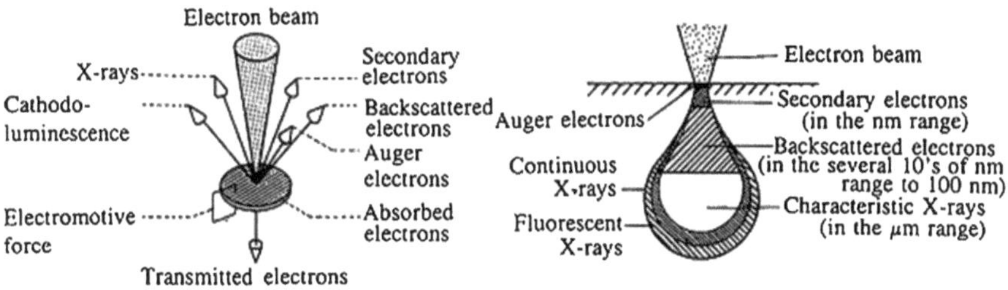

Fig. 3.14 Types of interactions of an electron beam with a specimen and the depth at which they occur. Image by permission of Jeol Ltd

specimen at 5× to 60×, although some can magnify up to 100×. High-power microscopes, on the other hand, typically magnify the specimen at 100× to 1,000×.

The use of an optical microscope is always the first step towards the characterisation of objects of artistic and archaeological interest. Based on that observation, the decisions about further methods that need to be employed are made.

3.2.3.2 The Scanning Electron Microscope

Scanning Electron Microscope (SEM) is often referred as "a tool to observe an invisible tiny object in a stereographic image with a magnified scale" and covers a wide range of magnification, roughly from 10× to 1,000,000×. Its most important advantages are the ease of magnification change, a large depth of field (depth of focus) and the stereographic image display.

Scanning electron microscopy has been extensively used for the material characterisation of objects of artistic and archaeological importance, especially in combination with energy dispersive X-ray microanalysis (SEM/EDS). The various kinds of signals that can be obtained by a SEM, the different types of information they carry, and their main applications are shown in Fig. 3.14 (Table 3.1).

In detail, as presented in Fig. 3.13, a typical SEM consists of: (a) an electron source (also called an electron gun) which produces an electron beam which is accelerated down the column; (b) a series of lenses (condenser and objective) controlling the diameter of the beam and focusing it on the specimen; (c) a series of apertures (micron-scale holes on a metal film) which the beam passes through and which affect properties of that beam; (d) controls for specimen position and orientation; (e) an area of beam/specimen interaction that generates several types of signals that can be detected and processed to produce an image or a spectra. The entire electron pathway is maintained at high level vacuum. The upper column is kept in higher vacuum than the specimen chamber.

The SEM technique requires electrically conductive specimens. Insulating specimens need to be covered by a thin (1–10 nm) conducting surface layer,

Table 3.1 SEM signals and respective information carried

Signal	Information carried
Secondary electrons	Topographical observation of surface
Backscattered electrons	Compositional observation of surface
X-rays	Elemental analysis of specimen
Transmitted electrons	Internal structure observation
Cathodoluminescence	Internal characteristics observation
Electromotive force	Internal characteristics observation

usually applied by sputtering with gold or carbon in an argon atmosphere. Other metals, such as Pt or Au–Pd alloys, may also be used.

When scanning a specimen's surface with a finely focused electron beam in both X- and Y- directions, secondary electrons and backscattered electrons are emitted, carrying information. These electrons, as well as photons in case an EDS detector is available, are detected by the instrument's detectors and the information is converted into an electric signal, amplified and then fed into an observation monitor as a light spot. The spot is produced in real time as the electron beam scans the specimen's surface, allowing the information emitted from the specimen surface to be simultaneously displayed as an image. The type of information obtained can be changed by switching the view mode. In this way, specific desired characteristics of a specimen's surface can be seen on the monitor in a magnified scale. The whole system is controlled by a computer, rendering the interaction of the user much more efficient.

The collection of X-ray photons produced from the interactions of the electrons of the beam with the specimen is another significant source of information that can be exploited when the SEM technique is associated with an energy dispersive X-ray spectrometer (EDS). By analysing the characteristic X-rays generated, the elements that constitute the specimen can be identified, so quantitative analyses and calculations are possible.

Mainly, three signals are measured in the SEM providing different information about the sample: secondary electrons, backscattered electrons and X-rays. Ultimately, image formation in a SEM is dependent on the acquisition of signals produced from the interaction of the specimen and the electron beam. These interactions can be broken down into two major categories:

1. those resulting in *elastic* collisions of the electron beam on the sample, where total kinetic energy of electrons after the encounter is equal to their total kinetic energy before the encounter; and
2. those resulting in *inelastic* collisions, where kinetic energy is not conserved throughout the encounter.

In addition to those signals that are utilised to form an image, a number of other signals are also produced when an electron beam strikes a sample (Fig. 3.14). We will be referring to the illumination beam as the "primary electron beam". The electrons that comprise this beam are thus referred to as *primary electrons*. Upon contacting the specimen surface, a number of changes is induced by the interaction of the primary electrons with the molecules and atoms contained in the sample. Upon contacting the surface of the specimen, most of the beam is not immediately bounced off in the way that light photons might be bounced off in a light dissecting microscope. The electrons with sufficient energy penetrate into the sample for some distance before they encounter an atomic particle with which they collide. In doing so, the primary electron beam produces what is known as a region of primary excitation. Because of its shape, this region is also known as the *tear-drop zone*, which is schematically presented in Fig. 3.14. A variety of signals are produced from this zone, and it is the size and shape of this zone that ultimately determines the maximum resolution of a given SEM working with a particular specimen. The various types of signals produced from the interaction of the primary beam with the specimen include secondary electron emission, backscatter electrons, Auger electron, characteristic X-rays and cathode luminescence.

The most widely utilised signal produced by the interaction of the primary electron beam with the sample is the secondary electron emission signal. A secondary electron is produced when an electron from the primary beam collides with an electron from a specimen atom and loses energy to it. This will serve to ionise the atom and in order to re-establish the proper charge ratio following this ionisation event, an electron may be emitted. Such electrons are referred to as *secondary electrons*. Secondary electrons are characterised from other electrons by having an energy <50 eV. This is by far the most common type of image produced by modern SEMs. It is most useful for examining surface structure and provides the best resolution image among the various scanning signals. Depending on the initial size of the primary beam and various other conditions (composition of sample, accelerating voltage, position of specimen relative to the detector) a secondary electron signal can resolve surface structures down to the order of 10 nm or better. The topographical image is dependent on how many of the secondary electrons actually reach the detector. Although an equivalent number of secondary electrons might be produced as a result of collisions between the primary electron beam and the specimen, secondary electrons that are prevented from reaching the detector will not contribute to the final image and these areas will appear as shadows or darker in contrast than those regions that have a clear electron path to the detector. One of the major reasons for sputter coating a non-conductive specimen is to increase the number of secondary electrons that are emitted from the sample. In order to detect the secondary electrons that are emitted from the specimen a specialised detector is required. This is accomplished by a complex device that first converts the energy of the secondary electrons into photons. It is referred to as a scintillator-photomultiplier detector. The principle component that achieves this is the scintillator. The photons that are produced then travel down a light pipe and are amplified into an electronic signal by way of a photocathode and photomultiplier. Thus, the signal produced can be used to control the intensity of brightness on the monitor in proportion to the number of photons originally produced. A photomultiplier tube converts the quantum energy contained within the photon into an electron by a process known as electron–hole replacement. This results in a cascade of electrons and eventually this amplified signal strikes the anode. The anode then sends this amplified electrical signal to further electrical amplifiers. Because secondary electrons are emitted from the specimen in an omnidirectional manner and possess relatively low energies, they must be in some way collected before they can be counted by the secondary electron detector. For this reason, the secondary electron detector is surrounded by a positively charged anode or Faraday cup or cage that has a potential charge on it in the neighbourhood of 200 V. This tends to draw in many of the secondary electrons towards the scintillator.

A second type of electrons, the backscattered electrons, is also produced when the specimen is irradiated with the primary electron beam. Together backscattered and secondary electrons contribute to the signal that reaches the scintillator and form what we refer to as the secondary electron image.

A backscatter electron is defined as one which has undergone a single or multiple scattering events and escapes with an energy greater than 50 eV. Backscattered electrons are produced as the result of elastic collisions with the atoms of the sample and usually retain about 80 % of their original energy. The number of backscattered electrons produced increases with increasing atomic number of the specimen. For this reason, a sample composed of two or more different elements differing significantly in their atomic numbers, will produce an image showing great contrast of the elements despite a uniform topology. Elements that are of a higher atomic number will produce more backscattered electrons and will therefore appear brighter than neighbouring elements. The region of the specimen from which backscattered electrons are produced is considerably larger than it is for secondary electrons, as seen in Fig. 3.14. For this reason, the resolution of a backscattered electron image is considerably less (1.0 μm) than it is for a secondary electron image (10 nm). Because of their greater energy, backscattered electrons can escape from much deeper regions of the sample than can secondary electrons, hence the larger region of excitation. By colliding with surrounding atoms of the specimen, some backscattered electrons can also produce X-rays, Auger electrons, Cathodoluminescence and even additional secondary electrons. The detector for backscattered electrons is similar to that used in the detection of secondary electrons in that both utilise a scintillator and photomultiplier design. The backscatter detector differs in that a biassed Faraday cage is not employed to attract the electrons. Only those electrons that travel in a straight path from the specimen to the detector go towards forming the backscattered image. So that enough electrons are collected to produce an image, many SEMs use multiple backscattered detectors positioned directly or nearly above the specimen. By using these detectors in pairs or individually, backscattered electrons can be used to produce a topographical image that differs from that produced by secondary electrons.

Another class of signals produced by the interaction of the primary electron beam with the specimen comes under the category of characteristic X-rays. When an electron from an inner atomic shell is displaced by colliding with a primary electron, it leaves a vacancy in that electron shell. In order to re-establish the proper balance in its orbitals following an ionisation event, an electron from an outer shell of the atom may "fall" into the inner shell and replace the spot vacated by the displaced electron. In doing so, this falling electron loses energy and this energy is referred to as X-radiation or X-rays. The SEM can be set up in such a way that the characteristic X-rays of a given element are detected and their position recorded or "mapped". These X-ray maps can be used to form an image of the sample that shows where atoms of a given element are localised. The resolution of these X-ray maps is on the order of greater than 1 μm.

A rather new usage of secondary electrons is employed in so-called *Environmental SEMs*. Unlike a conventional SEM, the environmental SEM is designed to image specimens that are not under vacuum. In fact, for an environmental SEM to function properly there must be air or some other gas molecules present in the specimen chamber. The way an environmental SEM works is by first generating and manipulating a primary beam in much the same way as in a conventional SEM. The primary beam then enters the specimen chamber through a pressure limiting aperture (PLA) that is situated beneath the final lens pole piece. This PLA allows the chamber to be kept at one pressure (e.g. 0.1 Atm) whilst the rest of the column is at a much higher vacuum (e.g. 10^{-6} Torr). The primary beam strikes the specimen and produces secondary and backscattered electrons in the same manner as a conventional SEM. The difference is that these secondary electrons then strike gas molecules in the specimen chamber, which in turn produce their own secondary electrons or "environmental electrons". This results in a cascading or propagation effect and greatly increases the amplitude of the signal. It is all of these electrons that are then used as signals by the detector positioned near the final aperture. Because of this unique design, wet or even uncoated living specimens can be imaged in a SEM. There are, however, some significant drawbacks.

3.2.3.3 Transmission Electron Microscope

Transmission electron microscopy (TEM) can provide useful data for the characterisation of the pictorial layers due to the fact that individual, ultrathin sections of the paint sample can be studied. Ultimately, TEM provides information about local structure, by imaging defects such as dislocations, about the average structure, by using diffraction to identify crystal class and lattice parameters and also chemical composition, almost simultaneously. For the art conservation community, TEM serves in the aspects of identifying the minerals present in the sample and also, determining its crystalline structure, combined with the morphological analysis of each layer.

TEM exploits the very small wavelengths of high-energy electrons to probe solids at the atomic scale [2]. However, the use of a TEM microscope requires a highly skilled user and the interpretation of the information gained calls for a good understanding of the processes occurring in the microscope and ultimately, the structure of materials.

A transmission electron microscope generally consists of three parts (Fig. 3.13): an illumination system, the sample stage and objective lens and a projection system.

The illumination system is an electron source located at the top of the TEM column. So, like in SEM systems, an electron gun is the source of electrons that are afterwards accelerated to high energies (typically 100–400 keV) and then focused towards the sample by a set of condenser lenses and apertures. The source is carefully chosen so that the rates of electrons incident on the sample per unit area and leaving the source per unit solid angle (brightness) are maximised. Thus, the maximum amount of information can be extracted from each feature of the sample.

The manipulation of the electron beam is the key to getting information from the sample. This is achieved using electromagnetic lenses. A typical TEM uses a system of two condenser lenses to control the beam incident on the sample. The first lens demagnetises the source, either to increase the brightness or decrease the area of the specimen that is exposed to the beam. A second lens with an aperture above it controls the convergence angle of the beam at the specimen. By using a large number of finely adjustable lenses acting in series, it becomes possible to reduce the effects of spherical aberration drastically. With the computing power available today it is possible to adjust the lenses simultaneously to find the optimal combination of strengths. This has made it possible to construct aberration-corrected microscopes with a resolution better than 1 Å.

In the central section of the microscope, the electron beam interacts with the specimen and the resulting electrons are gathered and focused, ready for further magnification of the desired images. At the sample stage, as the electrons collide with the sample, they may be scattered by several mechanisms, changing the trajectory of the electrons. Scattering may be either *elastic* (conserving energy) or *inelastic* (with energy dissipated as heat). The needed information about the sample can be obtained by measuring the changes to the electrons passing through the sample, either by measuring the angle that they have been scattered through, that is by studying diffraction patterns or images, or by measuring the amount of energy lost.

The specimen is inserted into the path of the electrons. To achieve optimal image resolution, the specimen must be extremely thin, a few nanometers to be exact. This is to minimise multiple scattering of the electrons which decreases the number of electrons detected making it more difficult to deduce information about the sample. It is necessary to align the specimen very accurately with the electron beam to achieve the required images. Common specimen holders allow rotation about two horizontal axes, along with lateral movement.

The objective or intermediate lens collects the electrons leaving the specimen and forms a diffraction pattern in the back focal plane of the lens, and an image of the specimen in the image plane. In conventional TEM, there is the option of magnifying either the image of the sample formed by the objective lens, or the diffraction pattern. The ease with which the user can move between the two modes (imaging mode and diffraction mode) is one of the key advantages of TEM.

In order to view an image, the user should focus the intermediate lens onto the image plane of the objective lens. To view a diffraction pattern, the lens is refocused onto the back focal plane of the lens, where the diffraction pattern of the sample forms and the image may be magnified for viewing on the screen.

After the electrons have passed through the specimen and have been scattered to varying degrees, the information they carry is converted into a macroscopic image. The simplest way of doing this is by simply magnifying the diffraction pattern or image formed of the sample until it reaches the required size for analysis. This is the basis of *conventional TEM* (CTEM). Alternatively, if a very fine beam of electrons is rastered across the sample, the amount of scattering from each point may be measured separately and successively, and an image gradually built up. This technique, requiring no lenses after the specimen, is called *scanning TEM* (STEM).

The projection system magnifies the images or diffraction patterns formed by the specimen and focuses images in the plane of the screen, where the electron density is converted into light-optical images for the user to see. The effects of spherical and chromatic aberrations in the lenses are quite significant at this stage. Especially chromatic aberrations are much more pronounced, since the interactions of the electrons with the sample often absorb energy, so the beam of electrons passing through the projector lenses contains electrons of a much wider range of energies.

Beneath all the lenses there is a phosphorescent screen that glows when it is struck by electrons, displaying the image or diffraction pattern. Differences in the degree of absorption result in the contrast of the image, implying that the information contained in an electron micrograph is solely due to the difference in the flux of electrons through each point in the image. That contrast may result from mass absorption, diffraction or phase differences. The screen is viewed through a lead-glass window.

To obtain lattice images, a large objective aperture has to be selected that allows many beams including the direct beam to pass. The image is formed by the interference of the diffracted beams with the direct beam (phase contrast). If the point resolution of the microscope is sufficiently high and a suitable crystalline sample is oriented along a zone axis, then *high-resolution TEM* (HRTEM) images are obtained. In many cases, the atomic structure of a specimen can directly be investigated by HRTEM. The incident parallel electron beam, ideally a plane wave, interacts elastically while passing through the specimen, and the resulting modulations of its phase and amplitude are present in the electron wave leaving the specimen. Thus, the wave exiting the object contains the information about the object structure. Unfortunately, the objective lens is not ideal but exhibits aberrations that reduce image quality. However, in many cases, the direct observation of atomic-scale structures is feasible, making HRTEM a powerful tool for almost any case study.

In the following, the advantages and limitations of SEM/EDS are presented in four case studies

3.2.4 Evaluation of the Method as Used in Conservation Science

Since its invention and development in the 1930s and 1940s by E. Ruska and M. von Ardenne, the electron microscope (EM) has become an irreplaceable tool for the material characterisation of objects of artistic and archaeological importance.

Cooperation between science and art is taking place providing a scientific base for the gradual development of art history, archaeology, authentication, conservation and restoration of our cultural heritage. Objects made of stone, ceramics, glass, metal as well as pigments in wall paintings have to be examined in order to get answers in questions like: what is the chemical composition, the origin or the authenticity of an artefact, the present state of works of art and the deterioration status.

Under these circumstances, it seems appropriate to present a short overview of the applicability of electron microscopy to the analysis of works of art. In this chapter, four different case studies will be examined; wall paintings and pigments analysis, icons on leather support, coins and archaeological ceramics.

3.2.4.1 SEM–EDS, OM for Pigment Analysis in Wall Paintings: Samples from the Monastery of Mileseva

From the concentrations of the basic elements with SEM–EDS it is possible to identify the different pigment grains contained in each color layer, and to characterise the different grains that give the tone of each color layer. In some cases, (e.g. carbon black), it is not possible to verify the pigments by SEM, and another method, such as optical microscopy or micro-Raman spectroscopy, must be used. In the present case [3], SEM–EDS microscopy confirmed the use of the fresco technique for the construction of the wall painting, identifying calcium in all measurements on the substrate. Optical microscopy allowed the examination of grain shapes in the sample section of the blue color. It consisted of two different layers of almost the same thickness. The first layer contains black and white grains and since the SEM–EDS analysis detected only calcium, they must be from calcium carbonate. From their irregular shape, is deduced that this is made of "wood-coal". The analysis of the upper layer by SEM using energy dispersive spectroscopy detected Ca, Si, Na, Cl and a small percentage of S. From the inorganic blue colours, only Lapis Lazuli contains sulphur into the crystals and for this reason its colour is blue [4]. The variety of the elements and especially the presence of S and Cl guide us to the estimation that the pigment is natural Lapis Lazuli.

Element concentration analysis via EDS ascertained that the green grains are green earth, since K, Al and Si was detected. Among mineral pigments, only green earth contains the above elements, while the absence of Cu excludes the existence of malachite or atacamite.

SEM and OM detected the occurrence of three layers in a gold sample obtained from the area of the angel's halo. A metallic compact layer of argent, an intermediate organic one and third a thin surface layer of gold (Fig. 3.15). Since microscopy is not adequate for the complete identification of organic materials, an analysis using another technique is necessary.

Deterioration status of archaeological materials is of great significance in conservation science and microscopy is a very useful tool for a close observation and identification. In wall paintings two main types of deterioration exist: (a) the alteration of calcite ($CaCO_3$) —which is in high amount in wall paintings—to gypsum

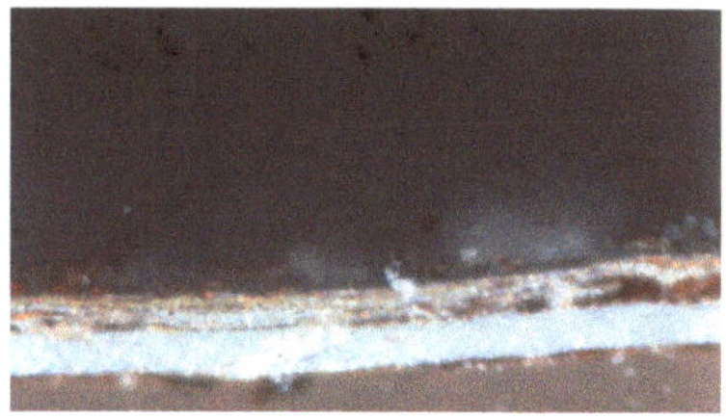

Fig. 3.15 Optical photograph of the gold sample section

Fig. 3.16 Sample of wall-painting affected by deterioration

($CaSO_4 .2H_2O$) due probably to acid air pollution (sulphur oxides); (b) biodeterioration by lichens which apart from mechanical damages also cause chemical damage by the excretion of oxalic acid. In Fig. 3.16 a picture from an affected surface is presented. SEM–EDS analysis (Fig. 3.17) revealed the presence of gypsum.

3.2.4.2 Stratigraphy and Pigment Identification with SEM–EDS and OM Cross-Sectional Analysis. A Post-Byzantine Icon of St Nicholas Painted on a Leather Support

Sample cross-sections may be very simple, containing just one or two colour layers, or very complex, as in the case of over-painting when 10 or more layers may be detected. Furthermore most pigment layers contain a mixture of different pigments species, which means that every grain of pigment must be analysed separately. Electron microscopy is the ideal method for an in-depth research like this.

In the present case [5], electron microscopy is the tool to point out the technological knowledge used for the preparation of an icon or an artefact on a leather support. The icon of St Nicholas is painted on a single piece of leather measuring 56.5×33 cm^2, and was found in the Northern Greek (Macedonian) city of Kastoria (Fig. 3.18). Samples (which measured less than 1–2 mm^2) were removed from the edges of cracks and damaged areas for cross-sectional analysis.

SEM–EDS elemental microanalysis determined the stratigraphy of the icon (Fig. 3.19). The painter covered the whole surface of the leather with a silver leaf in order to create a smooth and durable surface. The leaf was attached to the leather with animal glue (identified with FTIR), while the same organic material was used as a primer between the silver leaf and the pigment layers. In all the

Fig. 3.17 Gypsum as seen via SEM

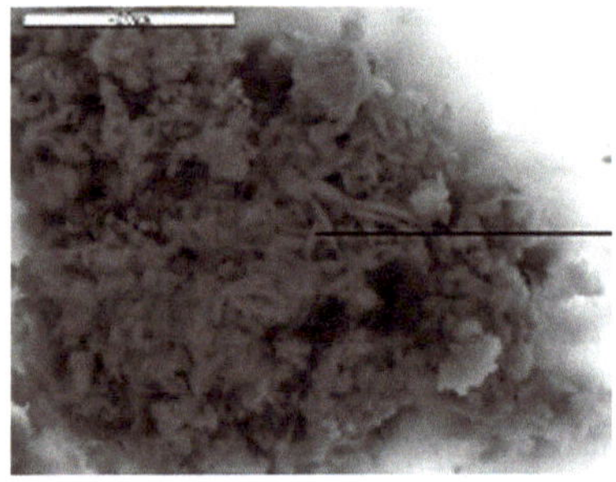

Fig. 3.18 St Nicholas icon

samples, a thin layer of lead white paint was observed immediately above the animal glue primer. Lead white confirmed by SEM–EDS microanalysis, showing nothing but Pb with no more than a trace (<1 % by weight) of Ca. This layer serves the same purpose as the white substratum of gesso on wood panels. Except where smalt is used, the other pigment layers are relatively thin, for better transparency, since the binder used was egg yolk, which is less transparent than other binders. An example of pigment identification with SEM–EDS is sample 4 (Figs. 3.20, 3.21) taken from the forehead area and having a warm red tone.

Optical microscope examination of its cross-section revealed three distinct color layers. An initial white layer of 15–20 μm thickness is followed by a red layer—about 15 μm thick—composed predominately of grains of red pigment ranging in size from 1–5 μm, with white grains apparently constituting the minor component. However, SEM–EDS microanalysis showed a Pb content of 19 %, with Fe being second at 15.4 %. Other elements present were Si (3.7 %), Ca (2.8 %), Al (2.1 %)

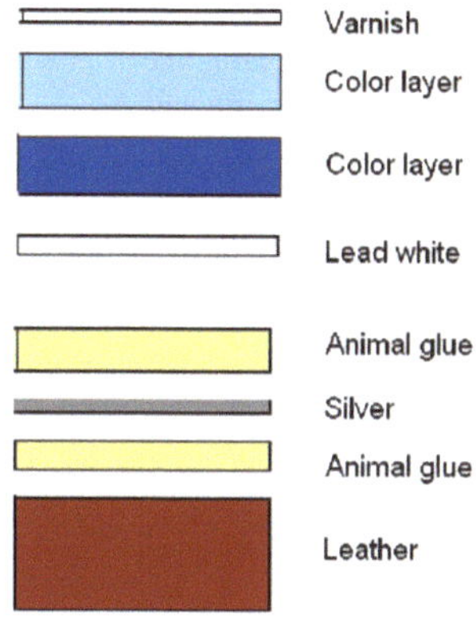

Fig. 3.19 Stratigraphy of the icon of St Nicholas

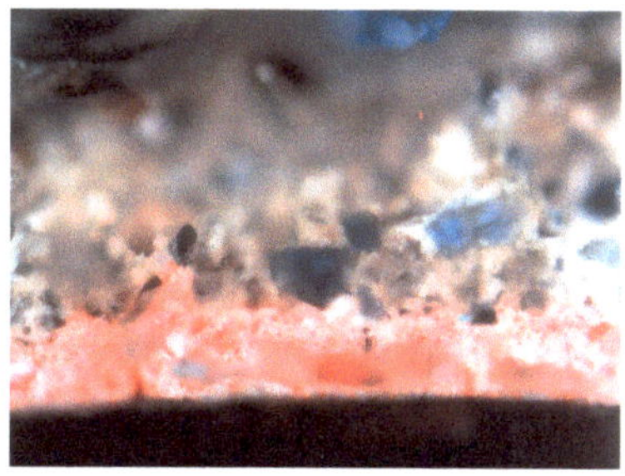

Fig. 3.20 Optical image for sample 4

and a small amount of Mg. The predominant element in the paler red upper layer was Pb (>27 %), followed by Ca (6.6 %), Fe (6 %), Si (2.8 %), Al (0.8 %) and tiny amounts of Mg and K (close to the limits of the analytical method).

3.2.4.3 Microstructure, Morphology and Chemical Composition Determination of Silver Coins with Microscopy

Microscopy is a very powerful tool supporting archaeological research in every field, including numismatics. In this case study, physical and chemical methods were used in order to assist the determination of the composition and the microstructure of ancient coins. As a result, information provided on the alloy employed, the minting and metallurgical technology of the coinage era, (e.g. the era the coin was minted). In addition, information can be obtained concerning the remelting of previously struck coins, since alterations of the microstructure and especially the grain size as well as the existing crystal defects could be indicative of processes such as annealing and quenching. It is also possible to correlate variations in alloy composition with ores so as to determine the origin of different coins, by taking into account trace elements exclusively present in every ore.

Coins of King Monounios were studied [6]. These coins belong to the Kreshpan hoard (third century BC) which was founded in 1982 near the village of Kreshpan,

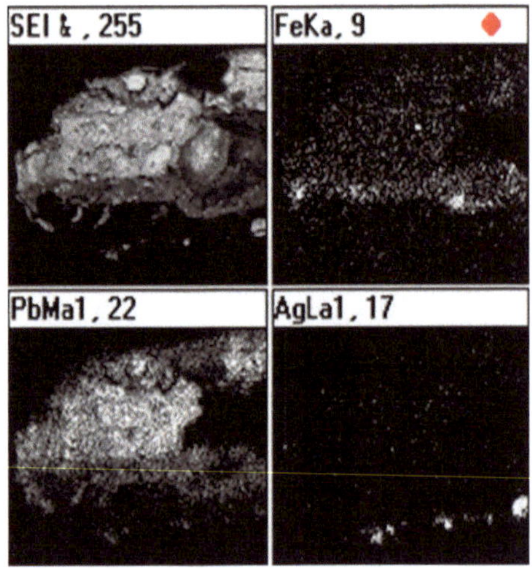

Fig. 3.21 SEM images for sample 4, revealing the presence of Fe, Pb and Ag

located in southern Albania, about 30 km in the southeast of the ancient Greek colony of Apollonia.

For the examination with optical and scanning electron microscopy (SEM), small sections of the samples were cut using a low-speed saw. SEM associated with EDS gave a preliminary estimation of the chemical composition and revealed varying average stoichiometrical compositions of the coins. Besides silver, which is the main component of all coins, copper and gold were also detected.

More information was obtained via SEM micrographs of the coins. In Figs. 3.22, 3.23 is presented the microstructure of coin 7922: In this coin there are no inclusions. The grain boundaries are visible, while twins at high density and slip lines within the grains are observed. Furthermore, the study of crystal defects such as dislocations and twins, which are critical for the determination of the minting method, could be accomplished with precision only with TEM. The presence of micro twins was unambiguously verified with TEM (Fig. 3.24). The formation of twins could be ascribed to hot-working or cold-working and annealing, because the mechanical twinning in face centred cubic (FCC) structured metals is rare, while annealing twins are often formed. Furthermore, the presence of slip lines (arrow 1 in Fig. 3.23) and the bending of certain twins (arrow in Fig. 3.23) are probably due to further cold working after cooling.

Hence, the minting method of this coin could have involved the usage of a manual pressing tool. A piece of silver with the necessary weight after being heated up to glowing was placed between the stamped surfaces of the tool and pressure was applied to form the symbols of the issuing city. Probably the coin was pressed again after cooling (Fig. 3.22).

Fig. 3.22 Optical image of the silver coin #7922

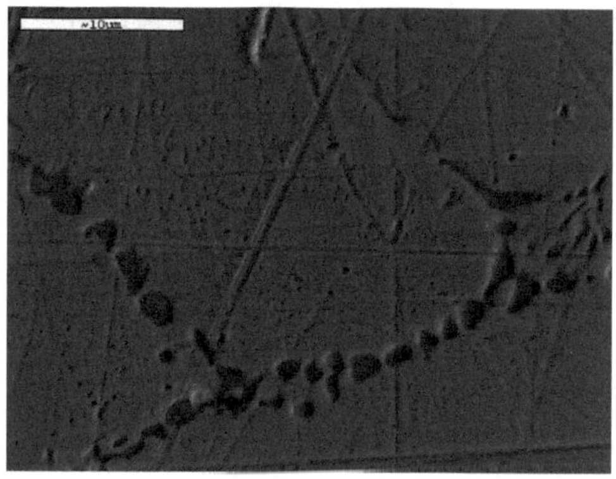

Fig. 3.23 SEM micrograph of the coin's surface

3.2.4.4 SEM, SEM–EDS Analysis in Archaeological Ceramics

Applications of scanning electron microscopy to pottery and ceramics are manifold. The impact of SEM and X-ray microprobe analysis on ceramic studies has been reviewed by Froh [7], who mentions mainly three types of information that can be obtained from SEM: information on the raw materials used, information on firing procedures and information on surface decoration, i.e. the application and nature of slips and glazes.

Most applications made use of polished sections, often sections perpendicular to the surface. SEM has the advantage of better resolution than optical thin section microscopy under a polarising microscope, but has the disadvantage that the identification of individual mineral phases is more difficult. SEM using backscattered electrons is therefore often preferred to secondary electron micrographs, because of its sensitivity to the mean atomic number of individual phase. Information on the firing temperature can be obtained from the presence or absence of vitrification.

Figure 3.25 shows the secondary electron micrograph of the cross-section of a ceramic masterpiece from Vergina (Greece). This micrograph shows the

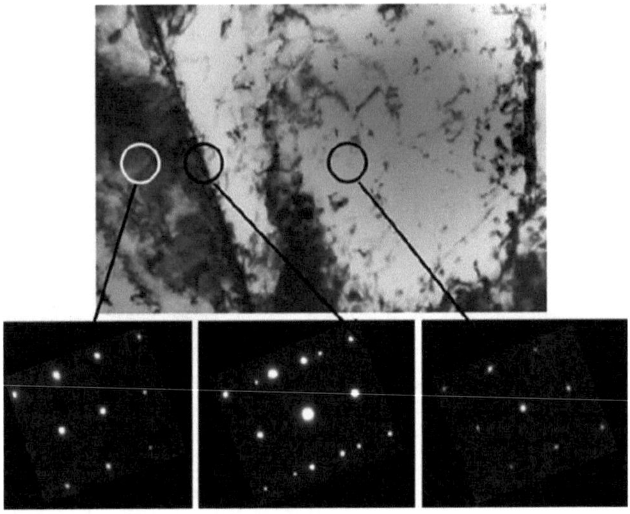

Fig. 3.24 Coin 7922: Bright field TEM micrograph and related electron diffraction patterns of a twin. Inset 1 (*left*) reveals the presence of slip lines, while inset 2 (*middle*) indicates the bending of twins

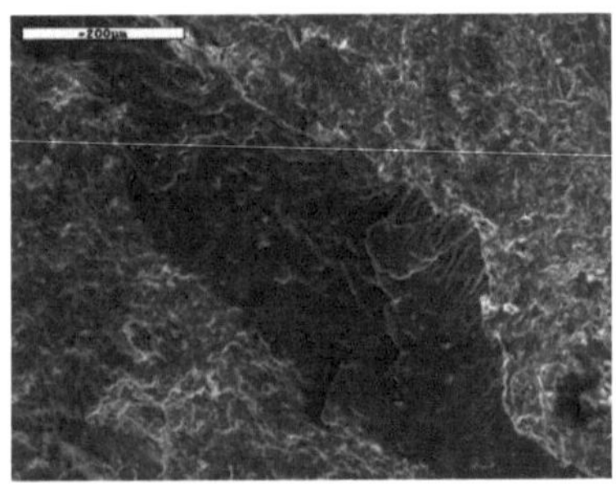

Fig. 3.25 SEM cross-section image of a ceramic sherd from Vergina showing the clay matrix and crystalline inclusions

morphology of the fine grained clay matrix and some rather large inclusions, but it is difficult to identify the nature of the latter, because they exhibit no idiomorphic crystal faces. The inclusions can, however, be identified with the help of EDX. Peaks of Si and O reveal the presence of quartz, while the inclusion identified as a feldspar additionally exhibits the lines of Na, Al and faintly K. Most probably, it is a sodium feldspar ($Na[AlSi_3O_8]$). The micrograph of Fig. 3.26 shows the glaze from the external surface of the ceramic with the pores formed from the firing process, which is indicative of the kiln temperature.

Studies of surfaces, glazes and slips are one of the most promising fields of application of SEM in ceramic studies, especially in the authentication of pottery. Ceramics with glossy black surfaces and replicas very similar to the ancient finds were examined. Both the ancient and the replica pieces have shiny black surfaces that were supposed to be caused by carbon deposits formed during the reducing firing. No slip was applied to the ancient black ware pottery. In the production of the replica pottery, no slip was applied either, but the surface of the dried vessels

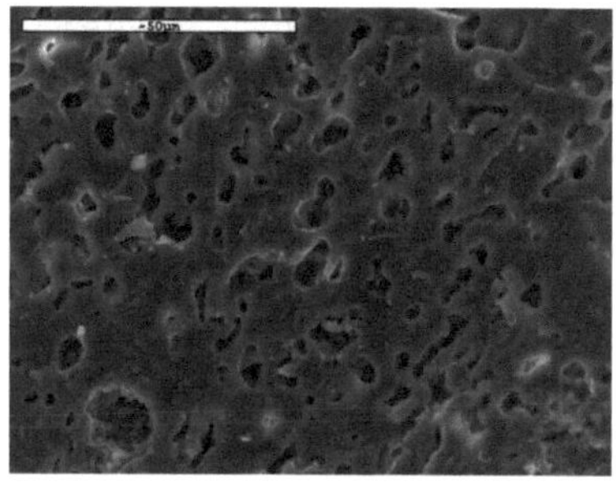

Fig. 3.26 SEM image from the ceramic glaze indicative of the firing process

was very carefully burnished before firing. Furthermore, replicas exhibit a continuous, smooth surface layer on top of the fractured clay matrix. The surface of the ancient sherd is also rather smooth, but rougher than that of the replica although visually both sherds appear glossy black.

3.2.5 Conclusions

Microscopic methods are widely applied for the identification and documentation of the preservation state of archaeological materials. Scanning electron microscopy, especially in combination with energy dispersive X-ray microanalysis, has contributed extensively to the analysis of objects of cultural heritage and furthermore has been an indispensable tool in delineating the degradation processes of ancient as well as modern materials in art and archaeology. However, the results presented imply that scientific investigations should be carried out by a combination of several analytical techniques. There are many cases where microscopy has provided some information for identification purposes in order to exclude some other options, but on many occasions microscopy alone is not adequate for the complete identification of an object and its deterioration (e.g. characterisation of organic materials). In addition, it should always be kept in mind that the so-called non-destructive or non-invasive methods should be preferred to minimise the amount of sample material that has to be taken from an object. Finally, the evaluation and interpretation of the results should be carried out as an interdisciplinary cooperation between the analytical chemist and the art historian, archaeologist or conservator-restorer prior to the analytical procedure.

3.3 Infrared and Raman Spectroscopy

Reiner Salzer

Abstract Infrared (IR) and Raman spectroscopy have a high potential for characterisation of material. Extensive series of wet chemical analysis may be substituted by a single spectroscopic measurement followed by detailed chemometric data evaluation. Topics of this chapter are: (i) basics of IR and Raman spectroscopy, (ii) the registration of "correct" spectra, and (iii) spectra evaluation. Dedicated applications in the area of conservation science are collected in separate chapters. The infrared (IR) spectrum is often called the fingerprint of a substance. An IR spectrum identifies a substance like a human fingerprint. Due to their origin the features of an IR spectrum are bands, not peaks. They indicate vibrations within the molecular framework. Such vibrations are excited by irradiation with infrared light. The infrared spectral range joins the red end of the visible range, it extends from 780 to 1 mm wavelength. Radiation in this spectral range is of low energy, it does not harm material. Commercial IR spectrometers are available since 1945. They dominated the field of structural analysis (identification) of molecular substances until NMR spectrometers became affordable. Nowadays small and easy-to-use IR spectrometers are abundantly available in laboratories [8]. Raman spectroscopy is complementary to IR spectroscopy. It is named after its discoverer, the Indian Nobel Laureate Sir C. V. Raman. Molecular vibrations are excited here by irradiation with intense visible or near IR radiation. The excitation mechanism is different from the excitation of vibrations by IR light, which results in some well-defined differences between IR and Raman spectra of the same sample. For this reason IR and Raman spectra are complementary, not identical. The complementarity of IR and Raman spectra provides additional information about molecular properties of the sample. The need for intense light sources hampered Raman spectroscopy until the advent of lasers. Today a range of miniaturised lasers and convenient fibre-probes are available, which permit the construction of small portable Raman spectrometers for difficult field applications.

3.3.1 Origin of IR and Raman Spectra

We may imagine a molecule as a geometrically arranged system of tiny mass pieces—the atoms. These mass pieces are kept in equilibrium position by connecting springs—the chemical binding forces. A system consisting of n mass pieces disposes of $3n-6$ (linear molecules: $3n-5$) *vibrational degrees of freedom*. The corresponding vibrations are called *normal vibrations*, i.e. the vibrational movements of the respective *oscillators* are fully independent of each other and cannot be displayed as linear combination of other vibrations (normal vibrations are orthogonal to each other) [9].

The *frequency* of a normal vibration is determined by the atomic masses and the binding forces. A vibration can be excited by the oscillating field of electromagnetic radiation if the normal vibration induces a change in the molecular dipole moment (change in size or direction) or a change in the polarisability of the molecule. A change of the molecular dipole moment is required to induce IR transitions, a change of the molecular polarisability in order to excite Raman transitions. Whether or not a normal vibration involves changes in dipole moment or polarisability depends on the *symmetry properties* of the molecule. Some symmetry properties may cause mutual compensations of bond dipole changes of particular vibrations. Such vibrations do not appear in the spectrum, they are IR inactive or Raman inactive. The simplest example is the linear molecule CO_2:

symmetric stretching vibration ν_s:

$$O \leftarrow C \rightarrow O \quad \text{IR inactive} \quad \text{Raman active}$$

antisymmetric stretching vibration ν_{as}:

$$O \leftarrow C \leftarrow O \quad \text{IR active} \quad \text{Raman inactive}$$

deformation vibration δ:

$$O{-}C{-}O \quad \text{IR active} \quad \text{Raman inactive}$$

Stretching vibrations are characterised by changing bond lengths. Adjacent bonds of identical type may oscillate in phase (*symmetric stretch*, see CO_2 above) or out of phase (*antisymmetric stretch—not asymmetric!*). *Deformation vibrations* cause changes in bond angles. Deformations vibrate at lower frequencies than stretches. Deformations can be subdivided into *scissoring, rocking, wagging* and *twisting* vibrations. The most common abbreviations are str for stretching and def for deformation vibrations.

In case of highly symmetric molecules, individual normal vibrations may have identical frequencies and thus form a single absorption band (*degenerate vibrations*, e.g. the two orthogonal deformation vibrations of CO_2 above). The appearance of degenerate vibrations often results in simpler spectra for symmetric molecules compared to spectra of molecules with the same number of atoms but less symmetric arrangement. In case of molecules with centre of symmetry the mutual exclusion principle is in effect: "IR active vibrations are Raman inactive and vice versa".

Every active, non-degenerate normal vibration exhibits its own absorption band, the so-called *fundamental* (Fig. 3.27). In case of oscillators comprising very different atoms the vibration may be more anharmonic. *Anharmonic oscillators* give rise to *overtones* and *combination* vibrations in addition to fundamentals (Fig. 3.28). The latter are usually distinctly more intense than the former. Accidental proximity of overtone/combination levels and stretching levels may result in resonances, which are accompanied by altered intensity distributions (Fig. 3.28).

The entirety of all features yields the *spectrum*. Abscissa units of IR or Raman spectra are *wavenumbers* (cm^{-1}). The ordinate scale of IR spectra from collections is usually transmittance, preferred for sample identification due to the overestimation of smaller spectral features. For quantitative evaluations the ordinate scale of IR spectra must be transformed from *transmittance* to *absorbance* (*Beer Lambert law*), because only absorbance values are linearly related to concentrations. The ordinate of Raman spectra (now usually *intensity* or *photon counts*) is always linearly related to concentrations.

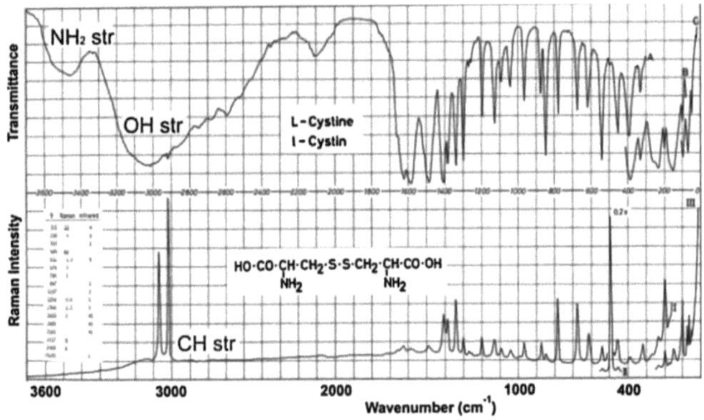

Fig. 3.27 IR spectrum (*top*) and Raman spectrum (*bottom*) of L-Cystine (from [10], with permission). The rich structured region below 1,300 cm^{-1} is called fingerprint region. Identical spectra in this region indicate identical samples. The characteristic bands above 1,300 cm^{-1} permit quick screening tests. Polar groups show more pronounced bands in their IR spectra, non-polar groups in their Raman spectra. Structural units with strong inter- or intramolecular interactions like –OH or $–NO_2$ groups give rise to very broad bands

Each individual spectral band is characterised by three parameters:
frequency or position, depends on the vibrating masses and the bond strengths;
intensity (band height as rough approximation), depends on the size of the changes in molecular dipole moment or polarisability during the vibration;
band shape (width at half height as rough approximation), depends on the surroundings of the oscillator.

Raman spectra provide an additional parameter, the *depolarisation ratio* ς. When an isotropic molecule (spherical electron cloud) is irradiated by polarised laser light, the excited Raman light maintains the polarisation direction of the initial radiation. No Raman intensity is measured in the direction perpendicular to the initial polarisation direction. In case of anisotropic molecules, the non-spherical electron cloud causes vectorial splitting of the excited Raman light into parallel and perpendicularly polarised components $I_\perp$ and $I_\parallel$ (Fig. 3.29). The depolarisation ratios $= I_\perp/I_\parallel$ of a non-symmetric vibrational mode equals 0.75, and that of the other modes <0.75. The depolarisation ratio permits easy assignment of a band to a symmetric or a non-symmetric vibration.

The vibrational spectra of two molecules carrying the same atoms but in different order (*isomers*) must be different because the individual oscillators are different. If particular atoms are replaced by *isotopes* the positions of the corresponding bands are shifted according to the isotopic mass difference. Polar molecules or groups usually give stronger IR signals, non-polar molecules or groups usually give stronger Raman signals. This different behaviour is a result of the different excitation schemes for IR and Raman spectra. It provides access to extra information about the molecular structure of the sample.

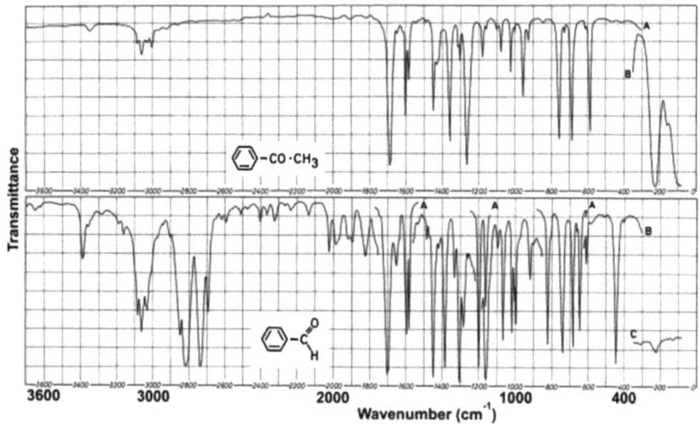

Fig. 3.28 IR spectra of acetophenone (*top*) and benzaldehyde (*bottom*) (from [10], with permission). The small feature below 3,400 cm^{-1} is the first harmonic of the strong C = O stretching fundamental of acetophenone at 1,685 cm^{-1}. This harmonic must not be confused with a –NH2 stretching fundamental in the 3,400 cm^{-1} range. Benzaldehyde has less C–H bonds than acetophenone but exhibits distinctly stronger C–H str features. The reason is a harmonic of the aldehydic C–H def in accidental proximity of the aldehydic C–H str fundamental. Both vibrations coupled particularly strong (Fermi resonance). This enormous resonant intensity enhancement is a very convenient indicator for aldehyde groups

IR and Raman spectra of gases are characterised by closely spaced sharp signals, the so-called *rotational fine-structure*. These sharp features broaden and merge into a composite band upon increasing intermolecular interaction caused by increasing pressure, by more polar solvents, by condensation or solidification. Very strong interactions—like hydrogen bonds—may not only result in band broadening but also shifting of the band position (Fig. 3.30).

IR and Raman spectra complement each other during sample identification, but they may be used independently. Complementarity provides additional certainty.

According to their physical origin, features in IR and Raman spectra are bands, not peaks.

Stretching vibrations like that one in CO_2 are symmetric or antisymmetric, by no means asymmetric.

It is of crucial importance to estimate intensity, height and width of an IR band with the ordinate scale in absorbance, not in transmittance.

Chemically equivalent groups within the same molecule exhibit a common absorption band (e.g., >CH_2 in hydrocarbons). Groups with heteroatoms (e.g., = C–O, –COOH, –NO_2) exhibit *characteristic bands*, whose frequency is largely independent of the rest of the molecule (*group frequencies*). The observation of a particular group frequency is a confident indicator for the presence of the corresponding structural group in the sample. Slight shifts of the group frequency indicate finest alterations in the binding between the structural group and the remainder of the molecule (different substituents, different geometries). The

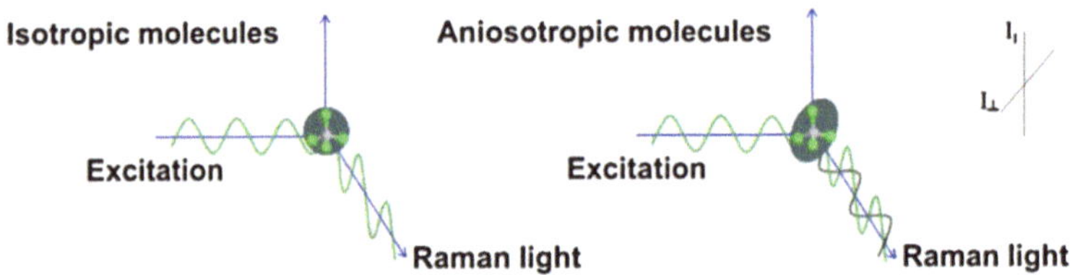

Fig. 3.29 The depolarisation ratio is a unique parameter of a Raman band. Isotropic molecules preserve the polarisation direction of the excitation radiation, anisotropic molecules excite Raman light of parallel and perpendicular polarisations

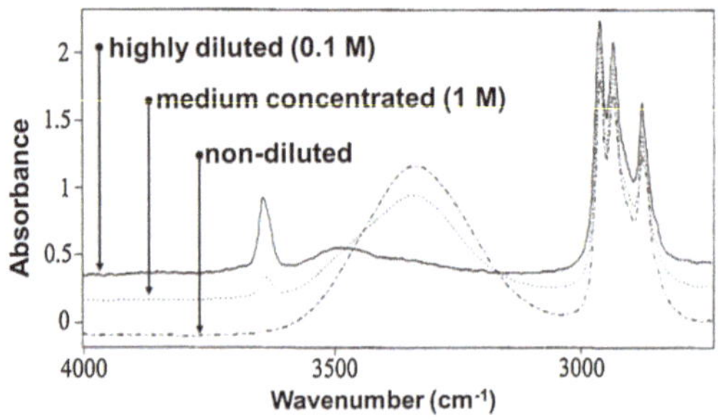

Fig. 3.30 Polar –OH groups may form hydrogen bonds, which result in redistribution of electron density, i.e. reduced O–H bond strength and hence reduced O–H oscillator frequency. In the non-diluted case all *n*-butanol molecules are in close proximity, and all –OH groups are engaged in hydrogen bonds. The distance between *n*-butanol molecules grows upon dilution by CCl_4. More and more –OH groups remain isolated, indicated by the rising size of the band of the OH monomers at 3650 cm^{-1}. The intensity ratio between monomer and H-bonded bands provides quantitative assessment of the interaction strength

most frequently used group frequencies of organic molecules are located between 4,000 and 1,300 cm^{-1}, whereas frequencies of the molecular skeleton are found in the so-called *fingerprint region* from 1,300 to 700 cm^{-1}. If heteroatoms are involved or in case of inorganic compounds group frequencies usually appear below 1,400 cm^{-1} [11].

Vibrational spectra are a kind of molecular spectra because chemical bonds are key constituents of any molecular oscillator. Multi atomic ions like CO_3^{2-} contain bonds as well, they exhibit characteristic IR and Raman spectra. In case of inorganic ions the oscillating masses are usually higher, hence the corresponding vibrational frequencies are lower. Moreover, ions are charged. The polar environment may result in broadened bands, in particular in case of IR spectra. Group frequencies of ions are often slightly shifted by the polar field of their counter ions. Such slight shifts permit the *identification of minerals* from the group frequency of one of the ionic partners (Fig. 3.31). Already crystal seeds or nanocrystals can be identified, whereas X-ray diffraction requires a distinctly developed crystal lattice.

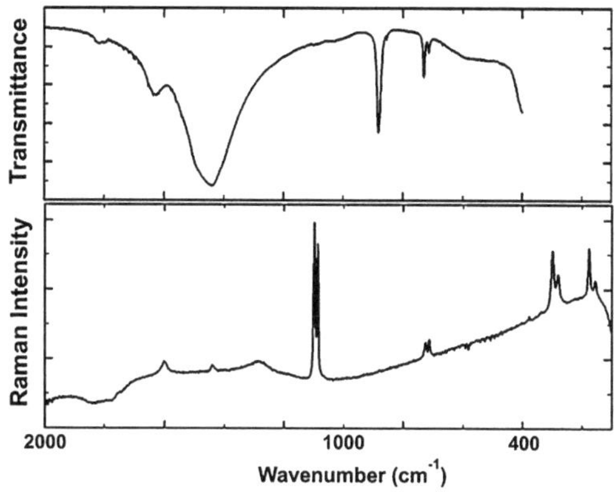

Fig. 3.31 IR spectrum (*top*) and Raman spectrum of limestone (origin: Acropolis/Greece). The very sharp Raman features indicate presence or absence of closely related minerals: calcite ($CaCO_3$) is indicated by its band at 1,086.5 cm^{-1}, dolomite ($CaCO_3{*}MgCO_3$) by its band at 1,098.6 cm^{-1}. There is no indication of magnesite ($MgCO_3$) at 1,095.1 cm^{-1}

Vibrational bands exhibit identical positions in IR and in Raman spectra, even though the overall shape of the two spectra looks different because of the different excitation schemes.

IR spectroscopy is superior for polar compounds, Raman spectroscopy for non-polar compounds.

Raman spectroscopy permits the identification of microcrystalline samples, even of X-ray amorphous samples.

The fingerprint capability of IR and Raman spectra provides easy identification of compounds from reference spectra.

Vibrational spectra usually do not permit the complete constitutional analysis of an unknown molecule, but the presence or absence of a particular constituent or structural group based on its group frequency can often be achieved. This forms the basis for reliable, quick *screening tests* even under field conditions.

3.3.2 Registration of "Correct" Spectra

It is a common misconception that the performance of the analytical instrument—here the spectrometer—determines the quality of the measured data, expressed as *accuracy and reproducibility*. Ninety-five percent of all errors occur before the sample is inserted in the spectrometer, 5 % of all errors are contributed after the spectrum was measured, during evaluation and interpretation. The dominating error contributions before insertion of the sample into the spectrometer arise during the initial steps of the analytical process: fuzzy description of the research target, errors during sample collection, inappropriate transfer to the laboratory, wrong sample preparation.

Conventional *monochromators* are widely replaced in current spectrometers: (i) by *interferometers* in case of infrared (Fourier transform or FT-IR spectrometers) and (ii) by *polychromators* in case of Raman spectrometers. Regardless the internal processes in these instruments, the user needs to adjust just two parameters for the spectral analysis, *spectral resolution* and number of accumulations. The

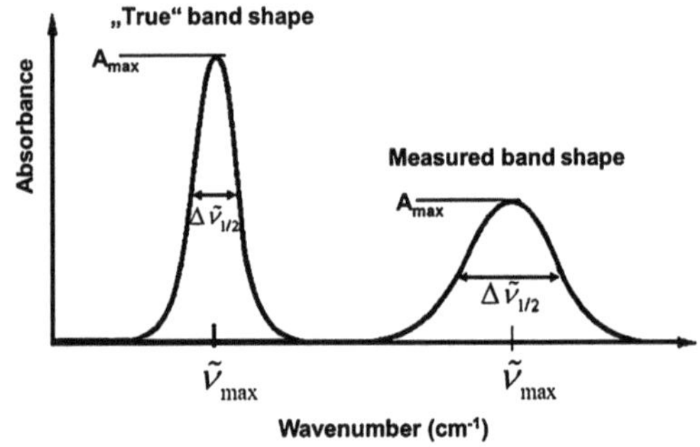

Fig. 3.32 Broadening of the registered band shape due to insufficient spectral resolution

correct spectral resolution should be smaller than the widths of the smallest spectral band to be measured. If a too large value was chosen for the spectral resolution, narrow spectral bands shall become broadened (Fig. 3.32). Band broadening may result in merging of spectral bands, i.e. in loosing spectral information. The second parameter, number of accumulations of scans, controls the *signal-to-noise (S/N) ratio* in the measured spectrum. Too low *S/N* may prevent small spectral features from being identified. *S/N* depends on the square root of the number of scans *n*:

$$S/N \sim n^{1/2}.$$

The number of accumulations is limited by the required measurement time: the measurement time grows linearly with *n*, *S/N* only as square root. The initial *S/N* depends on experimental conditions and on the chosen resolution. Lower spectral resolution improves *S/N*, hence spectral resolution and *S/N* have to be optimised with respect to each other.

The operation of IR spectrometers is much simpler than that one of Raman spectrometers. A particular feature of Raman spectroscopy is the *excitation wavelength*. Until recently a green laser line was preferred. Its short wavelength ensures high intensities of the Raman bands. Unfortunately, very strong *fluorescence* may be excited at the same time, which covers all Raman signals. Modern Raman spectrometers employ lasers emitting in the NIR, where fluorescence can hardly be excited. The reduced Raman intensity upon NIR excitation is compensated by new detectors of enhanced sensitivity.

Errors are usually caused by inappropriate sample preparation, not be spectrometer malfunction.

Two registration parameters are important: spectral resolution and number of accumulations.

Excitation of Raman spectra by NIR lasers usually avoids superimposition by strong fluorescence.

Raman spectroscopy hardly requires *sample preparation*, because the exciting laser light can directly be shone on the sample spot. In contrast, IR spectroscopy usually requires dedicated sample preparation (Fig. 3.33) [12].

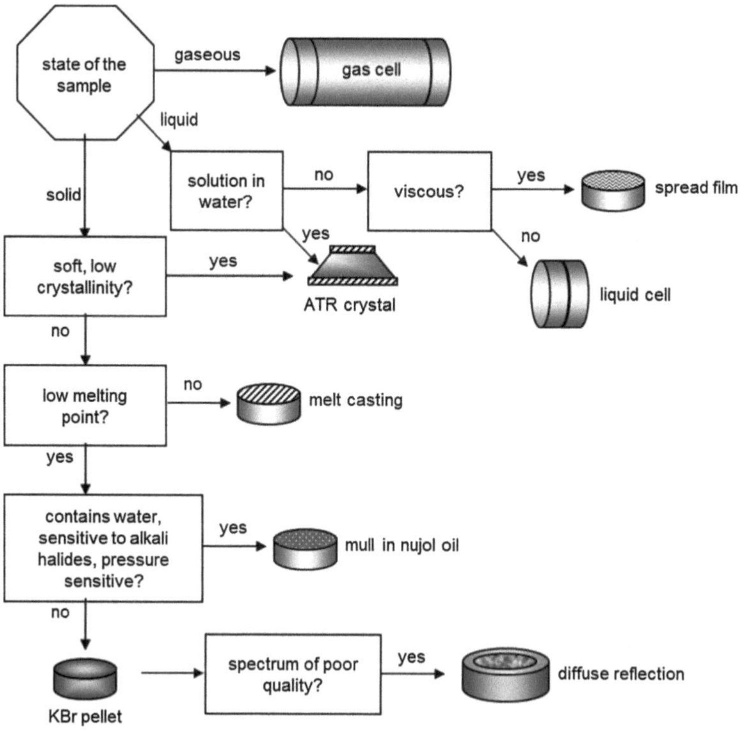

Fig. 3.33 Sample preparation techniques in IR spectroscopy (from [12], with permission)

Sample preparation in IR spectroscopy is needed because the amount of light absorbed by the sample cannot be measured directly. Instead, we measure the amount of incident light (I_0) and the remaining light intensity after the sample (transmitted (I_T), reflected (I_R) or scattered (I_S)). As only one of the parameters (I_T, I_R or I_S) can be measured in a particular experiment, the remaining two have to be zero. This is the goal of sample preparation: a sample measured in transmission must be clear (no *scatter losses*) and have a flat surface normal to I_0 (no *reflection losses*), a sample measured in diffuse reflection must not transmit light nor have reflection losses at its front surface.

Scatter losses occur in cases of inappropriate selection of cell windows or mulling agents. This shall be illustrated for a series of inorganic oxides, which were prepared as pellets using either KBr of TlBr as mulling agent (Fig. 3.34). The *refractive indicesn* of the oxides covers the range between 1.4 and 3.0. The refractive indices of the mulling agents are $n = 1.55$ for KBr and $n = 2.37$ for TlBr. The pellets show the highest transmission (least scatter losses, least measurement errors) when the refractive indices of the oxide and the mulling agent coincide (Al_2O_3/KBr and HgO/TlBr), whereas the pellets made of components of

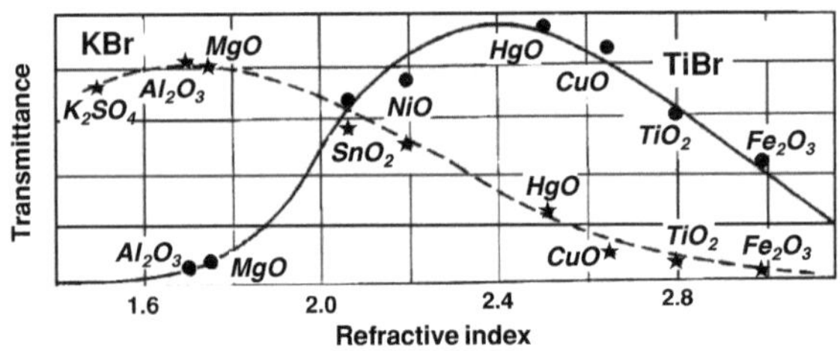

Fig. 3.34 A series of metal oxides prepared as pellets with either KBr or TlBr as mulling agent. The oxides are arranged according to their refractive indices *n* on the abscissa and the IR transmittance of the pellets on the ordinate. The pellets embedded in KBr are connected by the broken line, the pellets in TlBr by the solid line

very different refractive indices become almost transparent (Al_2O_3/TlBr and Fe_2O_3/KBr). Similar scatter losses are also observed for liquid samples when the refractive index of the window material of the liquid cell does not match the refractive index of the sample.

Sampling preparation involves appropriate changes in the physical state of the sample (e.g., melting or grinding) or dilution by a solvent, dispergent or mulling agent. These substances often contribute their own features to the measured IR spectrum. If these features are very strong, they practically absorb all light and thus all information around their locations ("dead" spectral regions).

Weaker features can be compensated by inserting a reference cell R parallel to the sample cell S in the spectrometer's sample compartment. The measured transmission spectrum T is than automatically calculated from the measured intensities: $I_T = I_R/I_S$. This procedure requires an extremely delicate adjustment of the layer thickness of the reference cells. Otherwise the measured spectrum may be much deteriorated by in compensations or even by newly introduced impurities (Fig. 3.35).

Attenuated total reflection (ATR) gained much popularity in IR spectroscopy. It permits easy measurement of liquids (also aqueous), of very viscous samples and even of solids (optical flat or soft surface). Perfect optical contact is required between the ATR crystal and the sample, because the IR light must penetrate the interface to the sample upon reflection inside the ATR crystal. A perfect optical contact presupposed, the penetration depth of the IR light into the sample depends on its wavelength. Long wavelengths penetrate deeper, hence the spectral bands at the long-wavelength (low-frequency) side of the IR spectrum are always over-emphasised (Fig. 3.36). The spectrometer software usually provides a correction algorithm, which render ATR spectra comparable to transmission spectra.

The corrected ATR spectrum in Fig. 3.36 exhibits very noisy portions (above 3,200 cm^{-1} and above 2,000 $cm^{-1)}$. These features are not due to electronic noise but to in compensation of atmospheric water vapour. Accessories like the ATR unit distort the balance between sample and reference beams and make the water residues appear. The removal of water residues from the IR spectra usually requires purging of the spectrometer with dry air. The necessary equipment is a common part of IR spectrometers.

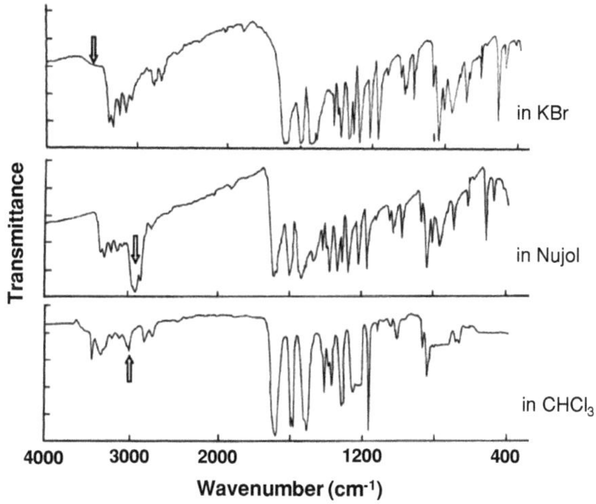

Fig. 3.35 Influence of sample preparation on the measured IR spectrum: An organic solid (p-acetamidobenzaldehyde) prepared as KBr pellet (*top*), dispersed in Nujol (*mid*) and dissolved in $CHCl_3$. Note the distinct differences in the spectral region between 3,600 and 2,400 cm^{-1}. The KBr spectrum shows the stretching bands most clearly but introduces water impurities (broad band at the high frequency wing) and causes stay light losses (distinct slope of the spectral background above 2,000 cm^{-1}). Nujol introduces a dominating CH str band below 3,000 cm^{-1} and a stray light slope. $CHCl_3$ exhibits a CH str in compensation below 3,000 cm^{-1} and extinguishes most of the bands in the region ("dead" region)

IR spectroscopy is experimentally simpler than Raman spectroscopy but requires sample preparation.

IR and Raman *microscopes* often do not require sample preparation and permit spatial resolution down to the diffraction limit.

Window material and mulling agents are chosen according to their transmission range and their refractive index.

The compensation of solvent bands requires particular care in order to avoid severe distortions of the measured IR spectrum.

The ATR technique is very convenient but requires mathematical compensation of varying penetration depths.

3.3.3 Evaluation of Spectra

Modern spectrometers are able to produce thousands of spectra in a few seconds. This applies in particular to spectroscopic imaging [13]. Such huge numbers of spectra can only be evaluated by automatised routines. The corresponding results are only trustworthily if sophisticated tests for the correctness of automatised evaluations are implemented. Implementations of this kind are dedicated to the particular experiment and require experienced spectroscopists.

The information content of a spectrum is collected during the measurement. After the registration of the spectrum is completed, the information content cannot

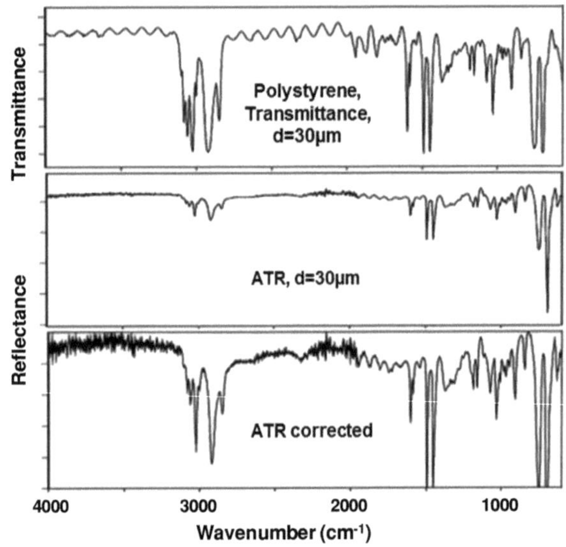

Fig. 3.36 The ATR spectrum of polystyrene (*mid*) shows identical band locations as the transmission spectrum (*top*) but the band intensities are severely distorted due to the varying penetration depths in the ATR configuration. A mathematical function can be used for correction (*bottom*)

anymore be raised, but it can easily be reduced. Some algorithms help to present the results more clearly. This does not mean the information content is raised. Some procedures are reversible, but some change (reduce) the information content irreversibly. Noncritical application of easily accessible software poses a severe thread to the quality of the investigations.

Spectra Treatment

Common tasks for spectra treatment may be: smoothing of noisy spectral regions, removal of a sloping spectral background, narrowing of overlapping spectral bands. All these problems can be solved by the Savitzky-Golay algorithm. This convenient and versatile approach is based on a linear equation with coefficients:

$$O_t = k_1 O_{r;1} + k_2 O_{r;2} + \ldots + k_n O_{r;n}$$

O_t is the smoothed ordinate value. k_n are coefficients, which depend on the treatment task. $O_{r;\,n}$ are n registered subsequent ordinate values, the so-called data window. n is always odd. In case of smoothing, all ordinate values inside the window are added. The sum is normalised by appropriate coefficients. The signal is not altered during this process but the noise is reduced according to its statistical nature. The wider the window the stronger smoothing (noise reduction) is achieved. The calculated value O_t replaces the middle point in the window, the window is shifted by one point and the procedure repeated until the other end of the spectrum is reached. Unfortunately, the linear smoothing function flattens spectral bands as well (Fig. 3.37). This may introduce severe error in quantitative evaluations.

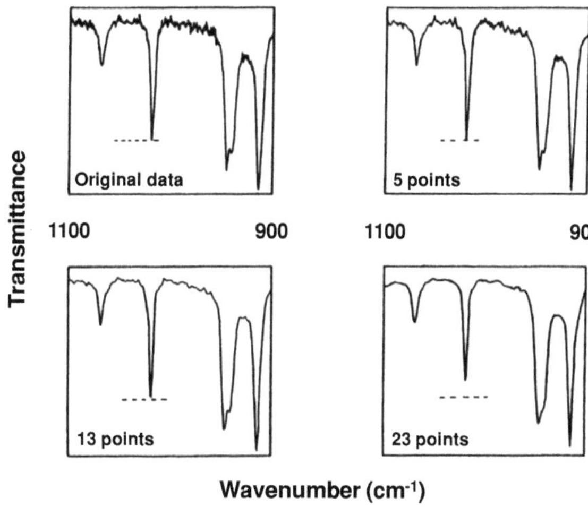

Fig. 3.37 Smoothing of noisy spectral data. The broader the data window the more noise is removed. At the same time spectral bands are flattened as well. The dashed line indicates the decrease in band height

Smoothing broadens the band, its height decreases accordingly.
The area under the envelope of a broadened band remains constant.
The smoothing window should not be wider than the half-width of the band.

By appropriate selection of their coefficients the Savitzky-Golay equation may be used to calculate the first, second, etc., derivative without any increase in computing expenses. Of particular interest in IR and Raman spectroscopy is the second derivative. Assumed a spectrum could be described by a set of quadratic equations. The spectral background would mainly be covered by the absolute and linear terms of these equations. These terms (and the corresponding background contributions) would be lost after calculating the second derivative. Even a difficult quadratic background would be transformed into a simple offset. Moreover, second derivatives of band-shaped features are narrower than the initial bands (Fig. 3.38). No assumptions are required for this procedure. It is a valuable tool to resolve overlaid spectral features. The sharpening by derivation works most effective for the narrowest features in the spectrum. By far the narrowest feature is noise, hence derivation enhances the noise (lowers *S*/*N*).

Qualitative Evaluation

A compound is considered identified when its IR spectrum and a reference spectrum are fully identical (fingerprint capability). The same applies to Raman spectra, but for historic reasons data banks hold much more IR spectra than Raman spectra. If no reference spectrum is available, the interpretation of a spectrum is mainly based on the identification of characteristic bands. Important characteristic bands of organic molecules are found above 1,300 cm^{-1} (Fig. 3.39). The situation for inorganic substances is more complicated (correlation tables in [14–16]). Structural groups with single bonds or heavy atoms (most inorganic compounds) appear in the fingerprint region below 1,300 cm^{-1}. The abundance of bands in this region may render the identification characteristic bands difficult (cf. Fig. 3.27).

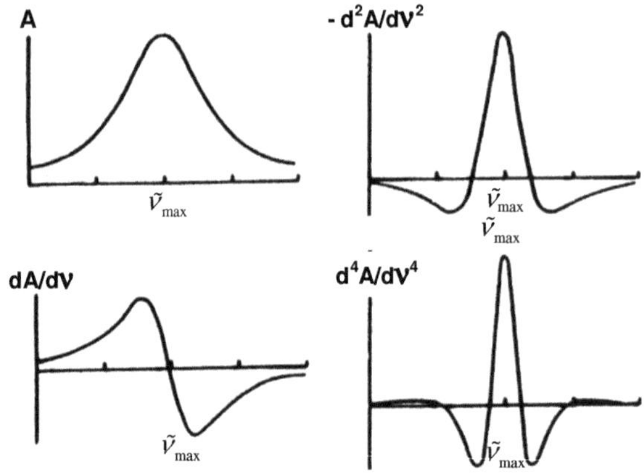

Fig. 3.38 The original band-shaped curve (*top left*) and its subsequent derivatives. The first derivative (*bottom left*) has a zero-crossing at the position of the maximum of the original curve. It can be used for automatic location of band maxima. The second derivative (*top right*) recovers the band shape and is distinctly narrower than the original curve. Overlapping features can be resolved this way. The fourth derivative is even narrower but requires original data of much higher *S/N* (like in UV spectra). Distorted wings of the derivatives, so-called side lobes, disturb the final result

The interpretation of IR or Raman spectra based on their characteristic bands follows a well-established procedure:

Note how the sample was run: physical state, thickness, solvent.
Mark the positions of strong solvent bands or of mulling agent (dead areas).
Start at the high-frequency end of the spectrum and work downwards.
Concentrate on the intense bands (usually due to polar groups).
Do not try to account for all the bands (many of them are fingerprint bands).
The absence of a band is often as informative as its presence.

The interpretation must not be restricted to band positions only. Valuable additional information is usually available or can be obtained:

In the spectrum: bandwidths, depolarisation ratio
About the sample: a polymer, a detergent etc.
Elemental analysis: no $-SO_2$—vibration without the presence sulphur
The degree of unsaturation can be calculated from the elemental analysis.
The molecular weight can be obtained from mass spectra.
Other kinds of spectra provide supplementary information (NMR, UV).

Spectra of mixtures with many overlapping bands may require dedicated mathematical procedures to separate the component bands. The calculation of derivatives (cf. Fig. 3.38) is a convenient approach, no assumptions are required. In contrast, band fitting routines require important assumptions. Band fitting routines should only be used by experienced spectroscopists.

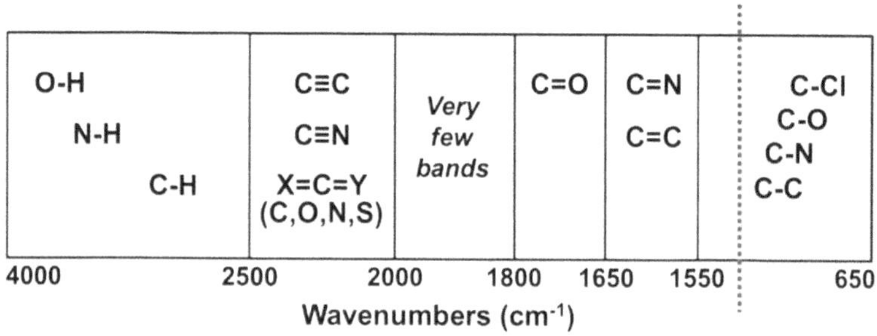

Fig. 3.39 Location of important characteristic bands of organic structural groups

Contemporary procedures for spectra evaluation can be subdivided into two groups. Unsupervised procedures are employed for explorative data analysis or for empirical investigation of samples with no additional information available, e.g. cluster analysis (CA) and principal component analysis (PCA) (Fig. 3.40). Supervised procedures are used to assign new objects to already established classes (available additional knowledge), e.g. discriminant analysis (DA) and SIMCA. All these procedures nowadays are absolutely essential, in particular when large amounts of data like in spectroscopic imaging have to be dealt with.

3.4 Solid State NMR

D. Capitani, A. Spinella and E. Caponetti

3.4.1 Introduction

Nuclear magnetic resonance (NMR) is one of the most powerful and versatile techniques in many scientific fields. It can be applied to a wide variety of samples, both in liquid and in solid phase. Recently NMR technique has also been developed to investigate soft matter.

In this chapter the physical principles of NMR, a brief description of the instrument features and an evaluation of NMR technique in the field of cultural heritage, will be presented.

As conservation science requires nondestructive and minimally invasive investigation methods, only solid state NMR (ss-NMR) technique will be deepened. This because liquid state NMR, even if more established, is a destructive technique in which the sample must be dissolved in a solvent.

The NMR phenomenon consists in the absorption of energy in the range of radio frequencies by atomic nuclei subjected to a static magnetic field $\boldsymbol{B}_0$.

The analysis of an NMR spectrum can provide many types of information such as the number and type of chemical species in a molecule, rather as infrared

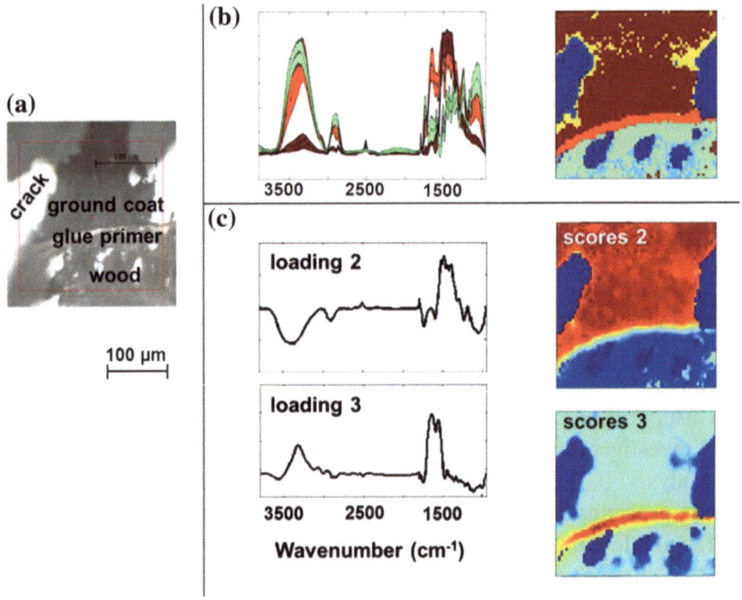

Fig. 3.40 Evaluation of spectroscopic images by chemometric procedures. **a** Cross-section of a painted art object (size 267 × 267 μm^2) A FT-IR image of 64 × 64 pixels was measured. Each pixel constitutes a complete FT-IR spectrum. **b** CA of the spectroscopic image. The spectra are clustered (indicated by color codes) according to their similarity. The color codes are then transferred to the position of the spectrum in the image. **c** PCA of the same data set. The loading plots indicate spectral positions of importance. The score plot reveals the importance of the respective loading at a particular location (*red*–high importance, *blue*–low importance) [17]

spectroscopy is used to identify functional groups. Furthermore, NMR can be used to elucidate molecular structure, to study mixtures of analytes, and to investigate molecular motions in liquid and solid samples.

Although NMR spectroscopy has enormously contributed to chemical knowledge, there are two reasons why ss-NMR technique has developed only in the last three decades: the first one is that NMR for many years has almost exclusively been a technique used for structural determination of molecules in liquid state in the field of organic chemistry. The second reason is that ss-NMR requires spectrometer settings such as more powerful amplifiers, devices for the high speed spinning of the sample and multinuclear probes that are different from the ones used for liquid state NMR. These problems have been overcome with the improvement of the NMR technology and the availability of multinuclear probes that are able to investigate most of the nuclei of the periodic table. As a consequence ss-NMR has become a powerful technique used in a wide range of applications such as catalysis [18], soil investigation [19], polymers [20], material science [21], supramolecular chemistry [22]. A case study regarding ss-NMR application to the analysis of paper and parchment collected from a sixteenth century book will be presented in another chapter.

3.4.2 Principles and Concepts

The physical NMR phenomenon depends on the fact that many nuclei possess an angular momentum called nuclear spin. When a sample is placed in a magnet, nuclear spins generate a bulk macroscopic magnetisation. The simplest experiment consists in perturbing the system from the equilibrium applying a proper radio-frequency pulse and then monitoring the response of the system to the applied perturbation. The so-called NMR-active nuclei are those nuclei having a nuclear spin I $\neq$ 0. The number N of energy levels generated by Zeeman effect for a certain nucleus in the presence of a static magnetic field $\boldsymbol{B}_0$ depends on the nuclear spin I and can be calculated using the following equation:

$$N = 2I + 1 \tag{3.1}$$

Eq. 3.1. The number of nuclear energy levels calculated from the nuclear spin I

Because NMR signals are observed when transitions among energy levels occur, it is evident that nuclei with $I = 0$ do not give rise to any NMR signal. To simplify the discussion, in the following only nuclei with $I = ½$ will be considered. These nuclei generate only two energy levels, namely $m_I = +\ 1/2$ and $m_I = -1/2$, where m_I is the spin quantum number. The energy difference ΔE between the two levels depends on the applied magnetic field $\boldsymbol{B}_0$ and it is described by Eq. 3.2:

$$\Delta E = \gamma \hbar B_0 = \hbar \nu_L \tag{3.2}$$

Eq. 3.2. Energy difference between the two levels generated by Zeeman effect for a nucleus with $I = 1/2$. γ is the gyromagnetic ratio and it is a constant characteristic for each nucleus, η is the Plank constant divided by 2π and ν_L is the Larmor frequency specific to each nuclear species.

The frequency ν_L known as Larmor frequency, is specific to each nuclear species. In the absence of magnetic field, all nuclear magnetic moments $\boldsymbol{\mu}_i$ are randomly oriented. Since nuclei possess a magnetic moment, when a strong external magnetic field is applied along the z-axis, they align either against or with it, with a small excess of nuclei (population excess) aligned with the field.

As a result a bulk magnetisation $M = \sum_{i=1}^{n} \mu_i$ aligned with the applied field is generated. However, the only presence of $\boldsymbol{B}_0$ is not sufficient to observe the NMR signal. This point may be easily explained calculating the ratio between the populations n_1 and n_2 of the two energy levels (E_1 and E_2) which is determined by the Boltzmann distribution. This ratio is obtained from Eq. 3.3:

$$\frac{n_1}{n_2} = e^{-\frac{\Delta E}{kT}} \cong 1 + \frac{\mu B_0}{kT} \tag{3.3}$$

Eq. 3.3. The Boltzmann distribution: n_1/n_2 is the ratio between the population of the two levels, $\Delta E = E_1 - E_2$ is the energy difference between the two levels, T is the absolute temperature, μ is the nuclear magnetic moment, k is the Boltzmann constant and $\boldsymbol{B}_0$ is the static applied magnetic field.

Because the energy difference between the two levels $\Delta E = -\mu B_0$ at normal temperature is much lower than the thermal energy kT, it is evident that the ratio n_1/n_2 is just slightly higher than 1. Under these circumstances the population difference between the two energy levels is too low, thus preventing the observation of the NMR signal. In order to observe the signal the population difference must be increased applying a proper radio frequency pulse, in other words the system must be perturbed from the equilibrium.

In a semi-classical description, nuclear magnetic moments in a static magnetic field $\boldsymbol{B}_0$ applied along the z-axis, precess at the Larmor frequency ν_L which is proportional to $\boldsymbol{B}_0$, see Eq. 3.4.

$$\nu_L = \gamma B_0 \tag{3.4}$$

Eq. 3.4: The Larmor frequency dependence from the magnetic field strength $\boldsymbol{B}_0$; γ is the gyromagnetic ratio.

The system is perturbed from equilibrium applying along x- or y-axis, a radio-frequency field $\boldsymbol{B}_1$, which oscillates at the proper radio-frequency ν_1. When ν_1 is equal to the Larmor frequency, the resonance condition occurs. This is the reason why the NMR spectrum gives information for a single nuclear species at once. At the resonance the field $\boldsymbol{B}_1$ interacts with the magnetisation $\boldsymbol{M}$ to produce a torque which moves the magnetisation towards the xy plane. Since the precession frequency about the static magnetic field $\boldsymbol{B}_0$ is equal to the rotating $\boldsymbol{B}_1$ frequency, the magnetisation will remain perpendicular to the $\boldsymbol{B}_1$ field component. To summarise, the magnetisation will precess about $\boldsymbol{B}_I$ and, at the same time, it will precess about $\boldsymbol{B}_0$. The $\boldsymbol{B}_1$ field is applied as a pulse duration t_p which usually lasts for a few microseconds. The angle θ (flip or tip angle) through which the magnetisation is tipped from the z-axis, is easily calculated: $\theta = \gamma \boldsymbol{B}_1 t_p$. Each duration t_p of the radio-frequency pulse produces a particular tip angle. The intensity of the detected signal depends on the chosen tip angle. As an example, the signal will be maximum for a tip angle of $\pi/2$, or it will be null at π.

As the radiofrequency pulse is switched off, nuclei relax back to equilibrium generating a signal called free induction decay (FID). The frequency domain spectrum is obtained by applying a Fourier transformation to the FID, see Fig. 3.41.

It is worth noticing that, after perturbing a system with a radio-frequency pulse sequence, the system will return back to equilibrium through a process called "relaxation" characterised by a time constant known as "relaxation time". Relaxation times are parameters tightly related to the properties of the investigated material.

The main source of chemical information in high-resolution NMR spectroscopy is the chemical shift which is a very sensitive probe of the chemical environment of each resonating nucleus. This property is due to the fact that, in presence of $\boldsymbol{B}_0$, the electrons surrounding a nucleus, generate a small magnetic field opposite to $\boldsymbol{B}_0$

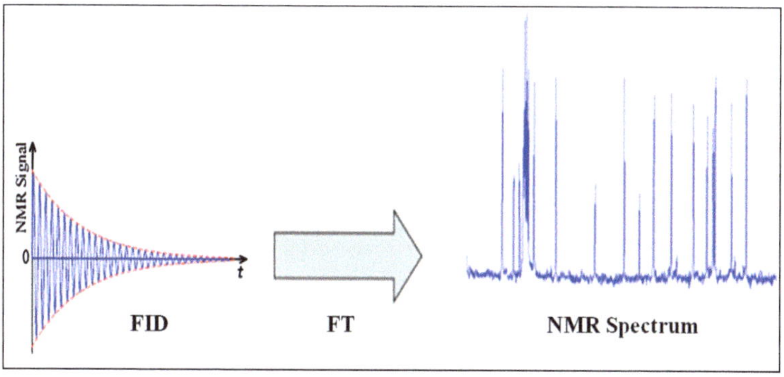

Fig. 3.41 By applying a Fourier transformation to the signal in the time domain (FID), the NMR spectrum in the frequency domain is obtained

which shields the nucleus, the higher the electron density surrounding the nucleus, the higher the shielding effect. Thus, $\boldsymbol{B}_0$ is not the effective magnetic field at which nuclei are subjected but it is slightly changed by the additional magnetic field generated by the electron currents. As a consequence, nuclei of the same species in different chemical environments resonate at slightly different frequencies. This allows one to obtain structural information from the position of the signals in the spectrum. As an example, the ^{13}C NMR spectrum of ethylbenzene is reported in Fig. 3.42. Note that different carbon nuclei in the molecule give signals at different positions, i.e. different chemical shifts.

Other important structural information can be obtained by investigating the coupling constants accounting for the coupling between nuclei through 1, 2 and 3 bonds. In addition to one-dimensional NMR spectroscopy, two dimensional approaches have been developed for the determination of the structure of complex molecules like proteins. A detailed discussion of these techniques that goes beyond the scope of this chapter can be found in specialised books [23, 24].

In liquid state, the Brownian motion of molecules averages to zero the dipolar interactions among nuclei, besides the position of lines in the spectrum does not depend on the orientation of the molecules on respect to the applied magnetic field. Therefore, spectra of liquid samples show sharp and well-resolved signals.

In the solid state, the dipolar interactions among nuclei are not averaged to zero, causing a marked enlargement of signals. Furthermore, in the solid state the position of signals depends on the orientation of molecules on respect to $\boldsymbol{B}_0$. As a consequence, each single crystal gives rise to a distribution of lines (powder spectrum) which are very close to each other, causing a further enlargement of the line width (chemical shift anisotropy). In the case of nuclei with spin higher than ½, the interaction between the nuclear quadrupolar electric momentum and the electric field gradient generated by electrons, is another source of line broadening, however, the case of quadrupolar nuclei will not be discussed. Another problem is that experiments on solid samples may be very time-consuming as nuclear relaxation times may be very long. To overcome these difficulties, three techniques

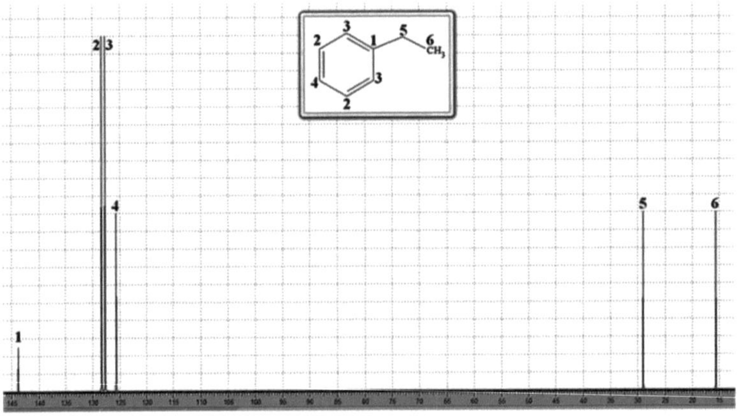

Fig. 3.42 ^{13}C NMR spectrum of ethylbenzene

of acquisition have been developed, namely magic angle spinning, dipolar decoupling and cross polarisation. The first two techniques average the dipolar interactions among nuclei and the chemical shift anisotropy effect, while the third one allows the obtainment of spectra of low abundant (dilute) nuclei in a reasonable time.

Many terms of Hamiltonian which cause line broadening in solid samples, involve the geometric factor $(3\cos^2\theta-1)$ which accounts for the orientation of the shielding/dipolar tensor on respect to the applied magnetic field $\boldsymbol{B}_0$. It is easy to demonstrate that by spinning the sample at the magic angle, i.e. 54.74°, the geometric factor $(3\cos^2\theta-1)$ is averaged to zero on the NMR timescale. Therefore, this technique is routinely used to remove the effects of chemical shift anisotropy and to assist in the removal of dipolar coupling.

The *magic angle spinning* technique involves the high speed spinning of the powdered solid sample packed in a rotor at the *magicangle* with respect to the direction of the static magnetic field, see Fig. 3.43.

The *dipolar decoupling* technique involves the irradiation of abundant nuclei (such as protons) having dipolar interaction with dilute nuclei (such as carbons) that must be observed. The high power decoupling thus allows the removal of the dipolar broadening to obtain a spectrum with much more sharp lines.

The *cross polarisation* technique is applied to dilute nuclei such as ^{13}C or ^{29}Si characterised by a low natural abundance (1.1 % for ^{13}C and 4.7 % for ^{29}Si) and low gyromagnetic ratio, and thus having a very weak observable net magnetisation. The NMR experiments on dilute nuclei are very time-consuming since they require a high number of scans. The CP technique consists in transferring the polarisation from abundant to dilute nuclei. The overall effect of the CP technique is to enhance the signal to noise ratio (*S/N*).

The polarisation transfer between nuclei with different Larmor frequencies is obtained by varying the radio-frequency pulses ($\boldsymbol{B}_1$) of the two nuclear species allowing the matching of their energy levels. This matching is called Hartman-

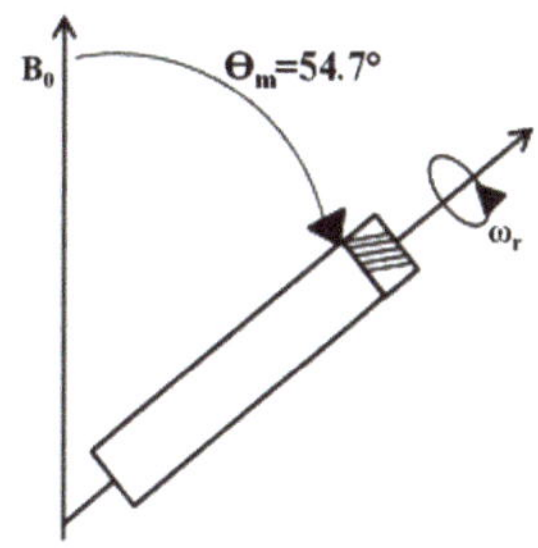

Fig. 3.43 The magic angle spinning of a solid sample

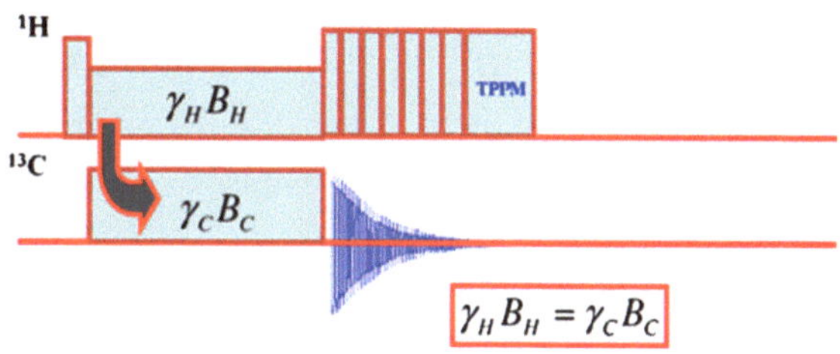

Fig. 3.44 Cross polarisation pulse sequence. The equation describing the Hartman-Hahn condition is reported in the insert

Hahn condition [25]. As an example, in the case of proton and carbon systems, the Hartman-Hahn condition is: $\gamma_C \boldsymbol{B}_{1C} = \gamma_H \boldsymbol{B}_{1H}$, where γ_C and γ_H are the carbon and proton gyromagnetic ratios, respectively. Since γ_H is four times γ_C the match occurs when the strength of the applied radio-frequency field $\boldsymbol{B}_{1C}$ is four times the strength of the applied proton field $\boldsymbol{B}_{1H}$. As a consequence, there is an enhancement of the dilute spin signal intensity by as much as the ratio of gyromagnetic ratios of the abundant and dilute spin. In the case of proton and carbon nuclei, because $\gamma_H/\gamma_c = 4$, the enhancement factor is 4. It is important to note that the Hartman-Hahn condition (i.e. the contact between the two nuclear species) must be maintained for a time (*contact time*τ_C) that allows the maximisation of the NMR signal. On the other hand, τ_C cannot exceed a certain value, depending on the analysed sample, above which relaxation processes occur thus causing a loss of the signal intensity. A representation of the CP pulse sequence is reported in Fig. 3.44.

In the presence of cross-polarisation, because polarisation is being transferred from protons to carbons, the shorter T_1 relaxation time of protons dictates the recycle delay for signal averaging. As a result, the time of acquisition of carbon spectra is markedly shortened on respect to carbon spectra acquired without cross-polarising.

3.4.3 The NMR Spectrometer

In modern high field NMR spectrometers the static magnetic field $\boldsymbol{B}_0$ is generated by a superconducting magnet (Fig. 3.45).

The magnet is constituted by an outer stainless steel or aluminium dewar which contains liquid nitrogen and an inner dewar containing the superconducting coil (1) immersed in liquid helium in order to reach the very low temperature needed to make

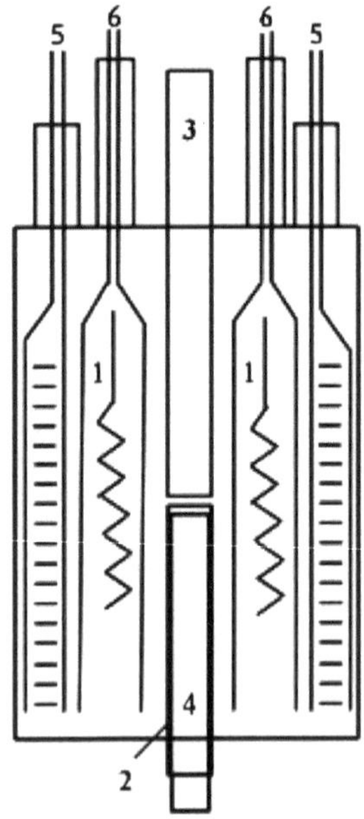

Fig. 3.45 Internal section of an NMR superconducting magnet: *1* Magnet coils and liquid He dewar; *2* Shim coils system; *3* Spinner assembly; *4* Probehead; *5* Liquid N_2 ports; *6* Liquid He ports

the material constituting the coils superconducting. A set of shim coils providing a room-temperature homogeneity adjustment is inside a room-temperature bore (2), a spinner assembly is equipped with the system for spinning the NMR sample holder (3). The probe-head (4) is introduced into the magnet from the bottom and is connected to at least three cables providing the ^{2}H lock, ^{1}H and ^{Y}X frequencies. In the case of solid state dedicated probeheads the lock channel is not necessary. The probehead transmits the radio-frequency energy to the sample and also detects the signal from the sample which is positioned within the rf coil of the probehead itself. The rf coil needs to be tuned for the specific sample to be analysed. Another important part of the spectrometer is the cabinet which is shown in Fig. 3.46.

The computer controls the radio-frequency and pulses generation. Typically three radiofrequency channels, namely the observed, the lock and the decoupling channel, are present inside the cabinet. The frequencies, after the amplification are transmitted to the probehead. As the output from the sample is small the output stage consists of a radio receiver able to amplify the signal. Then the signal is digitised and fed into the computer memory.

3.4.4 Evaluation of ss-NMR as Used in Conservation Science

A chemical characterisation of materials associated with archaeological, historical and artistic objects, completes the information obtained through the study of historical documents, supplying information on their nature and the modifications that might have occurred due to manufacturing processes and/or natural ageing.

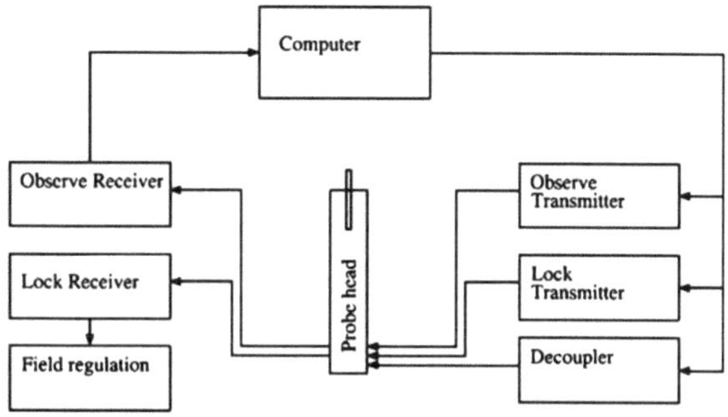

Fig. 3.46 Diagram of a NMR cabinet

Recently, in this field the ss-NMR technique started having an important role even if some drawbacks must be taken into consideration. The main drawback is that 50–80 mg of material is required to perform the analysis when using the standard 4 mm rotor to contain the sample. Nevertheless, the method is nondestructive and the sample can be recovered after the NMR experiment and investigated with different techniques. This is possible because the sample does not require any treatment before the acquisition. Furthermore, the quantity of information obtained from the same sample by means of an NMR experiment is often a sufficient reason for planning this kind of analyses. Besides, the use of 4 mm rotors with an internal insert to reduce the volume, or the use of 2.5 mm rotors, permits to lower the quantity of sample needed for the analysis (10–12) mg.

Solid state NMR is a powerful technique in the field of conservation science. First of all the NMR observables are specific for the chemical element under investigation and are sensitive to distances in short and medium ranges. Besides, even in the solid state, the study of the cross-polarisation dynamics process in many cases allows the obtainment of quantitative results.

A very interesting aspect is that ss-NMR can be used in the investigation of amorphous materials such as glasses and amorphous polymers used as conservation materials. As an example, the network structure in fluoride-containing aluminosilicate glasses has been studied through a quantitative determination of the coordination aluminum sites using ^{27}Al MAS NMR and the triple-quantum magic angle spinning (3QMAS) technique [26].

Some examples of ss-NMR application in the field of cultural heritage are reported in the following in order to highlight its potentiality.

The characterisation of stone material coming from the Cheops pyramid was performed by ^{29}Si and the ^{27}Al ss-NMR spectroscopy [27]. In particular, the hypothesis that a clay binder was added to stone material in order to obtain some sort of concrete was studied.

ss-NMR technique was applied in the UNESCO World Heritage site of Guanajuato City Main Church (Central Mexico) in which building stones from seventeenth century were analysed by ^{29}Si and ^{27}Al MAS NMR in order to determine the degradation degree and to plan preservation and restoration interventions [28].

An interesting ss-NMR application is represented by a rubber analysis. This natural polymeric material is slightly soluble and slightly volatile, so that techniques such as gas chromatography (GC–MS) or liquid chromatography (LC–MS) cannot be utilised. It was possible to demonstrate that the old Mesoamerican populations used a process similar to the modern vulcanisation by comparing ^{13}C ss-NMR spectra of samples from an ancient manufacturing with the spectrum of a modern rubber [29]. Recently ^{23}Na and ^{1}H solid state NMR has been used to study the salt quantity in the Statue of a Soldier eroded by salt weathering in La Rochelle (France). By quantitative (NMR) measurements of the amount of both Na and H ions in the solution inside a material, the salt concentration was determined. Furthermore, information on the pore size distribution in the manufacture was obtained by ^{1}H and ^{23}Na NMR experiment [30].

The study of the interactions between archaeological wooden samples and polyethylene glycol (PEG) was reported. Through ^{13}C ss-NMR it was possible to visualise the degradation extent in archaeological wood even after the conservation process. In addition, the investigation of the cross-polarisation dynamic allowed one to demonstrate that almost 30 % of PEG was in close molecular interaction with lignin [31].

As the ss-NMR spectra can be used as a fingerprint of samples constituted by a complex mixture of analytes, analytical procedures can be applied through the use of standards and reference materials. This is the case of the geographic origin assessment of several amber samples through the use of ^{13}C ss-NMR spectroscopy [32].

The ss-NMR can nowadays be applied to characterise artistic, archaeological and historical objects. The interpretation of the obtained data, together with the study of historical documents, is used to obtain information about the geographical origin and the use of the objects under investigation. Therefore, the integration of different research areas will allow to disseminate and to present in exhibitions not only the object itself, but also the technique used to create it and the description of its constituent materials. This new approach can be useful to reveal how the techniques of production have evolved over time.

References

1. JEOL Ltd (2007) Invitation to the SEM world. service advanced technology, JEOL, Japan
2. Williams DB, Barry Carter C (2004) Transmission electron microscopy: a textbook for Materials science. Springer, New York
3. Zorba T, Pavlidou E, Stanojlovic M, Bikiaris D, Paraskevopoulos KM, Nikolic V, Nikolic PM (2006) Technique and palette of XIIIth century painting in the monastery of Mileseva. Appl Phys A 83:719–725
4. Clark RJH, Dines TJ, Kurmoo M (1983) On the nature of the sulfur chromophores in ultramarine blue, green, violet, and pink and of the selenium chromophore in ultramarine selenium: characterization of radical anions by electronic and resonance Raman spectroscopy and the determination of their excited-state geometries. Inorg Chem 22:2766
5. Ganitis V, Pavlidou E, Zorba F, Paraskevopoulos KM, Bikiaris D (2004) A Post-Byzantine Icon of St Nicholas painted on a leather support microanalysis and characterisation of technique. J Cult Heritage 5:349–360

6. Pistofidis N, Vourlias G, Pavlidou E, Dilo T, Civici N, Stamati F, Gjongecaj S, Prifti I, Bilani O, Stergioudis G, Polychroniades EK (2006) On the comparative study of three silver coins of the third century BC minted in Korkyra, Dyrrachion and by the Illyrian king Monounios. Appl Phys A 83:637–642
7. Froh J (2004) Archaeological ceramics studied by scanning electron microscopy. Hyperfine Interact 154:159–176
8. Griffiths PR, de Haseth JA (2007) Fourier transform infrared spectrometry. Wiley, Hoboken
9. Skoog DA, Holler FJ, Nieman TA (1998) Principles of instrumental analysis. Brooks Cole, Philadelphia
10. Schrader B (1989) Ramn/infrared atlas of organic compounds. VCH, Weinheim
11. Mayo DW, Miller FA, Hannah RW (2004) Course notes on the interpretation of infrared and Raman spectra. Wiley, Hoboken
12. Gauglitz G, Vo-Dinh T (2003) Handbook of spectroscopy. Wiley-VCH, Weinheim
13. Salzer R, Siesler HW (2009) Infrared and Raman spectroscopic imaging. Wiley-VCH, Weinheim
14. Miller FA, Wilkins CH (1952) Infrared spectra and characteristic frequencies of inorganic ions. Analyt Chem 24:1253–1294
15. Nyquist RA, Kagel RO (1971) Infrared spectra of inorganic compounds. Academic Press, New York
16. Degen IA, Newman GA (1993) Raman spectra of inorganic ions. Spectrochim Acta 49A:859–887
17. Gruchow F (2007) IR-Spektroskopische Untersuchungen am Holzträger bemalter historischer Kunstobjekte. Thesis, TU Dresden
18. La Parola V, Longo A, Venezia AM, Spinella A, Caponetti E (2010) Interaction of gold with co-condensed and grafted HMS-SH silica: A 29Si 1H CP-MAS NMR spectroscopy, XRD, XPS and Au LIII EXAFS StudyEur. J Inorg Chem 23:3628–3635
19. Conte P, Smejkalovà D, Piccolo A, Spaccini R (2008) Evaluation of the factors affecting direct polarization solid state 31P-NMR spectroscopy of bulk soils Eur. J Soil Sci 59:584–591
20. Hedesiu C, Demco DE, Kleppinger R, Vanden Poel G, Gijsbers W, Bluemich B, Remerie K, Litvinov VM (2007) Effect of temperature and annealing on the phase composition, molecular mobility and the thickness of domains in isotactic polypropylene studied by proton solid state NMR. SAXS and DSC Macromol 40:3977–3989
21. Alemany LB (2007) Solid state NMR analysis of fluorinated single-walled carbon nanotubes: assessing the extent of fluorination Chem. Mater 19:735
22. Hoffmann A, Sebastiani D, Sugiono E, Yun S, Kim KS, Spiess HW, Schnell I (2004) Solvent molecules trapped in supramolecular organic nanotubes: a combined solid-State NMR and DFT study chem. Phys Lett 388:164–169
23. Becker ED (2000) High resolution NMR theory and chemical applications, Academic Press, Waltham ISBN: 0-12-084662-4
24. Macomber RS (1998) A complete introduction to modern NMR spectroscopy J. Wiley & Sons ISBN: 0-471-15736-8
25. Hartmann SR, Hahn EL (1962) Nuclear double resonance in the rotating frame. Phys Rev 128:2042–2053
26. Stebbins JF, Kroeker S, Lee SK, Kiczenski TJ (2000) Quantification of five- and six-coordinated aluminum in aluminosilicate and fluoride-containing glasses by high-field, high-resolution 27Al NMR. J Non-Cryst Solids 275:1–6
27. Demortier G (2004) PIXE, PIGE and NMR study of the masonry of the pyramid of Cheops Giza. Nucl Instr Meth Phys Res B 226:98–109
28. Cervantes J, Mendoza-Diaz G, Alvarez-Gasca DE, Martinez-Richa A (1999) Applications of 29Si and 27Al magic angle spinning nuclear magnetic resonance of the building materials of historical monuments. Solid State Nucl Mag 13:263–269
29. Hosler D, Burkett SL, Tarkanian MJ (1999) Prehistoric polymers: rubber processing in ancient Mesoamerica. Science 284:1988–1991
30. Rijniers LA (2004) PhD thesis, Technical university, Eindhoven

31. Bardet M, Gerbaud G, Trân QK, Hediger S (2007) Study of interactions between polyethylene glycol and archaeological wood components by C13 high-resolution solid-state CP-MAS NMR. J Archaeol Sci 34:1670–1676
32. Lambert JB Frye JS Lee TA Welch CJ Poinar GO (1989) Analysis of mexican amber by carbon-13 NMR spectroscopy, in archaeological chemistry IV (ed Allen RO), 381–388 Advances in Chemistry Ser. 220, Am Chem Soc, Washington

Chapter 4
Identification Techniques II

Giovanni Ettore Gigante, Stefano Ridolfi, Michele A. Floriano, Eugenio Caponetti, Lorenzo Gontrani, Ruggero Caminiti, Maria Luisa Saladino, Delia Chillura Martino, Nick Schiavon, Cristina Dias Barrocas, Teresa Ferreira and K. Chrysafis

---

G. E. Gigante
Department of SBAI, Sapienza University of Rome, Rome, Italy
e-mail: giovanni.gigante@uniroma1.it

S. Ridolfi
ArsMensurae, Rome, Italy

M. A. Floriano · E. Caponetti · M. L. Saladino · D. C. Martino
Dipartimento di Chimica "S. Cannizzaro", Università degli Studi di Palermo,
Parco d'Orleans II, Viale delle Scienze pad. 17, 90128 Palermo, Italy
e-mail: scatter@unipa.it

E. Caponetti (✉)
Centro Grandi Apparecchiature—UniNetLab, Università degli Studi di Palermo,
Via F. Marini 14, 90128 Palermo, Italy

L. Gontrani
Istituto di Struttura della Materia, Consiglio Nazionale delle Ricerche, Area della Ricerca di Tor Vergata, Via del Fosso del Cavaliere, 100, 00133 Rome, Italy

R. Caminiti
Dipartimento di Chimica, Università di Roma, 'La Sapienza', P. le Aldo Moro 5,
00185 Rome, Italy

N. Schiavon
Evora Geophysics Centre and Hercules Laboratory for the Study and Conservation of Cultural Heritage, University of Evora, Largo Marqués do Marialva 8,
7000-809 Evora, Portugal
e-mail: schiavon@uevora.pt

C. D. Barrocas · T. Ferreira
Evora Chemistry Centre and Hercules Laboratory for the Study and Conservation of Cultural Heritage, University of Evora, Evora, Portugal

K. Chrysafis
Department of Physics, School of Science, Aristotle University of Thessaloniki,
54124 Thessaloniki, Greece
e-mail: hrisafis@physics.auth.gr

E. A. Varella (ed.), *Conservation Science for the Cultural Heritage*,
Lecture Notes in Chemistry 79, DOI: 10.1007/978-3-642-30985-4_4,

4.1 X-Ray Techniques and X-Ray Fluorescence with Portable Systems

Giovanni Ettore Gigante and Stefano Ridolfi

4.1.1 The Nature of X-Rays and Their Interactions

The X-rays are very useful in the characterisation of works of arts for several reasons:

- an excellent penetration in many materials constituting the artefacts;
- basically simple interaction mechanisms that facilitate the development of techniques that have good specificity for some diagnostic tasks (such as X-ray fluorescence (XRF) for inorganic pigments);
- sources and detectors that can be miniaturised to set up portable or mobile systems (if the X-rays' intensities are low).

The X-rays are an electromagnetic radiation with a very short wavelength (0.01–1 nm) produced in the interaction with targets of high energy charged particles (essentially electrons). They also can be produced in the de-excitation of inner shells of atoms (in this case they are monochromatic) and in the deviation of very high energy electrons beams as happens in the synchrotron radiation. The most important mechanisms of interaction of X-rays are:

1. photoionisation of atoms;
2. elastic and non-elastic scattering.

The photoionisation is an interaction with inner electrons of atoms leading to the creation of a vacancy in the shell and the subsequent emission of a photon of energy equal to the difference of binding energies of electrons that fill the vacancy and that of previously occupying the vacancy. This interaction is also called 'photoelectric' because the electron extracted by his core gains an energy equal to the difference of that of the impinging photon and the initial binding energy of the electron. With determined probability, the process of de-excitation of the atoms can happen with the production of electrons of given energies, called Auger electrons. In order to create a vacancy in a given shell, an X-ray of energy is needed higher than the binding energy of the electron, therefore the probability of interaction between photons and electrons is maximum near the shell edge and decrease quickly with the increase of the photon energy. This is the mechanism that promotes the development of many X-ray techniques either spectroscopic or imaging.

The scattering of X-ray with electrons is characterised by two main mechanisms:

1. elastic (or coherent) scattering where the scattered photon keeps the energy of the primary;
2. non-elastic scattering (Compton effect) where the primary photon energy is subdivided between a scattered photon and the electron.

In the elastic scattering, the direction of the secondary electron has memory of the position of the interacting electrons, thus a regular frame of atoms produce an enhanced emission of coherently scattered photons at particular angles: this is the background of the X-ray diffraction (XRD). This technique reconstructs the structure of a molecule (or of a crystal) by the pattern of coherently emitted photon for the interaction with the sample of a well-collimated monochromatic X-ray beam.

Elastic and non-elastic scattering are complementary interactions producing secondary photons that are anisotropically distributed with the elastically scattered photons practically only in the forward direction. The photons scattered in all directions, origin of the background that drastically limits the sensitivity of the X-ray techniques and are the main hindrance in the development of them.

4.1.2 Radiography

The very small refraction index of the X-rays is at the base of imaging techniques. The starting experiment to define the quantities describing the interaction between photons and mass is the absorption of a well-collimated X-ray beam by a homogeneous target of thickness x:

$$I = I_0 e^{\frac{\mu}{\rho}\rho x} \tag{4.1}$$

where ρ is the mass density and μ is the linear absorption coefficient [1]. So, μ/ρ is the mass absorption coefficient. In order to have a synthetic view on the penetration capacity of the X-ray radiation, it is useful to remember that the Mean Free Path (MFP) of photons in a material is $1/\mu$ and in a material thickness of 2 MFP, 86.5 % of incoming photons are absorbed. In Fig. 4.1, the MFPs of X-rays in different material are shown.

The interactions that occur on the electrons are due to the high energy of the X-rays, thus the total mass absorption coefficient is the sum of element coefficients constituting the material

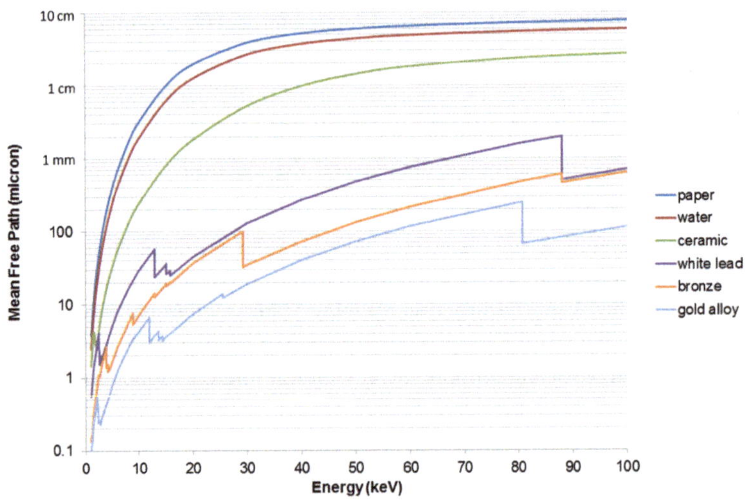

Fig. 4.1 Mean Free Paths of X-ray photons in some materials

$$\left(\frac{\mu}{\rho}\right)_{\text{tot}} = \sum_i \left(\frac{\mu}{\rho}\right)_i w_i \tag{4.2}$$

being w_i the weight fractions of the elements i.

The different absorption of X-rays in materials constituting an artefact is the basics of radiograph that consists on a single projection of a point X-ray source onto a detector area (see Fig. 4.2). In the case of a painting the inorganic pigments containing heavy elements, as for example the white lead or Biacca absorb more photons rather than, for example, white titanium pigment. Adjusting the X-ray tube supply voltage and filtering the beam, it is possible to select and use photons falling in an opportune energy range. Choosing proper photon energy ranges makes it possible to cross the artefact and, simultaneously, increase the contrast in the image. Looking at Fig. 4.1, it can be understood that in order to cross a painting having an overall thickness in the range of some millimetres (centimetres in the case of tables) an energy of 20–30 keV can be sufficient while in the case of a bronze statue characterised by an alloy thickness of few millimetres higher photon energies are needed.

One of the most interesting things of X-ray Radiography on paintings is the capability of seeing beneath the surface. The "pentimento" of the painter is most of the times evident. In the Fig. 4.3 on the right, it is clear the existence of two faces of Saint Paul, the former (hidden) is higher and bigger, while the latter (visible) is smaller and lower. Even the hand of Saint Peter underwent variations. It is easily understood the presence of two thumbs.

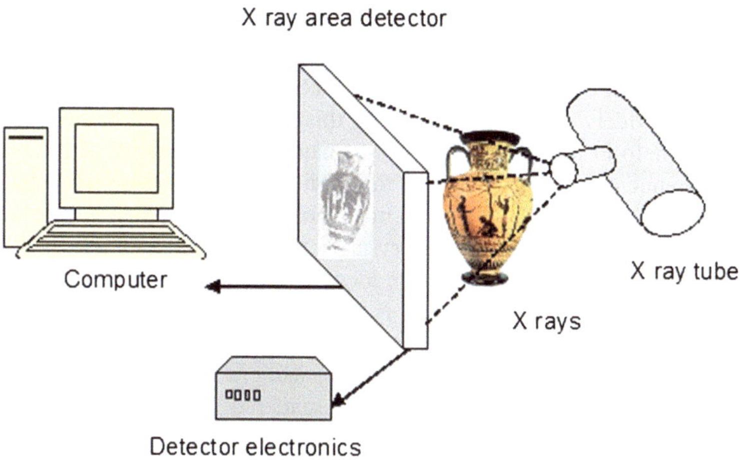

Fig. 4.2 Scheme of a digital radiography set-up

Fig. 4.3 Particular of a digital radiography of Ludovico Stern (1708–1777) painting of Saints Peter and Paul in Rome (38 kV—5 mA—120 s)

4.1.3 X-Ray Diffraction

With X-ray diffraction, it is possible to detect the molecular and crystal structure of materials constituents; therefore it is a very useful tool when a speciation is needed. During analysing the corrosion products of a metal artefact, sometime it is necessary to identify the individual crystals formed in order, for example, to determine whether the degradation process is active or not. Another example is the ochre that is widely used as pigments, in this case only the detection of crystal species can help in finding the exact provenance or authenticity of the pigment. The most frequently used technique in the studies on ancient materials is the X-ray powder diffraction, commonly used in many analytical laboratory, with whom it is possible to identify the presence of a crystal or a molecule in a sample by detecting the series of peaks that allow a precise identification of crystalline species, in the acquired spectrum. The physical basis is in the coherent scattering of X-ray and in the local interference of scattered waves where the electrons belong to atoms arranged in a lattice. In this case, there are, using a monochromatic impinging radiation, angles in which the coherently scattered photons are preferentially directed.

4.1.4 X-Ray Fluorescence

XRF is probably, currently, the most widely used technique in the analysis of cultural heritage both in the laboratory as in the "in situ" modalities. The reasons for this is the extreme simplicity of the technique and the continuous development of specific analytical procedures in which the XRF plays a role of protagonist, as in the case of characterisation of glass and ceramics with laboratory spectrometers or of pigments in a painted surface and metal alloys in statues, or a secondary role in the first and fast approach to the characterisation of the material. The reasons for this success are in the multielemental and non-destructive capabilities of the technique and in its simplicity of use.

The underlying physics is in the excitation of electrons belonging to the inner shells of atoms by X-ray photons of convenient energy and the following de-excitation process with the emission of X-ray having energies that are characteristic for each emitting element. The de-excitation process is independent from the excitation one; therefore, it is always the same also in the case of excitation with charged particle or other kind of high energy radiation. This is the profound reason why the energy dispersive spectroscopy (EDS), Particle (or Proton) Induced X-Ray Emission (PIXE)and XRF spectra are so similar. Each element emits a set of lines that are related to the internal transitions of the electron starting from the outer shells, for example the strongest lines in an X-ray spectrum are the K-lines originating by the transition of electrons of shells L and M toward the K one and they are called K-lines.

An XRF spectrometer mainly consists of an excitation source, a detector with electronics and a multichannel analyser with its chain for data processing. There are two modalities of detection of X-ray, with the use of a monochromator [wavelength dispersive (WD) mode] or using a detector capable to determine the energy of counted photons [energy-dispersive (ED) mode]. Both the modalities are

Table 4.1 Characteristics of small-size, portable X-ray tubes according to the element or elements to be analysed

Element or groups of element to be analysed	Anode material	Anode voltage (KV)	Current (mA)
Phosphorus, sulphur, chlorine	Calcium (K-lines)	8–10	0.1–1
Phosphorus, sulphur, chlorine	Palladium (L-lines) or silver (L-lines)	8–10	
Chlorine, potassium, calcium	Titanium (K-lines)	10	0.1–1
From potassium to yttrium (K-lines) and from cadmium to uranium (L-lines)	Molybdenum (K-lines)	30	0.1–1
From potassium to tin (K-lines) and from cadmium to uranium (L-lines)	Palladium (K-lines)	35	0.1–1
From potassium to tin (K-lines) and from terbium to uranium (L-lines)	Tungsten	35	0.1–1
From zinc to barium (K-lines) and from terbium to uranium (L-lines)	Tungsten	50	0.1–1

widely used being prevalent the firmer among the laboratory spectrometers and the latter in the field portable ones. This chapter is mostly focused on the use of "in situ" applications of XRF [2]. Recent technological developments have resulted in small, low-power, dedicated X-ray tubes and thermoelectrically cooled semiconductor detectors comprehensive of pulse height analyser, and thus many type of portable spectrometers featuring different capabilities and applicative results have become commercially available.

Concerning the excitation source, besides radioactive sources, the progress of technology has made available small-size, low-power X-ray tubes for Energy-Dispersive X-ray fluorescence (EDXRF). It should be considered that a 35–50 kV X-ray tube is adequate to the analysis of almost all elements of the periodic table, being able to excite K-lines of elements up to $Z = 60$ and L-lines of heavier elements.

Photon output from an X-ray tube of the type described in Table 4.1 depends on many parameters, such as kilo-voltage, current and anode material.

In the last decade, thermoelectrically cooled Silicon Drift Detector (SDD), Si-PIN and HgI_2 detectors progressively replaced by the portable apparatuses, the other detectors, such as nitrogen-cooled Si(Li) and HpGe-detectors.

The SDD, currently the most used, has a thickness of about 300–500 μm, which makes this detector useful up to about 30 keV, and an energy resolution of about 140 eV at 5.9 keV. The only current limitation of the new thermoelectrically cooled detectors is the reduced sensitive area, which limits the efficiency of the system.

Portable systems based on the use of X-ray tubes coupled to thermoelectrically cooled semiconductor detectors are now manufactured and supplied by many manufactures with characteristic that enhance the portability and the sensitive, according to the application (Fig. 4.4).

Fig. 4.4 Portable XRF system for the study of the "Gates of Paradise" by Lorenzo Ghiberti between 1426 and 1452, east door of the Baptistery of Florence

EDXRF technique can determine simultaneously with little or no sample preparation elements with concentrations ranging approximately from ppm to percentage levels, whether the sample is a solid, powder, liquid, thin film or coating. Materials that can be studied by using an EDXRF apparatus are different, for example:

- alloys (bronzes, brasses, and so on);
- jewellery, gold and silver alloys
- paintings of all type and frescos;
- ceramics;
- papers and parchments;
- enamels;
- stones and marbles of all type;
- building materials
- glasses;
- illuminated manuscripts.

In many cases, a qualitative (or semi quantitative) analysis is sufficient (for example in the case of paintings) and in others, a quantisation is required (for example in the case of alloys or ceramics).

EDXRF analysis generally involves an area of some mm^2, and a thickness between μm and fractions of mm (see Fig. 4.1). The analysis is, therefore, superficial and dependent on the surface conditions. It is therefore important to stress the fact that due to the reduced thickness analysed, EDXRF analysis is representative of the bulk composition only for homogeneous samples.

In the case of a non-destructive analysis, it is helpful to highlight the point in which the excitation beam impinge using a pointer (Fig. 4.5) In many cases, "capillary collimators" are employed, to focus the radiation into smaller areas of the order 10^{-2}–10^{-4} mm^2.

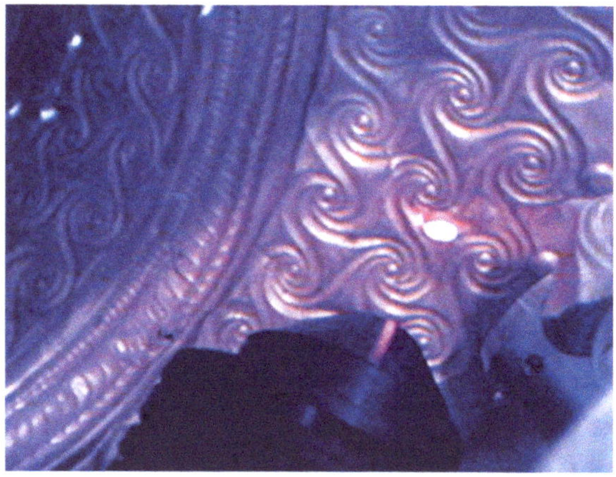

Fig. 4.5 An XRF measure using a laser pointer to highlight the sampled point, during the campaign of studies on Messenian gold in the 2002

4.1.4.1 Quantitative Analysis

As above mentioned, the analysed volume is from the surface to a variable depth depending on the energy of the XRF characteristic line of the different chemical elements. The detected counts in a given peak in a spectrum depend on several things that can be synthesised in the following equation

$$N_a = N_0 \varepsilon(E_a)\left(k\mu_{\mathrm{ph,a}}(E_a)\omega_{a,s}F_{a,s}p_l w_a\right)\frac{1-e^{-(\mu_0+\mu_1)d}}{(\mu_0+\mu_1)} \tag{4.3}$$

where:

N_0 is the number of exciting photons, only in the case of a monochromatic source
k is a geometrical term, depending on the source–sample and sample–detector distances;
$\varepsilon(E_a)$ is the intrinsic detector efficiency at the energy of the line of the element a;
$\mu_{\mathrm{ph,a}}(E_0)$ represents the photoelectric mass attenuation coefficient of element a at incident energy E_0;
$\omega_{a,s}$ is the fluorescent yield of element a for the shell $s = K, L, \ldots$;
$F_{a,s}$ is the fraction of photoelectric interactions occurring in the shell s;
p_l is the probability that the fluorescence event belong to the $l = a, b, g, \ldots$line of interest;
w_a is the weight fraction of element a;
d is the thickness of the sample in cm^2/g;
$\mu_t(E_0)$ and $\mu_t(E_a)$ represent the total mass attenuation coefficients of the sample at incident and fluorescent X-ray energy respectively.

In the four parts compounding the above equation, the first depends only on the source, the second only on the detector, the third and fourth both on the source and detector with a modelling of the self-absorption of the primary and fluorescent

Table 4.2 Depths of irradiated objects producing 90 % of the fluorescent radiation

Object	Fluorescent radiation of element	Depth involving 90 % of fluorescent radiation (μm)
Fresco	Sulphur (2.3 keV)	20
Fresco	Iron (6.4 keV)	300
Fresco	Tin (25.3 keV)	1000
Bronze	Copper (8 keV)	55
Bronze	Lead (10.5 keV)	18
Bronze	Tin (25.3 keV)	120
Gold	Copper (8 keV)	5
Gold	Gold (9.7 keV)	9
Gold	Silver (22 keV)	14

radiation in the sample. The first three parts are fixed in a given instrumental set-up while the fourth depends also on the sample density and composition. This is a problem because it is variable in a way that can be taken into account only knowing the results of the analysis. There are two theoretical models that can help in the XRF quantitative analysis: (i) thin and (ii) thick sample.

In these two specific situations, the Eq. 4.3 greatly simplifies.

For more information, we rely on specific publications [3, 4].

The majority of the objects generally appear to EDXRF analysis as "infinitely thick samples", in the sense that the thickness of the objects is much greater than the "radiation penetration". This is the general case of solids and liquids such as solutions, minerals, soil samples, artefacts like statues, columns, alloys and so on [5].

To give an idea of what "thick" means for the X-rays involved in XRF, the depths of irradiated objects producing 90 % of the fluorescent radiation collected by the detector are reported in the Table 4.2.

4.1.4.2 Stratified Samples and the Calculation of Thin Superficial Layers Thickness

Ancient materials are seldom homogeneous. This is a strong limitation for the quantitative analyses. In the case of presence of grains or of inclusions, the only solution is the use of micro-analytical methods, such as electron microscopy. Sometimes, there is a superficial coating or, as is the case of painted surface, a layered structure. In this case, the fluorescent radiation coming from the inner layers is partially or totally absorbed by the outer ones. This makes a real quantification practically impossible. For example, in the case of frescoes there is a thick layer of plaster over which thin layers containing the pigments were painted (with a thickness from fractions of mm to mm). Finally, over the pictorial layer there is, in many cases a thin layer due to pollution effects (tenths of μm) containing, for example, sulphur, typically in the form of gypsum. Recently, different methods have been developed that use the knowledge of the production technologies to verify the presence and characteristic of each layer. In the case in which a coating is present over a chemical element that responds in XRF, it is possible to determinate the thickness of the superficial coating by the change of the ratio of two peaks emitted by the element belonging to the bulk.

The ratio of two characteristic lines of one element is, in principle, constant, apart from the influence of self-absorption. It can be calculated in the case of a bulk sample:

$$\left(\frac{K_\alpha}{K_\beta}\right) = \left(\frac{K_\alpha}{K_\beta}\right)_0 \frac{\varepsilon(K_\alpha)\,(\mu_0+\mu_2)}{\varepsilon(K_\beta)\,(\mu_0+\mu_1)} \tag{4.4}$$

where K_α and K_β are the intensities of the two K-lines of an element and $(K_\alpha/K_\beta)_0$ is the tabulated value, valid for an infinitely thin sample. If a thin layer of thickness d is present, this ratio varies by a factor that can be calculated

$$\left(\frac{K_\alpha}{K_\beta}\right)' = \left(\frac{K_\alpha}{K_\beta}\right) e^{\left(\mu_2'-\mu_1'\right)d} \tag{4.5}$$

where μ_1 and μ_2 are the mass attenuation coefficients of elements of bulk, in the superficial layer of thickness d at the energies of K_α, K_β lines, respectively. Thus, it is possible to calculate the coating thickness. However, knowing (or hypothesising) the inside material composition it is possible to use any two peaks in order to estimate the superficial layer thickness. In this case in the Eq. 4.5 only the second term, which is replaced by the calculated peak ratio (measured) without the coating.

Some applications of this methodology have been developed for the detection of superficial layers in (i) painting and frescos, (ii) the determination of gilding thickness (as is the case of gold leaf in Byzantine icons, but also in Italian early renaissance period (e.g. Giotto), (iii) the identification of specific production techniques with the qualification of the surface (as is the case of Tumbaga in the pre-Columbian metallurgy in the Andes) [6].

As an example, in the case of Giotto's haloes in the chapel of the Scrovegni the complexity of the X-ray spectra puts in evidence of the presence of various pigment layers below the gold leaf. Applying the above equations, it was possible to reconstruct the stratigraphy shown in Fig. 4.6.

In the case of metal alloys, it is very useful to investigate (i) the gilding techniques—that have been extensively investigated by XRF and PIXE—and (ii) the qualification of prehistoric gold, for example of archaic Greece and pre-Columbian period. A systematic study of Peruvian Gold with portable XRF was recently carried out [7] using portable instruments. The finality was that to determine the technology (-ies) used and its (their) dissemination. It was possible to define that pre-Columbian used some peculiar technologies and the metallurgy in the area also showed periods of regression that are easily recognisable.

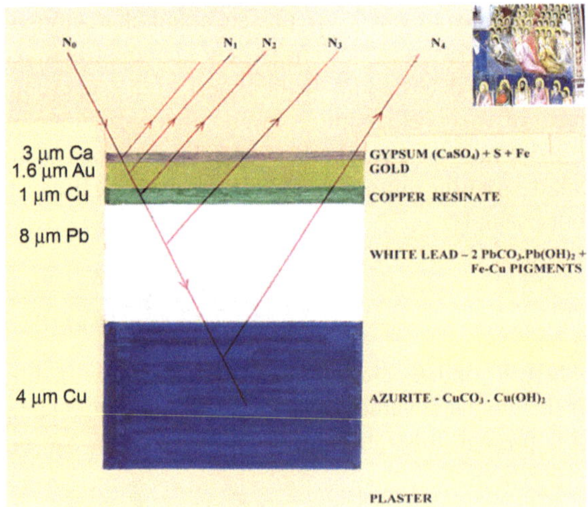

Fig. 4.6 Possible "reconstruction", through XRF measurements, of various layers present in a golden halos in the Scrovegni chapel by Giotto, Padua

4.1.4.3 Analysis and Surface Mapping of Low Atomic Number Elements with Portable Spectrometers

Deterioration of artefacts often starts from the surface by the action of chemical species, such as SO_x and Cl^-, which transform the materials composing the surface itself. Usually, outdoor monuments and works of art suffer more from these transformations but also in the indoor environment wrong ventilation and air pollution can produce similar problems.

In the last decade, the very small thickness of the beryllium window of new X-ray detectors and the developments of low noise electronics made it possible to use portable XRF systems in the analysis of elements with atomic numbers below calcium and above sodium. There are several considerations that have to be done to clarify this new possibility of use of portable XRF spectrometers. Only the surface or, better, the deteriorated part of it is analysable due to the very low energy of fluorescence photons that are related with the chemical elements present. The equipment can be very small, because it is possible to use a very low-power X-ray tube emitting photon of low energy; this improves the portability features (Fig. 4.7).

For example, on stone monuments the weathering deterioration process is caused mainly by human activity. One of the worst pollutants for stone monuments is therefore associated with sulphur compounds and, when all is said and done, with gypsum. Gypsum is a rather soluble compound that may be washed away by rain. It leaves the surface of the stone clean, but eroded and open to new corrosive processes.

Gypsum is produced by oxidation of sulphur during the combustion of substances, which contain this compound either as an impurity in combustion fossil or as a basic component. The oxidation of sulphur dioxide gives rise to a sulphur trioxide which after reacting with water, both in the liquid and gaseous states, gives rapid rise to sulphuric acid, the main cause of the acid-rain phenomenon.

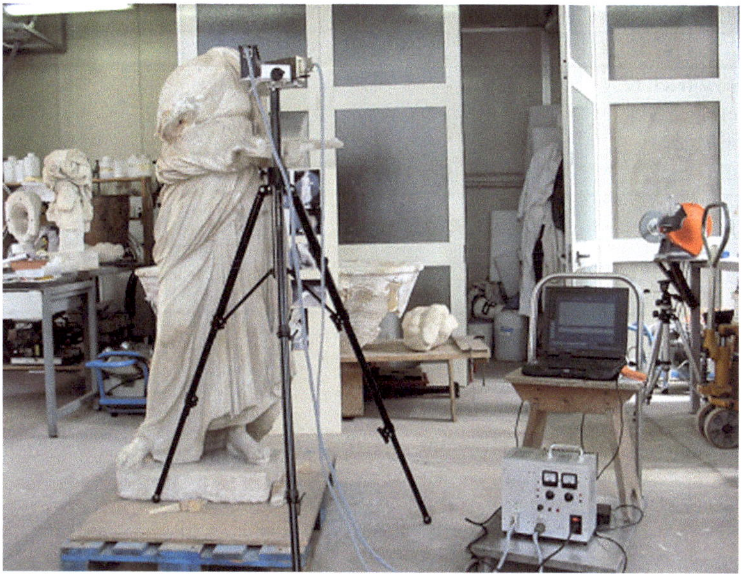

Fig. 4.7 Low energy XRF apparatus at work in the mapping of the ancient roman statue "Fanciulla di Anzio" in Rome

Once it is released into the air, sulphur dioxide remains suspended for a number of days undergoing transformational reactions, principally into sulphuric acid which precipitates with fog and acid rain, depositing itself on surfaces and monuments.

With a portable XRF equipment, it is possible to perform in situ investigations, mappings, registrations and evaluations of stone-degradation phenomena related to sulphate formation on stone and stone-like materials by quantifying sulphur concentrations directly associated with the presence of gypsum [8].

As an example, we show the analyses fulfilled on the ancient roman statue "Fanciulla di Anzio" and on the bronze monument of Sixtus IV by Antonio del Pollaiolo.

The "Fanciulla di Anzio" was found in the Imperial Villa known to belong to Nero, after a sea flood in 1878. In 1907, it was purchased by the Italian State and became a part of the National Roman Museum's collection at the Baths of Diocletian in Rome. The XRF analyses at low energy were made to support the restoration process. The presence of sulphur and other elements was used to ensure a proper cleaning of the surface.

The measurements carried out on areas of the statue which underwent cleaning yielded a sulphur count below the instrument's detection level, such as a recent fracture where a "new" surface came to light, while the measurements carried out on areas which were not cleaned yielded sulphur reading of 0.5 %.

In Table 4.3, we show the quantitative results of the measurements on the surface of the statue.

Table 4.3 Results on the statue surface on dirty or cleaned area

Table	S (%)	Cl (%)	K (%)
Body, new fracture	–	–	–
Body, cleaned area	–	–	–
Body, dirty area	0.44	0.23	0.64
Body, dirty area	0.5	0.5	1.08

Table 4.4 Results on the alloy composition of two of the panels

		Cu (%)	Sn (%)	Pb (%)	Ag (%)	Sb (%)	Fe (%)	Ni (%)	Zn (%)
Panel 1	Foil	86.7	9.1	2.4	0.1	1.4	0.3	–	–
	Body	85.9	11.7	2.3	0.1	–	–	–	–
	Leaf	88.0	9.4	2.5	0.1	–	–	–	–
Panel 2	Leaf	87.7	8.7	2.7	0.1	0.2	0.5	–	–
	Foil	82.2	13.1	1.9	0.1	2.3	0.4	–	–
	Foil	83.5	3.0	3.2	0.1	–	0.4	3.3	6.4
	Body	86.0	11.5	2.5	–	–	–	–	–

The cleaning of the surface was ended only when XRF did not detect any more presence of Sulphur and Chlorine.

The monument of Sixtus IV by Antonio del Pollaiolo, which was originally placed in the Choir Chapel in the Constantine Basilica of San Peter's in the Vatican, is now in the Treasure Museum of the Vatican. The masterpiece was commissioned by the Cardinal Giuliano della Rovere, nephew of Pope Sixtus, and later pope Giulio II.

The work was created between 1484, the year in which the Pope died, and 1493, the date inscribed on the monument together with the artist's signature.

The monument is composed of several pieces melted with a lost wax casting technique. By the feet, a large bronze epigraph in Latin describes the life of the Pope.

The monument of Sixtus IV has been undergoing restoration and XRF with high and low energy has been applied to determine the composition of the alloy of the different pieces and to map the distribution of low energy elements, like sulphur and chlorine, on the surface of the monument to aim the latter sampling for XRD.

To produce valuable XRF analyses on a bronze for bulk results, it is of paramount importance to eliminate the surface enrichments by scratching away a suitable layer of patina. Aside from all technical and theoretical matters, to do a good analysis, it is necessary to study a homogeneous amount of alloy.

In Table 4.4, part of the results of the quantitative analyses carried out on the alloy is shown.

The results suggest that the panels were produced in successive phases and welded together at the end of the process.

In Fig. 4.8, we show a moment of the analyses from a sky view.

Fig. 4.8 XRF instrumentation on the monument of Sixtus IV by Antonio del Pollaiolo, Vatican city

From a X-ray low energy point of view, all the surface of the statue was studied to map the elements S, Cl, Ar, K, Ca, Mn, Fe, Cu. In this case, it is necessary to eliminate all kinds of filters in front of the X-ray tube and detector, in order to maximise the flux of photons that excite the searched elements in the sample and to reduce the amount of photons absorbed by the filters. The system must be set as close as possible to the surface in order to reduce the path in air. The beryllium windows depth plays a role of great importance. Detectors with a beryllium window of 12 microns are now commercially available.

The systematic low energy EDXRF mapping allowed determining that K, Ca, Mn and Fe are elements involved in a surface treatment. In fact, there is a very good correlation between the results of such elements in all panels. About sulphur, its presence on the surface of a metal statue can be due to natural urban patinas that are composed of basic sulphates or artificial patination made of sulphides, yielding a surface of copper sulphides that give the bronze a brownish colour. Deposition of gypsum (calcium sulphate) is another possible source of sulphur. Since XRF is not able to give additional information of the chemical state of the sulphur, it is not possible to assess whether this element is part of a corrosion layer, a natural deposition or an artificially applied treatment. The importance of mapping the surface non-destructively with XRF is to have discrimination on areas on which to apply more invasive and expensive techniques such as XRD.

4.2 The Small Angle Scattering Technique: An Effective Tool in the Structural Investigation of Archaeologically Relevant Specimens.

Delia Chillura Martino, Michele A. Floriano and Eugenio Caponetti

Abstract An overview of applications of Small Angle X-ray Scattering (SAXS) techniques to topics of interest in the field of Cultural Heritage is presented. The basic concepts of the technique, a description of sources and laboratory instrumentation and some models and methods for data analysis are discussed. Applications of SAXS to various kinds of materials are reported to obtain information on the structure useful to shed light in some subjects such as preparation, physical treatment, traceability and degradation of materials.

4.2.1 Introduction

The development of new techniques and new instrumentation has generated an increasing interest in the physical investigation of artefacts in order to characterise their morphology, structure and composition. The number of scientific papers on this topic published either on highly specialised journals as well as in archaeometry journals is rapidly growing in the last years.

X-ray diffraction is a well-consolidated technique. In the last decade, neutron diffraction has become of interest due to its unique applications in the structural characterisation of marbles, ceramics, metals and alloys [9–17]. Several works have clearly demonstrated the prospective associated with the deep penetration (up to several centimetres) of thermal neutrons and the absence of damaging effects.

However, the above structural techniques cannot provide information on the presence of heterogeneities in a sample and on their size, shape and spatial arrangement. In this respect, it has been demonstrated that small angle scattering (SAS) techniques are the most useful ones. In fact, they probe portion of samples big enough to be assumed statistically representative of the whole and they do not require any particular manipulation of the sample. The main disadvantages are the need of complex instrumentations and the laborious data manipulation necessary to get structural information.

Indeed, these techniques have been applied in the field of Cultural Heritage to the investigation of a variety of materials with the aim to obtain details on the presence of porosity in degraded metals [18], on the structural characterisation of nanoparticles in glazed ceramics [19], on some features in petrified woods [20], on historical parchments [21]. In any case, the main application of SAS techniques is on the structural investigation of archaeological bones [22, 23].

The application of small angle neutron scattering (SANS) is limited to few particular cases [24, 25] that deserve deep investigation and peculiar experimental set-up probably because of the availability of neutrons only in large-scale facilities and the often difficult data analysis to get structural information. Due to the laboratory instrumentation that guarantees easiest availability of the techniques, in the following we will focus only on the use of the SAXS. The basic principles of SAXS and SANS methods are the same. The main difference lies in the kind of interaction between the two probes, X-rays or neutrons, and the specimens, in fact, the former interact with electrons and the latter with nuclei.

4.2.2 Theory

SAS is always observed when density in homogeneities exists in a sample. Since X-rays are primarily scattered by electrons, SAXS probes electron density in homogeneities.

The basic concept describing the scattering process can be summarised as follows. For simplicity, it is possible to refer to the diffraction process in a crystal. Any scattering process is accompanied by a reciprocity law, i.e. the distances among atom in a crystal (in the real space) correspond to the distances among planes indexed by the Miller indices in a crystal lattice (in the conjugated space). This implies an inverse relationship between distances and scattering angles. With typical X-ray wavelengths (0.1–1 nm), in the small angle regime, distances ranging between few and hundreds of nanometres are probed and, as a consequence, rather than observing atoms organisation within a crystal (as performed by wide angle X-ray scattering or XRD) one can evaluate size, shape and spatial arrangement of heterogeneities constituted by cluster of atoms that, thus, can be considered as scattering particles.

The electrons resonate with the frequency of the X-rays. As a consequence, each atom emits interfering coherent secondary waves. The interference is constructive when the phase difference fulfils the Bragg condition, and for large objects, the scattering maximum will be observed in the direction of the zero scattering angle, where all waves are exactly in phase. Incidentally, since the electron density is involved in the scattering process, the scattering intensity will depend on the atomic number of the elements contained in the sample.

A schematic diagram of the scattering process is shown in Fig. 4.9, where $\boldsymbol{k}$ is the incident beam wave vector, $\boldsymbol{k}'$ is the scattered beam wave vector, 2θ is scattering angle and $\boldsymbol{Q}$ is the scattering vector ($\boldsymbol{Q} = \boldsymbol{k} - \boldsymbol{k}'$).

For an elastic process

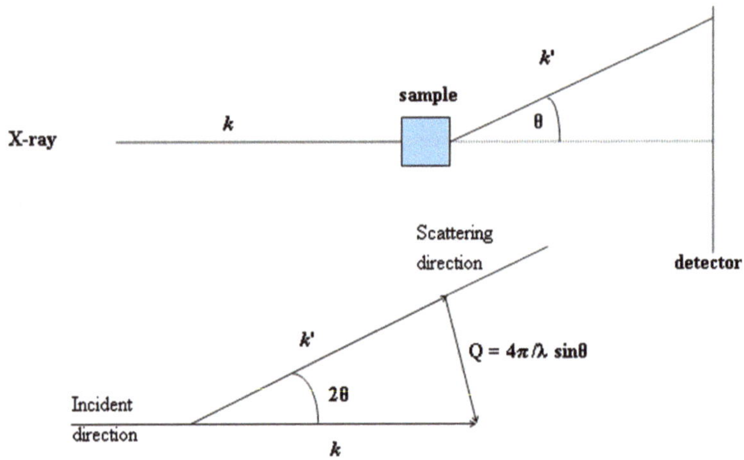

Fig. 4.9 Scheme of the scattering experiment

$$k = k' = 2\pi/\lambda \tag{4.6}$$

It follows that the scattering vector modulus is:

$$Q = \frac{4\pi}{\lambda}\sin\theta \tag{4.7}$$

The scattering of various particles can be treated at various degrees of detail and specific information on the size, shape, polydispersity, spatial arrangement and interaction among particles have been obtained since 1938, when Guinier postulated the scattering form spherical isolated particles [26]. Full details on theory and data analysis can be found in fundamental textbooks [27, 28].

Here, an introduction to some analysis methods applied to some topics in the field of Cultural Heritage will be given. Each of these methods applies in a well-defined region of the scattering pattern. Many other methods are used to get structural information from SAS data as can be inferred from specialised literature [27, 28], but their presentation is far from the aim of this paper, thus no further discussion will be given on them.

4.2.3 Instrumentation

The instrumentation used for small angle X-ray experiments can be, mainly, described as composed by a source, a sample holder and a detector. Obviously, many accessories are necessary for physical or instrumental reasons, like monochromators, slits to control the beam geometry, size and divergence, vacuum pumps, cooling systems, electronic control devices and so on.

4.2.3.1 Sources

The X-ray sources, both tubes and synchrotron, are key components for any type of material analysis by means of XRD, X-ray scattering, XRF or chemical or biological structure determination. The first developed source was the vacuum tube that allowed the improvement of instrumentation devoted to structural or medical investigation. Unfortunately, the flux per unit area of the radiation source per unit solid angle of the radiation cone per unit spectral bandwidth, i.e. the brilliance of emitted light is low, thus powerful tubes, like the rotating anode tube ensuring an higher brilliance, have been implemented. This permitted the diffusion of specialised laboratory equipments that allowed accurate structural investigations. In particular cases, especially when organic materials or fast processes have to be studied, it is necessary to use synchrotron radiation where the beam brilliance is several order of magnitude higher than the powerful laboratory sources.

Some samples that are of special interest in archaeometry scatter X-rays strongly, and essential information about the crystal habit can be obtained with conventional X-ray laboratory instrumentations.

X-ray tubes. In a vacuum tube, X-ray can be produced when accelerated electrons hit a target (anode). The most used X-ray laboratory source is the Coolidge tube where the electrons are produced by a tungsten filament (cathode) by thermionic effect. The emitted X-ray spectrum depends on the anode material and the applied accelerating voltage. The rotating anode is an improvement of the Coolidge tube. Because X-ray production is very inefficient (most of incident energy is converted to heat), the dissipation of heat at the focal spot is one of the main limitations on the electrical power which can be applied. By sweeping the anode past the focal spot, the heat load can be spread over a large area, greatly increasing the power rating. The anode consists of a disc with an annular target close to the edge.

Synchrotron source. The synchrotron radiation occurs when a charge moving at relativistic speed follows a curved trajectory. Since the first observation of visible light produced by accelerated charged particles at the General Electric Research Laboratory in Schenectady, New York, on April 24, 1947, synchrotron radiation in the X-ray and UV spectral regions has become a premier research tool for the study of matter and facilities around the world constantly evolve to provide this light over more useful forms. Synchrotron radiation sources provide high brilliance and highly parallel X-ray beams that can produce images from samples containing poor density contrast.

Synchrotrons are large-scale facilities that are constantly evolving. For this reason, in this chapter detailed information will not be given. The interested users can find information at http://www.lightsources.org.

4.2.3.2 An Example of Laboratory X-Ray Instrument

Among various laboratory instruments it will be described, as an example, the Bruker AXS Nanostar-U instrument, shown in Fig. 4.10, available at the Centro

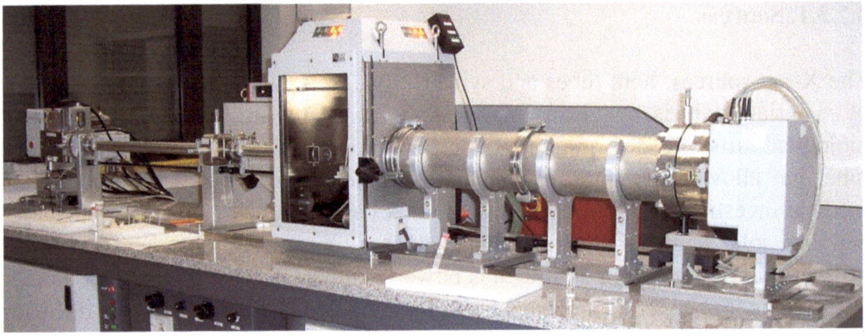

Fig. 4.10 Picture of the Bruker AXS Nanostar-U instrument

Grandi Apparecchiature of Palermo University. This instrument has a Cu rotating anode as source that can work at 40 kV and 18 mA. The X-ray beam is monochromatized at a wavelength $\lambda_{Cu} = 1.54178$ Å using two cross-coupled Göbel mirrors and is collimated using a series of three pinholes. The detector, a Hi-Star Bruker, is constituted by a 2D Multiwire Proportional Counter with a large imaging area (11.5 cm diameter). It is sensitive to the energy of X-ray used and can collect a data frame of 1,024 × 1,024 pixels with a pixel size of 105 μm.

The Nanostar is equipped with an extended chamber. Thanks to its generous size, it provides enough space to instal additional components for specialised handling of samples. Samples depending on their nature can be placed inside lime-glass capillary or directly mounted on the XY sample stage. This allows mounting of several samples simultaneously. The precise sample position is checked by software and then the bidimensional pattern is recorded.

This configuration provides advantageous results for structural investigation of samples of interest in the field of Cultural Heritage. In fact, fragments of an artefact can be directly inserted and eventually a map of the structure can be obtained. Data can be collected at room temperature or, when required, heated up to 150 °C. Temperature is controlled by a thermocouple and can be remotely changed (TLK-39 by Tecnologic SpA). The acquisition time depends on the property of the material under investigation and on the required statistic.

As an example, data acquired on a sample of archaeological bones from Iles de Porros, described in the following, are shown in Fig. 4.11. On the left is shown the pattern that has been acquired on a fragment of the sample, on the right is shown the pattern that has been acquired on the same sample after grinding. The former pattern result anisotropy as a consequence of an oriented arrangement of hydroxyapatite crystals in the fragment, it could be properly analysed to get information on the preferred orientation of crystals [29]. The latter pattern results isotropy indicating that in the powder, crystals are randomly distributed. In this case, by radially averaging, i.e. by an integration in annular rings, data are converted in the mono-dimensional scattering intensity versus the scattering vector $\boldsymbol{Q}$. The intensity is thus computed in arbitrary units after correction for the sample transmission and for the contribution of the empty capillary.

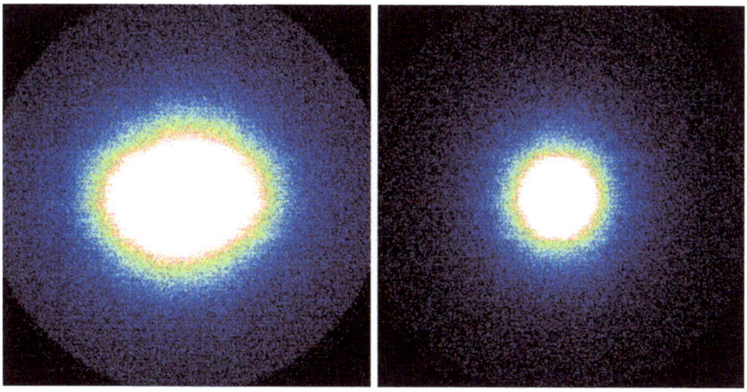

Fig. 4.11 The SAXS bidimensional patterns acquired on: a bone fragment (anisotropic, on the left); a bone powder (isotropic, on the right)

4.2.4 Models for Data Analysis and Some Applications

Once the intensity versus $\boldsymbol{Q}$ has been obtained, it can be compared to the computed on the basis of a model which depends on particle size, correlation lengths, shape, polydispersity, spatial arrangement and interaction among them. The model, based on some preliminary knowledge on the system allows to compute the whole scattering pattern. Alternatively, approximate analysis methods can be applied in well-defined regions of the scattering pattern.

Various models and methods have been reported in the literature. In the following, a short summary to some approximate analysis methods and to some models proposed to analyse SAXS pattern, will be given. For each model or approximate method a specific application will be reported. The attention will be focused first on the determination of structural detail of particles then on the structure of surfaces and interfaces. Finally, some models able to derive information on the particle size and the kind of interfaces will be presented.

Guinier approximation. In 1939, Guinier, in the limit of very small angle and for an assembly of non-interacting homogeneous particles, proposed the following expression:

$$I(\boldsymbol{Q}) = N\Delta\rho^2 V^2 \exp\left(-\boldsymbol{Q}^2 R_g^2/3\right) + C \tag{4.8}$$

where N is the particles number density, $\Delta\rho$ is the difference in the electron scattering density between the particles and the medium, V is the particle volume, Q is the scattering vector, R_g is the gyration radius and C is the background which includes the incoherent scattering of the sample and the instrumental noise. The scattering intensity expected for such a system is shown in Fig. 4.12 together with a pictorial representation of the model describing the system.

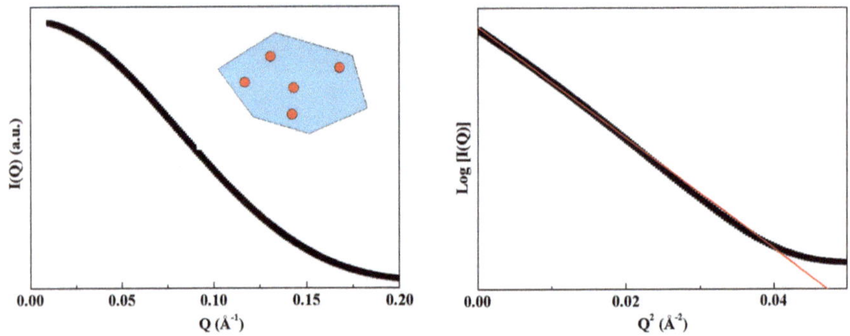

Fig. 4.12 Scattering intensity simulated for a system of homogeneous, monodisperse, spherical and non-interacting particle, *Left* I(Q) versus the scattering vector, Q. *Right* log(I(Q)) versus Q^2 (*Guinier* plot). In the *inset*, a pictorial representation for such a system is shown

By plotting the logarithm of the intensity versus Q^2 (Log $[I(\boldsymbol{Q})] = \text{cost} - \boldsymbol{Q}^2 R_g{}^2/3$) the R_g value can be evaluated from the slope of the linear fit of the experimental curve. R_g is defined as the mean square distance from centre of gravity, where the role of the "mass" is played by the electron density. In the Guinier formula, the only unknown parameter is R_g, and it holds surprisingly well in most cases. For an assembly of monodispersed spherical particles $R_g = R\sqrt{(3/5)}$ where R is the geometrical radius of the particle. Similar relations were developed for other particle shapes [27].

In the field of Cultural Heritage, the Guinier approximation has been applied to evaluate R_g of silver and copper nanoclusters in some Renaissance lustre decorated ceramics [19]. Lustre can be described as a thin film containing silver and copper nanocrystals dispersed in a silicium-rich matrix [30].

The above method has been, also, applied to characterise the size of a particulate deposited on a parchment sample. This investigation was only an aspect of a more general purpose, i.e. to get information on the possible collagen degradation induced on parchment as a consequence of particular cleaning processes like laser ablation [21].

Hosemann–Joerchel method. For systems composed by a polydisperse distribution of spheres the *Guinier* approximation does not hold: the graph of log($I(\boldsymbol{Q})$) versus $\boldsymbol{Q}^2$ is not linear. The simulated scattering pattern from a polydispersed distribution of spherical non-interacting particles is reported along with a pictorial representation of the system and the *Guinier* plot in Fig. 4.13.

The Hosemann–Joerchel method allows to evaluate the particles mean size, $\langle R \rangle$, and the polydispersity, $g_y = [(\langle R^2 \rangle - \langle R \rangle^2)/\langle R \rangle^2]$ assuming a Maxwellian distribution function for the radius R [31, 32].

By applying this method, experimental intensities are plotted as $Q^2\ I(Q)$ versus Q in order to obtain Q_m and Q_0. Q_m corresponds to the maximum of the curve and Q_0 is the abscissa at the point where the tangent, in the flexus point on the right side of the curve, intercept Q-axis as shown in Fig. 4.14.

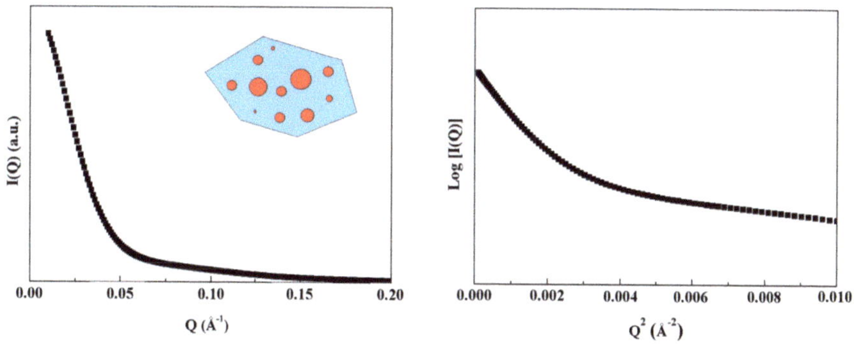

Fig. 4.13 Scattering intensity simulated for a system of homogeneous, polydisperse, spherical and non-interacting particle. *Left* I(Q) versus the scattering vector, Q; *right* log(I(Q)) versus Q^2 (*Guinier* plot). In the *inset*, a pictorial representation for such a system is shown

Fig. 4.14 Hosemann–Joerchel plot

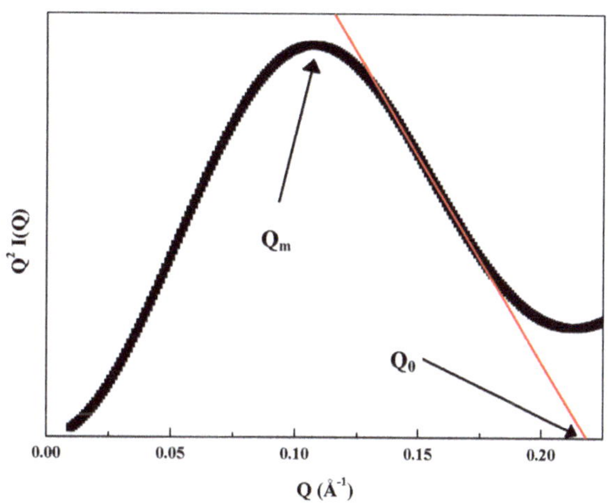

The ratio between Q_0 and Q_m gives the M^* parameter, which is related to $\langle R \rangle$ and g_y by the following formula

$$\langle R \rangle = \frac{0.96}{Q_m} \frac{9.6 - 2M^*}{\sqrt{7.5 - M^*}} \quad g_y = \sqrt{\frac{M^* - 2.1}{9.6 - 2M^*}} \tag{4.9}$$

and the value of the average interparticle distance, D, and the particle packing density, ε_y, can be obtained by

$$D = \frac{2\pi}{Q_m} \quad \varepsilon_y = \left(\frac{2r}{D}\right)^3 \frac{\pi}{6} \tag{4.10}$$

In the field of Cultural Heritage, the Hosemann–Joerchel method has been applied to evaluate the structure of silver and copper nanoclusters in lustre for which the Guinier approximation did not hold [19]. The knowledge of both

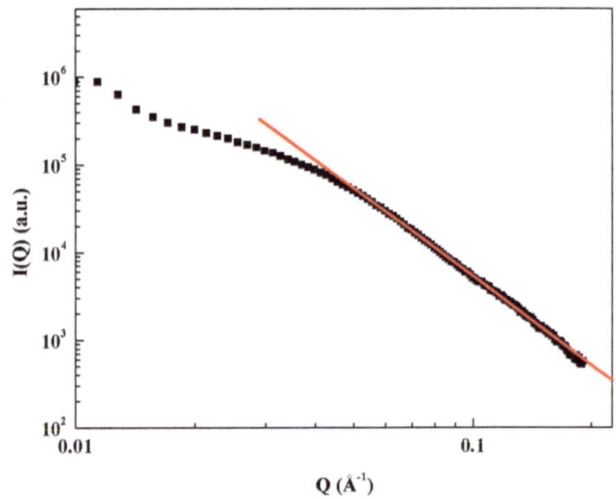

Fig. 4.15 Log–log plot of the scattering intensity I(Q) versus Q acquired for a sample of archaeological bone from the necropolis of S'Illot des Porros. The slope of the linear fit gives the exponent of Q in the power law in Eq. 4.11

composition and structure could help in understanding why recent glazed pottery is not comparable in quality to that achieved in the past.

Porod approximation. Porod in 1951 and Debye et al. in 1957 independently demonstrated that the slope of the log–log scattering curve depends on the fine structure of the particle surface. They derived a fourth power law which relates the scattering intensity in the high Q region of the SAS to the nature of the particle surface, S, by means of the following formula:

$$I(\boldsymbol{Q}) = 2\pi N \Delta\rho^2 S/\boldsymbol{Q}^4 + C \tag{4.11}$$

where the meaning of N, $\Delta\rho$ and C is the same as in Eq. 4.8.

This law holds not only for non-interacting particles, but also for densely packed particles as well as for non-particulate structures provided that a well-defined internal surface exists. For particles with a diffused interface, the $\boldsymbol{Q}$ power is different from 4.9. The surface can be described as diffused up to fractals depending on the exponent value derived from the slope of the linear fit as shown in Fig. 4.15.

Recently, the method has been applied in the field of Cultural heritage to some archaeological bones found in the necropolis of S'Illot des Porros, one of the most important prehistoric funerary sites of the Balearic Islands (Spain) [33] together with the method—described below—proposed by Fratzl, to obtain information about the alteration observed. It was concluded that the human bones were subjected to different temperatures accounting for the incineration processes involved.

The Porod approximation has been applied, together with the method described by Axelos et al. [34] to determine the morphology of interfaces and the pore size distribution in some samples of petrified wood [20]. The aim of the paper was to get some light on the mechanism of mineralization process of the organic objects, especially trees.

Fratzl method. This method was early proposed by Fratzl et al. [35] in 1992; later improved [36] in order to characterise the thickness and orientation of mineral particles in bones. In the model, the bone is approximated as a mixture of two phases—the mineral particles and the organic matrix. Assuming that the mineral phase is an assembly of individual slightly polydisperse crystals (needles or plates), the intensity was expressed as

$$I(\boldsymbol{Q}) = I_0 T^3 F(\boldsymbol{Q}T)[1 + G(\boldsymbol{Q})] \tag{4.12}$$

where I_0 evaluated at $\boldsymbol{Q} = 0$ depends on the mineral density in the sample, $T = 4$ V/S (being V the volume and S the average surface of the crystal) is a measure of the average size of the crystals, $F(\boldsymbol{Q})$ is a form factor characteristic for the shape of the mineral crystals, $G(\boldsymbol{Q})$ is the structure factor that reflects the degree of order of the crystals, e.g. for a randomly oriented ensemble of crystals $G(\boldsymbol{Q}) = 1$. If the two phases are separated by a sharp interface then the Porod law (Eq. 4.11) holds and T can be directly determined from $I(\boldsymbol{Q})$:

$$T = \left(\frac{4J}{\pi P}\right) \quad \text{were} \quad J = \int_0^\infty \boldsymbol{Q}^2 I(\boldsymbol{Q}) d\boldsymbol{Q} \quad \text{and} \quad P = \lim_{Q\to\infty} \boldsymbol{Q}^4 I(\boldsymbol{Q}) \tag{4.13}$$

This approach has been used in several works to investigate archaeological bones in order to understand the effect of postmortem combined with archaeological changes on the mean crystal thickness. Some processes, such as cooking or cremation, are expected to have a more global effect on crystal habit, whereas bacterial or hydrological changes are expected to produce changes in bone character at the micron length scale. The knowledge of changes in crystal habit, orientation and the statistical distribution of mean thickness can lead to the understanding of the archaeological remodelling that occurs at the molecular level and manifests itself at the level of macroscopic bone integrity [23, 24, 33, 37, 38]. *Beaucage method*. Beaucage in 1995, first, proposed a method, called the *Global unified scattering function* to analyse data for a system composed by related or unrelated levels of structures, i.e. for system composed by clusters of small primary particles. This method allows the simultaneous evaluation of the gyration radii of both primary and cluster particles and gives information on the kind of interface of both primary particles and clusters [39]. The method is based on the observation that an appropriate combination of *Guinier* (Eq. 4.8) and *Porod* (Eq. 4.11) approximations, in some cases, can successfully describe the experimental intensity:

$$I(\boldsymbol{Q}) = G \exp\left(\frac{-\boldsymbol{Q}^2 R_g^2}{3}\right) + B(\boldsymbol{Q}^*)^{-d}; \quad \boldsymbol{Q}^* = \frac{\boldsymbol{Q}}{\left[\operatorname{erf}\left(\boldsymbol{Q}R_g/6^{1/2}\right)\right]^3} \tag{4.14}$$

where $G = N\,\Delta\rho\,2V2$, $B = 2\pi\,N\,\Delta\rho\,2S$ and the meaning of N, $\Delta\rho$, V and S is the same as in Eqs. 4.8 and 4.9. R_g is the gyration radius, 'erf' is the error function accounting for the approximations introduced in the derivation of the law [39]. The exponent d is related to the fractal dimension.

Equation 4.15 is written for a single level of structure, such as a collection of spherical primary particles, but it can be extended to two levels of structures (i.e. a collection of primary particles that assemble to form clusters), as described by Beaucage [40] for mass fractal-aggregates.

In the field of Cultural Heritage, this approach has been used in order to evaluate both gyration radii and the fractal dimension for some pottery samples. The question was whether the size and surface characteristics of the aggregates of minerals are parameters sensitive to the firing technology used to produce the pottery [41]. In particular, the authors tested the possibility of determining the firing parameters and their correlation with the age of the finds, enhancing the knowledge on some technological choice made in producing the ceramics found in Cuma and Miseno. Moreover, the estimation of firing temperature of this ceramic typology could be useful to identify the production centres [42].

Teixeira method. Teixeira in 1988 modelled the scattering intensity as [43]:

$$I(\boldsymbol{Q}) = AP(\boldsymbol{Q}, r, d_s)S(\boldsymbol{Q}, r, d_m, R) \tag{4.15}$$

$$P(\boldsymbol{Q}, r, d_s) = \left(1 + \frac{\sqrt{2}}{3}\boldsymbol{Q}^2 r^2\right)^{\frac{d_s - 6}{2}} \tag{4.16}$$

$$S(\boldsymbol{Q}, r, d_m, R) = 1 + \frac{d_m \Gamma(d_m - 1)}{(\boldsymbol{Q}r)^{d_m}} \left(1 + \frac{1}{(\boldsymbol{Q}R)^2}\right)^{\frac{1 - d_m}{2}} \sin[(d_m - 1)\arctan(\boldsymbol{Q}R)] \tag{4.17}$$

where A is a constant dependent on the overall pore concentration and the chemical composition of the sample, $P(Q, r, d_s)$ is the form factor whose value depends on the shape of the individual scatterers and their surface fractal dimension, $S(Q, r, d_m, R)$ is the structure factor that reflects the degree of order in the system and mass fractal behaviour. In these equations, d_s is the surface fractal dimension, d_m is the mass fractal dimension and $\Gamma(d_m - 1)$ is the gamma function of $d_m - 1$. R and r are both correlation lengths, R gives the size of the fractal aggregates, and r is the size of the individual scatterers.

Teixeira method has been applied to some marbles. The provenance of stone object is of key importance to archaeology in so far as artistic, technological or commercial exchange patterns may be studied and correlated to historical events and social contact between cultures. It was demonstrated that the structural features at a mesoscopic level are strongly correlated to the metamorphic degree. Thus, it is believed that they can offer a fingerprint to trace the stone provenance. With this aim, various marble samples from the Mediterranean basin [24] and from Marble Canyon, Texas [25] were investigated. The scattering intensity analysis by means of Eq. 4.15 allowed to evaluate porosity, pore distributions (characterised by mass fractal dimensions) and grain surface morphology (characterised by surface fractal dimensions).

4.2.5 *Conclusions*

In this work, a short description of principles, instrumentation, and data analysis methods for SAXS and its application in the field of Cultural Heritage science was presented. The technique is widely used for structure investigation on various kinds of materials. Its application on the field of Cultural Heritage is rather new and, till date, rather exotic. With respect to the more traditional XRD it allows to evaluate size, polydispersity and spatial arrangement of particles in a sample. Moreover, it allows determining interfacial or surface structure. Data acquisition does not require any particular sample manipulation, but, surely data treatment requires a quite strong effort. This, together with the accessibility of high-flux instrumentation, accounts for the few SAXS application in the Cultural Heritage fields. Notwithstanding, the techniques reveal to be unique in order to get accurate and reliable structural details.

4.3 Energy-Dispersive X-Ray Diffraction in Cultural Heritage Science: The Winning Duo of Structural and Elemental Analysis

Lorenzo Gontrani, Ruggero Caminiti, Maria Luisa Saladino, Eugenio Caponetti and Delia Chillura Martino

Abstract The capabilities of Energy-Dispersive X-ray Diffraction (EDXD) in the field of Cultural Heritage are discussed. EDXD diffractometer, equipped with a white source, is particularly suited for the structural and elemental analysis of items having different nature. Given its compact and versatile design, the instrument allows to collect data from samples without tearing down of any portion and with no specific preparation. In the wide energy range sampled during the measurements, both X-ray fluorescence and diffraction features appear. In all cases, data are acquired in air and in a non-destructive way, and their collection is fast.

4.3.1 *Introduction: Why Energy-Dispersive X-Ray Diffraction?*

The spectroscopic techniques, employed for the investigation of various art works, based on X-ray radiation provide the opportunity to carry out analyses without sample damaging. Different kinds of X-ray analyses may be used such as XRD [44, 45], XRF [45, 46], PIXE [47] and X-ray photoelectron spectroscopy (XPS) [48]. Among these techniques, XRD represents a powerful non-destructive technique providing information on the structural features of the samples at the atomic level, although for a complete characterisation of the material the elemental composition is often desirable, since, for example, some artists may have used specific paints containing particular compounds. The detection of these components can help in tracing back the sample to a certain artist or time frame. Moreover, it may be essential to understand the elemental composition of the piece to find a means to prevent further decomposition for conservation purposes. XRF provides a valuable tool to obtain the sample bulk composition. This can be assessed since fluoresced X-rays are a fingerprint of the atoms from which they originate [49].

Structural and compositional features may be simultaneously extracted from EDXD experiments. This non-conventional diffractive technique has been generally used for the structural investigation of samples characterised by a short-range order [50]. For this reason, all systems of artistic and historical value with a low degree of crystallinity may potentially be studied by EDXD, though; the technique is not limited to non-crystalline systems only, as witnessed by some recent studies [51]. Therefore, since it provides XRF spectra in addition to XRD patterns, EDXD is a powerful tool that gathers the typical advantages of spectroscopic techniques, being completely non-destructive, with the added value of providing the structural features of the sample. Finally, the acquisition times are sensibly shorter than those typical of any other in-house apparatus being of the order of seconds [52].

4.3.2 Theory

Information about the atomic structure of a sample can be gained by observing the modulation of the intensity spectrum of a probe whose wavelength is comparable with the interatomic distances. If the probe is electromagnetic radiation, the wavelength is in the X-ray range.

The XRD theory due to Thomson and Debye (a completely classical approach) describes this modulation of the probe intensity spectrum, which arises because of interaction with the sample, as the effect of coherent superposition of waves elastically scattered by atomic core electrons that are regarded as small antennae scattering with a dipolar distribution. Despite the neglect of quantum effects, that can be anyhow added as perturbations by introducing some suitable corrections (e. g. Compton scattering), the theory is able to explain the features and the characteristics of diffraction patterns and to obtain the structural information they carry. The elastic contribution to the scattering is a function of the momentum $\Delta \mathbf{k}$ exchanged ("transferred momentum") by the radiation with the atomic tightly bound electrons. For a generic system, $\Delta \mathbf{k}$ is a vectorial quantity, having three components, and measurements of the intensity must be carried out by observing in turn each of them. However, for isotropic systems, the intensity depends only on the modulus of $\Delta \boldsymbol{k}$ and not on its orientation. In this case, a scalar quantity $\boldsymbol{Q}$ can be defined called the scattering parameter [45], which represents the modulus of $\Delta \boldsymbol{k}$ expressed in $\hbar$ units.

Diffraction patterns can then be drawn by plotting the scattered intensity as a function of Q.

The relation between q and the experimental quantities is:

$$\boldsymbol{Q} = 2k \sin\theta \tag{4.18}$$

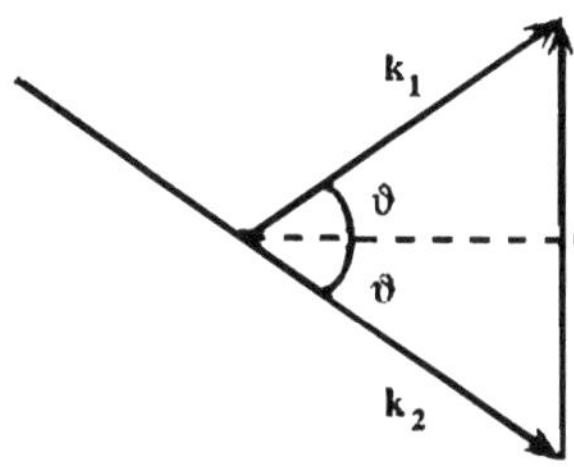

2θ being the total scattering angle and $\boldsymbol{k}$ the modulus of both wave vectors $\boldsymbol{k}_1$ and $\boldsymbol{k}_2$ of the incident and scattered radiation [46] respectively. In fact, by assuming that the scattering is elastic, the energy E of the radiation remains unchanged during the scattering process. Moreover, according to the dispersion relation of an electromagnetic radiation, $E = n\hbar c\boldsymbol{k}$, where n is the refraction index and c is the speed of light, and $\boldsymbol{k}$, the modulus of its wave vector (wavenumber) is proportional to its energy. Therefore, in the hypothesis of elastic scattering, the wave vector of radiation interacting with a sample can vary its direction but not its modulus; that is, the X-ray beam can be deflected but it maintains its wavelength unchanged. As E is proportional to $\boldsymbol{k}$ (Eq. 4.18), the following equation holds:

$$\boldsymbol{Q}(E, \theta) = \alpha E \sin\theta \tag{4.19}$$

where α is a constant (equal to $2/\hbar c \approx 1.01354$ Å^{-1} keV^{-1}) in the approximation that n at the typical energies of X-rays is always unity, regardless of the material forming the sample (usually $1 - n = 10^{-6}$, as an order of magnitude). The former relation is another way of expressing the well-known relation $\boldsymbol{Q} = (4\pi/\lambda \sin\theta)$, where λ is the radiation wavelength.

Summarising, two methods are available to perform a scan of q and to draw the scattered intensity profile as a function of it. In the first method, a monochromatic beam is used (the emission spectrum of a laboratory X-ray tube or of a synchrotron is filtered through a monochromator) and q-space is spanned with an angular scan [angle-dispersive X-ray diffraction (ADXD)], while the second one employs a continuous spectrum radiation, often called white by analogy with the visible light, for example the Bremsstrahlung of the same X-ray tube, the scattering angle being fixed (EDXD). Although it had been known for a long time, the second method became widely available only at the end of the 1960s [50, 53]. In fact, only at that time solid-state detectors with an acceptable resolution became available allowing the construction of a new kind of diffractometer.

Among the advantages of the use of EDXD to study disordered systems [49, 51], the most important are: (1) the intensity distributed in the entire white spectrum (EDXD) is, approximately, from one to two orders of magnitude higher than the intensity concentrated in the fluorescence line (ADXD), this implies a reduction in the acquisition time; (2) The apparatus remains fixed during data collection. This simplifies the experimental procedure, since the beam trajectory is fixed. In fact, unlike ADXD, no movements may compromise the alignment, inducing systematic errors. This is particularly important for the careful treatment of artistic samples. (3) Parallel collection of the spectrum points. While in the ADXD method spectrum points are collected in a sequential way, EDXD permits multiple acquisition—the points are collected simultaneously at each value of $\boldsymbol{Q}$ (or E; see Eq. 4.19).

Yet, some limitations must be pointed out: (1) Data processing is more complicated, because of the energy dependance of all involved quantities: a separate treatment of each energy component is required. (2) It is necessary to join various X-ray spectra.

In fact, according to Eq. 4.19, at a given angle, using a radiation having a continuous energy spectrum in the interval (E_{low}, E_{high}), the accessible $\boldsymbol{Q}$ range in the reciprocal space is $\Delta\boldsymbol{Q}\ (\theta) = \boldsymbol{Q}_{\text{high}} - \boldsymbol{Q}_{\text{low}} = \alpha\ (E_{\text{high}} - E_{\text{low}})\ \sin\theta$. Therefore, to cover the widest possible region of q-space, more than one measurement (i.e. more than one scattering angle, usually up to four-five) is needed. The various spectra must then be joined to reconstruct the whole diffraction pattern. (3) Decrease in the Q resolution. Besides angular resolution (shared by both ADXD and EDXD), in EDXD an additional electronic resolution term originated by the detector and by the electronic chain has to be taken into account. This is the most important reason why EDXD is preferred for non-crystalline systems: its broad peaks are less affected by the energy uncertainty than the sharp Bragg peaks of crystals. However, the versatility and rapidity of the technique makes a lot of non-conventional measurements otherwise not possible (e. g. time-resolved experiments, to follow phase transitions, etc.), so that the drawbacks are completely overcome.

As already mentioned, in addition to scattering, samples irradiated by X-rays can absorb and eventually re-emit a radiation (fluorescence). The fluorescence lines are detected and superimposed on the pure diffraction spectrum; thus, additive contributions represented by extra intensities at characteristic energies appear. Such lines that are generally removed in a diffraction experiment can be conversely very useful in the analysis of cultural heritage samples, since they yield the elemental composition (both qualitative and quantitative, if a suitable calibration is performed). It is worth noticing that the energy of a fluorescence emission is independent of the diffraction geometry (θ): this implies that the fluorescence lines do not vary with the angle. On the contrary, the diffraction peaks positions depend on both the energy and θ. This also implies that changing θ will result in the shift of diffraction peaks (ascribed to a given q value) to lower or higher energies.

Fig. 4.16 EDXD Diffractometer "at work" with a decorated eighteenth century Sicilian floor

4.3.3 Instrumentation and Data Treatment

The EDXD diffractometer shown in Fig. 4.16 operates in vertical θ/θ geometry and is equipped with an X-ray generator, a collimating system, two-step motors, and a solid-state detector connected via an electronic chain to a multichannel analyzer. The X-ray source is a Seifert tube (W target) operating at 55 kV and 45 mA whose Bremsstrahlung radiation is used whereas the detecting system is composed of a solid-state photon (HPGe) detector (Ortec) cooled by an X-cooler system and connected to a PC through ADCAM hardware.

Both the X-ray source and the detector can rotate, by an angle θ, around the focus where the volume of the sample to be investigated is placed. Alternative instrumental geometries are possible: in the $\theta/2\theta$ vertical geometry, the X-Ray tube remains fixed (X-axis), while the detector is moved vertically (Y-axis); the same configuration can be reproduced in the horizontal plane by replacing the step motors with a ring–nut system.

The measuring time for a given scattering angle, necessary to obtain data with a high signal-to-noise ratio, depends on the absorbance of the sample under investigation and ranges from thousands of seconds for the smallest angles (up to 8°) to hundreds of thousands of seconds for the largest ones (10–30°). Besides the scattering measurements, the "transmission" of the full and of the empty cell (i. e. at zero scattering angle) is recorded, and used to perform the necessary corrections for the sample absorbance. A couple of single angle measurements (that corresponds to diffraction spectra with $\boldsymbol{Q}$ range given by Eq. 4.18) may be sufficient to identify the fluorescence lines, and consequently to obtain the qualitative elemental composition of the sample, or to confront the diffraction peaks with those of known samples, to point out the presence of structural features in the specimen.

If more complete structural information is needed, the continuous spectrum in Q must be obtained, by joining the partial diffraction spectra recorded at more diffraction angles. Four measurements (at 1, 3, 8 and 24°) are enough to reconstruct the whole spectrum between 0 and 20 $Å^{-1}$. In the end, if desired, such spectrum could be Fourier transformed in space to point out specific structural correlations (i. e. hydrogen bond distances, as those occurring between proteins and painting dyes [54]).

4.3.4 *Examples of Simultaneous Detection of Diffraction and Fluorescence*

The presented technique has been successfully employed on several samples of interest in the field of Material Science. In the field of Cultural Heritage, information have been obtained on: composition and structure of layered samples (surface and cardboard of old photos), characterisation of buried samples, elemental composition of alloys of jewels, structural analysis of books, with particular regard to the ageing processes; [55] comparison of cellulose polymorphs in old and modern papers; [44] analysis of polychromies in ancient potteries from Pesaro and Faenza [56]

In the present paper, a study of an eighteenth century Sicilian finely decorated floor tile (see Fig. 4.16) from an ancient villa in Ribera (Agrigento, Sicily) is reported. The floor tile, in good overall conservation state, shows glossy front side stylized and coloured, reddish-white rear side. The aim of the investigation was to shed some light on the elemental composition of pigment used in the various decorated zones.

It is important to underline that all the experiments were performed with no removal of parts of the ancient tile, and with no modification of its properties.

Spectra at two angles were acquired to discriminate fluorescence from diffraction contribution. As an example, the spectrum of an orange coloured area is reported in Fig. 4.17. From the lower spectrum registered at 6.7° θ, several peaks were revealed. The correct attribution to emission fluorescence or diffraction is prevented unless, as previously said, the measure is repeated at a different angle. In fact, the upper spectrum acquired at 14.9° θ, shows a number of peaks falling at the same energy of some of the previous peaks. These peaks were thus attributed to the fluorescence emission of Pb, Sn, Ca, Fe and Sb. The unmatched peaks are diffraction ones. Measurements were repeated in various areas of the same colour and the spectra were superimposable.

Spectra acquired, by means of the same procedure, from the different coloured areas allowed to detect the emission lines of the elements listed in Table 4.5.

Pb and Sn are referred to the enamelled glassy surface, while Fe and Ca derive from minerals constituting the original argillaceous matrix. Fe engendered the reddish colour in the rear side of the tile. The simultaneous presence of Ca, Fe, Pb

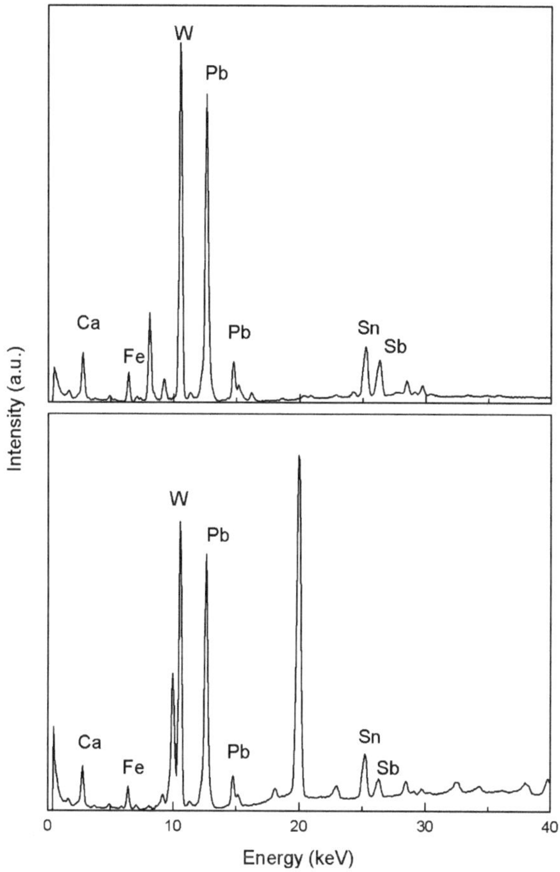

Fig. 4.17 EDXD spectra from *orange* coloured area: lower spectrum registered at 6.7° θ; upper spectrum registered at 14.9° θ. The fluorescence lines are indexed by the element symbols

and Sn, therefore, indicates that the analysed volume extends to a layer deeper than the thickness of the coloured layer.

The orange and yellow pigments are characterised by the Sb presence. It, as antimonates and mixed with Pb oxides in different ratios, accounts for the different shades.

The green pigment contains Cu, as the typical rameic compounds do.

The brownish/black colour evidences the Mn presence. It, as MnO_2, causes different brown shades [57–60].

In order to compare the diffraction spectra with those of the mineral compounds present either in the pigments and the ceramic body, the energy scale in the EDXD spectra was converted on a Q-scale by means of Eq. 4.19. As an example, data acquired at 6.7 and 14.9° θ for the orange pigment are reported in Fig. 4.18. The fluorescence lines have been manually removed in order to improve the readability of the graph.

The diffraction patterns of the Sb_2O_3*Sb_2S (orange) and $Pb_3(SbO_4)_2$ (yellow) are reported on the same ***Q***-scale at the bottom of Fig. 4.18. The comparison

Table 4.5 Elements revealed in the colour of the floor tile

Orange	Pb	Sn	Ca	Fe	Sb
Yellow	Pb	Sn	Ca	Fe	Sb
Green	Pb	Sn	Ca	Fe	Cu
Brownish/black	Pb	Sn	Ca	Fe	Mn
White	Pb	Sn	Ca	Fe	–

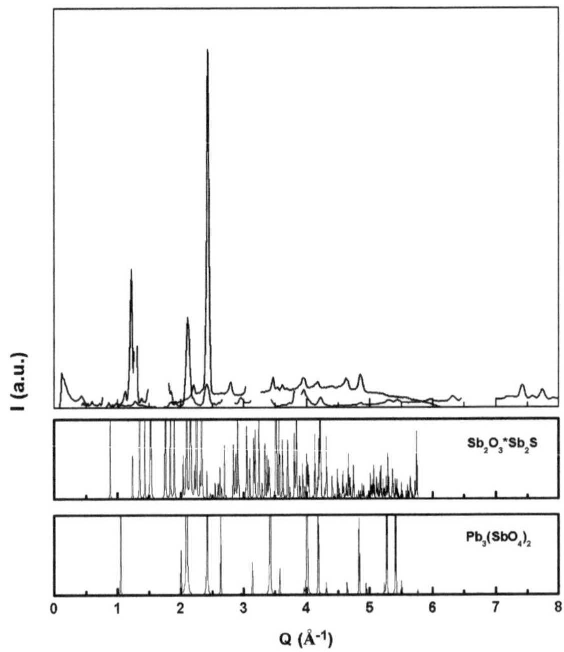

Fig. 4.18 EDXD spectra, intensity versus Q ($Q = \alpha E \sin\theta$), from *orange* coloured area. In the two graphs on the *bottom*, the diffraction patterns, on the same Q-scale, for Sb_2O_3*Sb_2S (*orange*) and $Pb_3(SbO_4)_2$ (*yellow*) are shown

between reference patterns and experimental spectra allows identifying some of the diffraction peaks thus further confirming the fluorescence findings.

Even if the study here discussed is purely qualitative, the presented approach is particular useful for the non-destructive investigation of decorated samples. It is worth noticing that the composition and the structure of pigments are strongly dependent from the raw material used and from the production methodologies.

4.3.5 Conclusions

In this work, the description of an Energy-Dispersive X-ray Diffractometer and its capabilities in the field of Cultural Heritage science is presented. The instrument proves to be well suited to get both compositional and structural information from a unique measurement. Its design turns to be particularly advantageous in the field of Cultural Heritage because it allows obtaining data from a very large variety of samples without any particular preparation and without sampling of any portion.

A study of a decorated floor tile has been presented. The analysis of the fluorescence emission from the decorated layers allowed identifying the elemental composition. In addition, the qualitative comparison of the diffraction peaks with those of reference patterns confirms the composition of the pigments, in terms of relative amounts of different minerals. The main advantage of this approach lies in the availability of both compositional and structural information by a unique measurement on the same area of the sample, thus avoiding variability of results deriving from a non-uniform distribution of components. Obviously, the method here presented can be developed in order to gain quantitative results.

4.4 Particle Induced X-Ray Emission Spectroscopy in Conservation Science

Nick Schiavon

4.4.1 PIXE: Introduction

In the often-interconnected archaeometry and conservation science research fields, acquiring data on the bulk and surface elemental composition of objects of art and archaeological artefacts constitutes an essential step in any methodological research approach. Geological provenance of raw materials, identification of ancient production techniques and reconstruction of commercial trade routes are just some of the topics in archaeology where advanced chemical analyses of materials play a crucial role in confirming (or refuting) hypotheses by historians and archaeologists. Similarly, identification of pigments in paintings, pottery and ancient manuscripts and/or characterisation of corrosion products and mechanisms affecting various cultural heritage materials (stone, glass, metal amongst others) are some of the topics in conservation where chemical data provides a much needed helping hand. In fact, it is fair to affirm that only by knowing the bulk and surface chemistry of materials a conservator may be in the position of selecting the most appropriate restoration products and methods to be applied to any given case study.

The range of analytical techniques available to scientists for the chemical characterisation of heritage materials is extremely vast. In this respect, spectroscopic analytical techniques, i.e. techniques that analyse the electromagnetic radiation absorbed, emitted or scattered by molecules or atoms in irradiated samples, now play a pivotal role in any analytical protocol in cultural heritage studies [61]. Spectroscopy-based techniques could be classified according either to the energy of the primary radiation used to "excite" the sample (linked to the radiation frequency and wavelength by the well-known photon energy equation $E = h\nu = hc/\lambda$, where E photon energy, n frequency, λ wavelength, h Plank's constant, c speed of light in vacuum) and/or to the effects that the electromagnetic radiation induce on the material irradiated. In Fourier Transform Infrared Spectroscopy (FT-IR), for example, a spectrum is obtained by irradiating a sample with infrared region (IR) radiation ($\lambda = 1$ mm to 750 nm) and by measuring what fraction of the incident radiation is absorbed at a particular energy. IR (and also visible) radiation is absorbed most efficiently when its frequency matches almost exactly the frequency of internal motions within a molecule. For radiation with frequencies in the IR, the relevant motions are the vibrations of atomic bonds within the molecules of inorganic and organic compounds, hence the name of vibrational spectroscopy as FT-IR (together with Raman) is often referred to. Other spectroscopic techniques exploit the manner in which more energetic radiation, i.e. the one falling within the X-ray range of the electromagnetic spectrum (λ in the range 0.003–3 nm and corresponding energy in the range 500–500,000 eV), interacts with and/or is emitted by matter. In this chapter, we will focus our attention on the spectroscopic X-ray-based technique called PIXE. PIXE belongs to the family of the so called Ion Beam Analysis (IBA) techniques which include also Rutherford Backscattered Spectroscopy (RBS) and Particle induced Gamma-ray Emission Spectroscopy (PIGE). In IBA, a beam of accelerated charged particles (ions) is used as primary radiation [61–65]. The ion beams required for IBA are usually produced in particle accelerator units at Nuclear Science Laboratories where ions are accelerated to energies in the range 0.2–4 MeV. In most cases, conservation scientists have to share these facilities with colleagues from other scientific disciplines. The PIXE technique, however, has now found so many applications in cultural heritage studies, that some conservation institutions have set up their own in-house IBA laboratory, the most famous one being the AGLAE accelerator at the C2RMF of the Louvre Museum in Paris (Fig. 4.19) [64].

Let us now look into more detail at the PIXE technique. For ease of reading, the chapter is subdivided into the first part outlining PIXE basic physical principles and methodology followed by a section focusing on its applications to the study of cultural heritage materials. Finally, a summary highlighting the pros and cons of the PIXE technique in Conservation science is presented.

4.4.2 PIXE: Principles, Practice and Recent Developments

The physical concept behind PIXE is basically similar to the one exploited in another spectroscopic technique, X-ray Fluorescence Spectroscopy (XRF, see Sect. 4.2), the main difference being the nature of the primary beam used for

Fig. 4.19 General view of the AGLAE (Accélérateur Grand Louvre pour l'Analyse Elémentaire) accelerator at the C2RMF of the Louvre Museum established in 1989

sample irradiation. In XRF, primary X-rays from an X-ray tube are used to excite secondary X-ray emission (fluorescence effect) from the sample. In IBA, the same effect is obtained by exposing the sample to a beam of highly accelerated charged particles, i.e. ions, most commonly protons H^+. Both in XRF and in PIXE, the secondary X-rays are produced by the ionising effects of the highly energetic protons; in other words, the ion beam creates electron "vacancies" in the inner atomic energy levels (the so called K, L and M shells) by ejecting one or more of its most tightly bound electrons out of the atom (Fig. 4.20). As a consequence, electrons from higher energy levels "jump down" almost simultaneously (within 10^{-15} s) to fill these vacancies; each one of these *electronic transitions* is responsible for the release of an energy equal to the difference (ΔE) in the electron binding energy between the outer shell (E_o) and the inner shell (E_i) (Fig. 4.20). ΔE is released through the emission of an X-ray photon with a discrete energy and wavelength.

As electron energy levels and ΔE are related to the charge of the nucleus (i.e. the atomic number—Z) and the λ of the emitted X-rays are, in turn, related to the same energy differences, dispersion of the emitted X-rays into distinct peaks, each associated with a particular electronic transition, in an energy spectrum using an X-ray detector allows the identification of the elements present in the samples (Fig. 4.21).

Each characteristic X-ray peak in the spectrum is conventionally named after the energy level from which the electron is ejected, K, L or M, while its intensity/area is proportional to the concentration of the corresponding element in the sample. Van de Graaff particle accelerators required in IBA to produce collimated energetic beams of particles (ions, protons) are extremely expensive. Notwith-

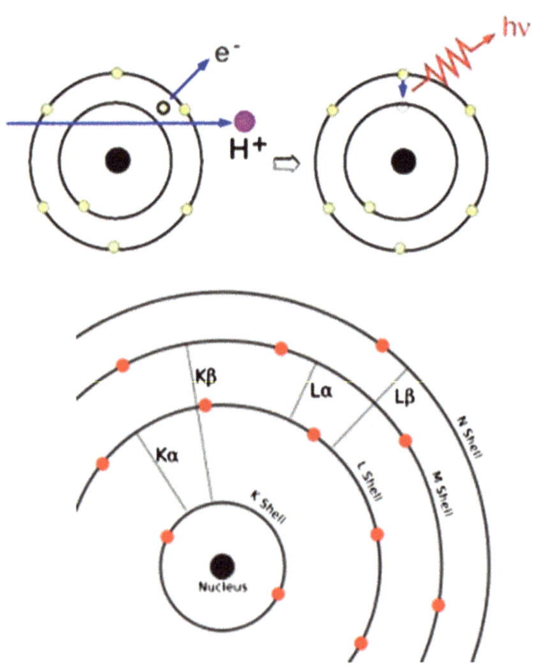

Fig. 4.20 Schematic view of proton atomic interactions and of electronic transitions between K, L and M atomic shells

standing this, the number of research institutions acquiring IBA facilities together with the rate of usage of these facilities by Conservation scientists is steadily increasing. This is true not only in Europe where the already mentioned AGLAE beamline at the Louvre, included in European Network Activities under the sixth and seventh Framework Programs of the EU through projects such as EU-ARTECH and CHARISMA, is routinely used by Cultural Heritage researchers throughout Europe but also in less economically developed countries [66]. In today's accelerators, proton beams of energies between 2 and 6 MeV can be produced with beam diameters commonly in the order of few micrometres (typically 10–30 μm). The emitted secondary X-rays are analysed by solid-state Si(Li) Energy-Dispersive detectors which discriminate between the various X-rays by their photon energy. The lowest energy detectable is around 1 keV so all elements with atomic number (*Z*) higher than 11 (Na) up to (with certain limitations) 92 (U) could be identified simultaneously. By varying the energy of the primary proton beam (usually between 1 and 3 MeV) and by using two detectors (one for low energy and the other for high energy X-rays), qualitative and quantitative data on major, minor and trace elements' concentrations can be obtained [67]. Depending on beam configuration, detection limits may reach 10–90 ppm (10,000 ppm = 1 %) for elements with *Z* between 20 and 60 and 100–300 ppm for $Z > 75$ [68]. The low detection limits of PIXE are one of the main reasons for PIXE popularity amongst increasing popularity in conservation science and archaeometry (see Chap. 5). The results of PIXE analyses are commonly displayed

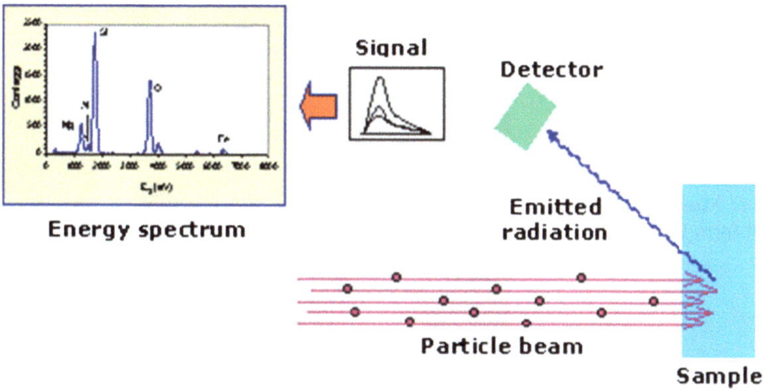

Fig. 4.21 Diagrammatic view of a typical PIXE analytical set-up and spectral output

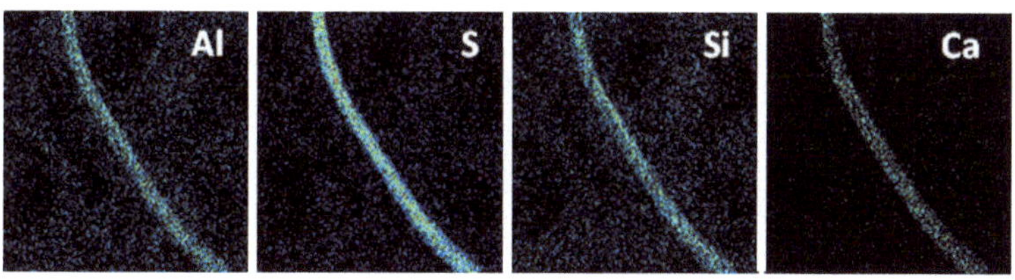

Fig. 4.22 μ-PIXE derived chemical map showing elemental distribution of selected elements (Si, Al, S, Ca) within tapestry fibres of eighteenth century rugs from Arraiolos, Portugal

as point analyses and/or chemical maps showing 2D concentration distributions of selected elements across an area (Fig. 4.22). Since protons have a very low penetration depth ($\approx$5-17 µm for an incident beam of 3 MeV protons, [68]), these maps usually refer to elemental data restricted only to the sample's most superficial layers. To overcome this problem, PIXE is often used in combination with "sister" IBA techniques such as PIGE and RBS. In PIGE and RBS, the relevant effects of the atomic interactions proton beam/sample are, respectively, the γ-rays emitted from nuclear processes (PIGE) and the protons backscattered from the sample after elastic collisions (RBS). A combined PIXE–PIGE approach provides an added value to the analysis for two main reasons: (a) PIGE is better suited than PIXE in detecting light elements (such as Be, Li, B, F); (b) γ-rays are produced deeper within samples (reaching for instance in gold objects about 27 µm in depth [68]) and, therefore, (c) PIGE data can be considered as bulk analysis while PIXE is primarily a surface sensitive technique [62].

Recent advances in IBA and PIXE technology has greatly expanded the potential of PIXE in conservation science. The introduction of microfocused beams (micro-PIXE), for instance, with beam dimensions reduced to even less

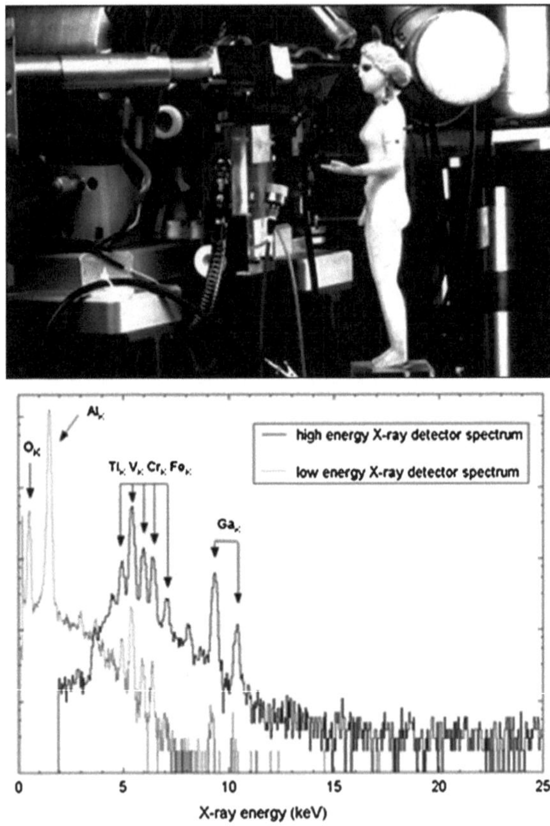

Fig. 4.23 PIXE set-up (*top*) and spectrum (*bottom*) of *red* rubies used as *inset* for the *left* eye in a statuette from ancient Babylonia. Analysis performed at the AGLAE Accelerator at the C2RMF of the Louvre Museum, Paris. Modified from Callegaro (2005)

than 1 μm, have brought dramatic improvements in the analytical spatial resolution pushing it down to a few hundred nanometers. This has made it possible to chemically resolve surface details of distinct composition in works of art such as multilayered pigmented surfaces or metal inclusions at a very fine scale [69–72]. But arguably the most important PIXE technical breakthrough with regards to its impact in cultural heritage studies has been the introduction of the *external* beam option (external PIXE), i.e. the possibility of analysing objects under ambient air atmosphere (utilising a flux of helium to minimise beam degradation in air) without the need to place samples under vacuum conditions [64] (Fig. 4.23).This option further upgrades PIXE's rating as a non-destructive technique inasmuch as: (a) the presence of air reduces the possibility of sample overheating under the ion beam; (b) vacuum related damages (such as dehydration processes in paper manuscripts) are avoided; (c) objects larger than the required size to fit vacuum chambers can be analysed avoiding the risks of damage related to sampling and/or transport procedures.

4.4.3 PIXE: Applications in Conservation Science

As already mentioned, chemical compositional analyses by PIXE are routinely performed on a variety of cultural heritage materials such as glass, metals, pigments, manuscripts, ancient inks, tapestry, glazed ceramics, ancient pottery, mortars, etc. [61, 65 and Refs. therein]. As already mentioned, only in a small number of cases, PIXE has been used alone [66, 72–75] as a multi-analytical approach has now become the rule of thumb in Conservation Science and Archaeometry [61]. PIXE, in fact, is often used in combination with techniques such as XRF, EPMA, SEM, Raman, AMS, ICP-MS and XRD as well as with other IBA techniques such as PIGE and RBS. Non-destructive PIXE–PIGE analytical combination, often in the micro-beam mode (micro-PIXE), has been extensively used, for instance, in the characterisation of historical glass objects examples of vitreous materials studied with PIXE–PIGE include glass beads from Archaic Greece [76], early glass from ancient Chinese dynasties [74], late Roman glass fragments [67, 77] and mediaeval stained glass windows [72]. PIXE–PIGE analysis in the external beam mode has proved particularly effective in identifying compositional non-homogeneities in bulk glass and in glass corrosion layers [67]. Degradation processes in archaeological glass objects otherwise displaying no visible alteration or corrosion features, for example, have been identified through PIXE/PIGE depth profiles of light elements (Na, Al, Si) concentrations across the corrosion layers [78–81].

PIXE–PIGE analytical combination has also been effective in the characterisation of ancient semiprecious stones and gemstones such as jade [82], ruby [83] or emeralds [84]. A fascinating case is represented by the PIXE study performed at the AGLAE lab on three red small spherical inserts (at eye and navel levels) found in an ancient mesopotamic statuette depicting the goddess Ishtar (Fig. 4.23) [83]. The PIXE analyses led to two important conclusions: the major element being Al, the red objects were confirmed as rubies (i.e. the mineral corundum, Al_2O_3) confirming them as the oldest ruby gemstones found in Middle-East; the PIXE trace elements (Ti, V, Cr, Fe, Ga) results proved effective in tracing the geological provenance of the rubies to sources in Burma several thousands of km away from Mesopotamia [83] and so providing evidence for the existence of commercial routes between Mesopotamia and South-Eastern Asia 2,000 years ago.

Many PIXE studies in the literature are concerned with the analysis of metals such as copper [73], aluminium [85] and gold [68, 71, 86, 87]. In the case of gold, for instance, PIXE combined with micro-SR-XRF has allowed the authentication of Dacian gold bracelets using Sn and Sb trace-element concentrations as discriminators [71].

Other interesting examples of successful PIXE application to Cultural Heritage material studies include the identification of pozzolanic material in Roman mortars [88] which could be extremely useful particularly to complement standard mortar analyses by OM, SEM+EDS, XRD and TG–DTA [89]; the reconstruction of Mediterranean trade routes during the Roman Empire by chemical fingerprinting of pottery mineralogical constituents in amphorae [66]; the compositional analyses of glazed polychrome layered porcelain, where the use of PIXE in the microbeam mode has allowed major, minor and trace-element characterisation of individual pigment layers which would have been not possible with conventional porcelain techniques such as SEM, XRF and XRD [90]; the identification of modern forgeries from authentic ancient porcelains [91]; the PIXE-based discovery of a previously unknown Central Mexico source for the obsidian slabs used by the Seville-based seventeenth century painter Murillo as the substrate for some of his famous paintings now stored in the Louvre Museum in Paris [92]. In the Conservation-related field of air pollution attack, on cultural heritage materials in urban environments, it is interesting to note how PIXE has also been used to complement other techniques such as SEM+EDS [93] to assess the particulate air pollution threat to paintings in outdoor and/or semiconfined environments [94].

The list of studies mentioned above does not intend, of course, to be exhaustive as it is only meant to provide a synthetic overview of the ever expanding use of the PIXE technique in the field of Conservation Science. The reader is referred to the bibliography provided at the end of this chapter for further study.

4.4.4 Why PIXE?

PIXE, as any other analytical technique, has its own array of advantages and shortfalls that have to be taken into account when the conservation scientist has to decide which one is the most appropriate analytical method to be applied to each particular case study [95]. Here is a summary of the main pros and cons:

Main advantages

- PIXE allows fast, multielemental qualitative and quantitative elemental characterisation for major, minor and trace light and heavy elements. 2D chemical mapping of element concentration distribution at the sample surface can be performed. The lightest measurable element depends on the ability to detect low energy X-rays. With a minimum X-ray energy set to 1 keV, PIXE can measure all elements starting from sodium (Na).
- PIXE is a technique with high sensitivity, i.e. low detection limits down to ppm levels, for most elements. This is due to the nearly absent background contribution to the PIXE induced X-ray emission (as compared with, for example XRF) and to the higher probability of collision between large protons and electrons.
- When used in the microbeam mode (micro-PIXE), PIXE analyses are characterised by a spatial resolution that could reach the nanometers scale. This allows the detection and analysis of compositionally distinct features within objects of art at a very fine scale.
- PIXE is a non-destructive technique. PIXE analyses, therefore, it has no need for sampling and/or for potentially damaging sample preparation procedure including chemical digestion and/or extraction procedures.

- When used in the external beam mode, PIXE analyses may be performed even on large and awkwardly shaped samples under ambient atmospheric pressure with no need to place samples under vacuum conditions that may damage delicate objects.

Main disadvantages

- Most Museum and Conservation laboratories cannot afford the particle accelerators required for IBA, the Louvre Museum in Paris being a most noticeable exception. In these cases, the transport of large Cultural Heritage items (statues, paintings, etc.) to IBA Laboratories to perform PIXE analyses represents often a rather difficult, if not impossible, task.
- PIXE analyses display all elements present in the analysed volume of the sample but do not tell the scientist in which chemical phase (i.e. oxides, sulphate, sulphides, organic compounds, etc.) the same elements are nor they provide information on the isotopic composition of materials which is often crucial in provenance analyses. Complementary techniques such as FT-IR, Raman spectroscopies, ICP-MS have to be implemented.
- Protons are characterised by a low penetration depth within the sample and therefore PIXE analyses provide elemental information primarily confined to the upper surface layers of the samples (1 μm depth in most cases). In the case of artefacts, showing outer corrosion patinas (very common in ancient bronze and glass objects) or a multilayered structure (finely stratified pigment layers on a painting, for instance) PIXE data alone are not suitable for obtaining bulk compositional data on the material. To obtain bulk data without the need for the invasive removal of the outer weathered layers in the artefact under investigation, PIXE has to be complemented with other techniques, such as PIGE and RBS which allow depth profiling data to be performed.

4.5 Atomic Absorption Spectroscopy in Conservation Science

Nick Schiavon, Cristina Dias Barrocas and Teresa Ferreira

4.5.1 AAS: Introduction

Spectroscopy deals with the study of the processes arising from the interaction between electromagnetic radiation of variable wavelengths and matter. The very birth of Spectroscopy can be dated back to 1672 with Newton's announcement to the Royal Society of London of the discovery of the solar electromagnetic spectrum, i.e. of the splitting of sunlight into various colours when passed through a prism. But it was only in the nineteenth century that important new findings were made thanks to the experimental work by scientists such as Wollaston, Fraunhofer, Brewster, Bunsen and Kirchhoff: amongst these new findings, of particular relevance was the discovery of the presence of black lines at well-defined and measurable wavelengths in the sun's spectrum. It was soon suggested that these lines were caused by absorption process of specific sunlight wavelengths in the atmosphere. In 1859, the famous Bunsen/Kirchhoff experiment demonstrated that the colours imparted to diffusion flames of alcohol by sodium salts (the typical "sodium yellow") occurred at the same wavelength of one of the sun's black lines. These effects are due to two phenomena occurring at the atomic level, i.e. atomic absorption and atomic emission of photons by elements in atoms. More specifically, they are related with the transfer of energy quanta from incident photons to the electrons in the atoms leading to the onset of electronic transitions from ground states to excited states and subsequent mission of quantized energy (photons) related to the difference between the atomic energy levels involved in the transitions. In other words, chemical species are able on one hand to emit radiation at specific wavelengths during relaxation from an excited state (induced, for instance, by thermal energy) and on the other absorb radiation at the same wavelengths during excitation by an incident radiation beam. The general relationships governing the electronic transitions between atomic levels and their energy differences (ΔE) is the one involving the well-known Plank's constant (h), i.e. $\Delta E = h\nu = hc/\lambda$ where ν and λ are, respectively frequency and wavelength of the absorbed/emitted photon and c is the speed of light in vacuum. The analytical potential of these atomic processes to provide information with good accuracy on the chemical nature of a variety of materials in several areas of application, including Archaeology and Conservation Science, forms the basis of Atomic Emission Spectroscopy (AES, also known as Optical Emission Spectroscopy—OES) and Atomic Absorption Spectroscopy (AAS). The latter technique is the subject of this chapter. In Atomic Spectroscopy, the material needs to be first decomposed into gaseous atoms in a flame, furnace or plasma. In AAS, the quantification of chemical species is based on the Beer's law, which relates directly the Absorbance ($A = \log [I_0/I]$ where I_0 is the intensity of the incident radiation and I is the intensity of the radiation exiting a predefinite layer of matter) to the concentration (c) of the absorber in the sample.

For ease of reading, the chapter is subdivided into a first part outlining AAS basic principles and methodology, highlighting the pros and cons of the AAS and comparing it with other elemental analytical techniques, followed by a section focusing on its Conservation Science applications to the study of cultural heritage materials.

4.5.2 AAS: Principles and Practice

The sample, usually in the liquid state, is atomized either by flame (FAAS: Fig. 4.24) or by electrothermal atomization (ETAAS: Fig. 4.25), the two most

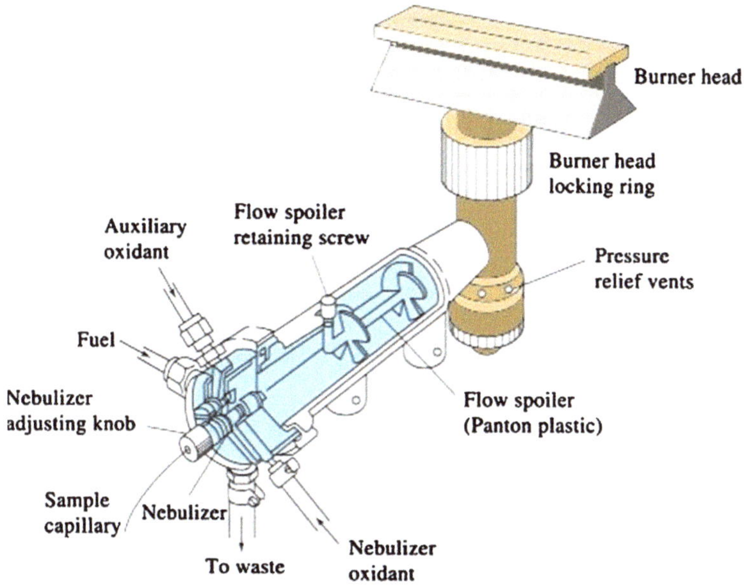

Fig. 4.24 Schematic diagram of a Flame Atomizer for FAAS

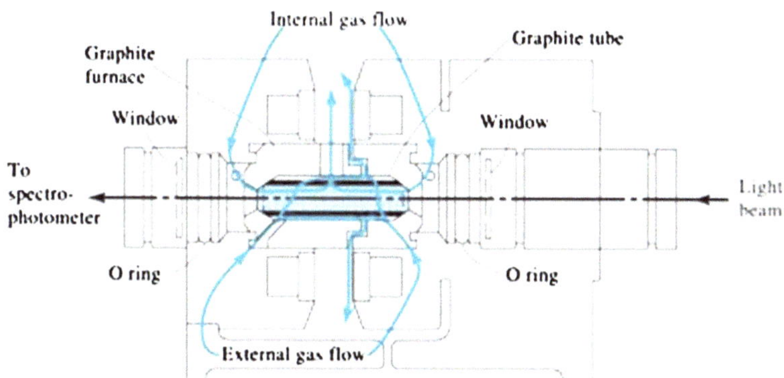

Fig. 4.25 Schematic diagram of an Electrothermal Atomizer for ETAAS

common methods of sample atomization, and radiation of suitable wavelength is shone on the atomic vapour formed. Some of the radiation is absorbed by the free atoms of the analyte while the remainder reaches the detector and it is used to calculate the concentration of the element in the sample. The technique is destructive, a feature that must be taken into account carefully when selecting this analytical procedure in Conservation Science case studies.

The basic components of an atomic absorption spectrometer comprise a radiation source, a sample holder (the atomizer unit which contains the gaseous

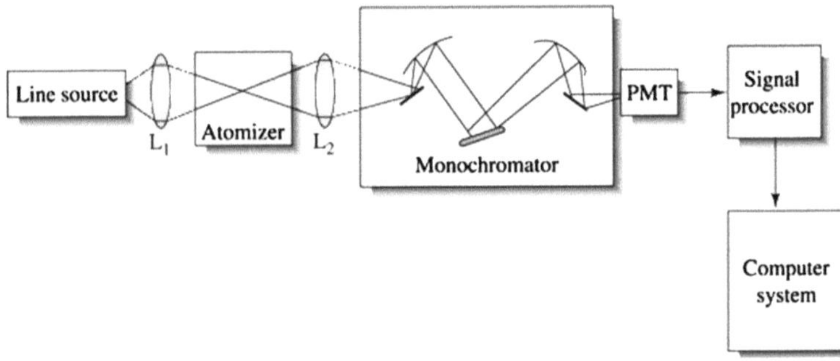

Fig. 4.26 General schematic diagram of an Atomic Absorption Spectrometer

atomized sample), a wavelength selector (usually a monochromator), a detector and a signal processor and readout (Fig. 4.26).

The most common radiation sources used in AAS are *hollow cathode lamps*. These lamps are made with a tungsten anode and a cylindrical cup-shaped cathode, made or coated of the element being studied, sealed in a glass tube filled with Ne or Ar at a pressure of 1–5 Torr. When an adequate potential is established between the anode and the cathode, the gas is ionised and the positive ions are accelerated towards the cathode, dislodging some of the metal atoms from the surface. Part of the sputtered metal atoms is excited and, as they return to the ground state, they emit characteristic radiation in the UV and visible regions. Usually, a different lamp is needed for each element being studied, making multi-element analysis by this technique not particularly time-effective. There are lamps for 65 elements, though some lamps with alloy cathodes can be used for multi-element analysis.

The main function of the *atomizer* is to produce ground state atoms of the analyte in the sample. This is the most critical step in the AAS procedure, since the sensitivity of the analysis depends on the effectiveness of the atomizer. The laminar flow burner is the most common type of flame atomizer, yielding a long narrow flame and creating a long path for the beam light to transverse. The liquid sample is aspirated up a capillary tube and converted into a molecular aerosol in the gas stream through a process called nebulization. A complex set of interconnected processes then occur in the flame leading to an atomic vapour. A quite common fuel–oxidant combination is acetylene–air mixture, providing a flame temperature of 2,400–2,700 K, while if a hotter flame is required for atomization of elements of thermally stable oxides (refractory elements) an acetylene and nitrous oxide mixture (2,900–3,100 K) must be used.

One of the most important advances of the AAS technique was the introduction of methods of sample atomization without the use of a flame (ETAAS: Fig. 4.25). Graphite furnace burners are electrically heated graphite or tantalum containers where a few microlitres of the liquefied sample are injected through a hole in an Argon atmosphere to prevent oxidation. The sample is first dried and then ashed at low temperature, before the temperature increases rapidly (up to 2,400 K) and atomizes the analyte. The light beam shines right above the furnace heater where the sample vapour is formed. This turns out to be a more efficient method of sample atomization, improving sensitivity up to 1,000 times for some elements (1–10 ppb against to 1–10 ppm) but with lower precision than FAAS method (5–10 % compared to 1–2 %). Although sample solutions are ordinarily used in furnaces, solid samples may also be analysed. Flame atomization is more reproducible than electrothermal atomization, but presents a lower sampling efficiency due to the larger sample volume needed (1–2 mL per analysis, as the majority of the sample flows down the drain) and the short residence time of the individual atoms in the optical path ($\approx 10^{-4}$ s).

In a hot gaseous medium, atoms are capable of absorbing radiation of characteristic wavelengths, enabling electronic transitions from the ground state to higher energy excited states. The most intense transition from the ground state to the first excited state (resonance transition) is the most widely used for analytical purposes because it is the most sensitive.

The non-absorbed radiation from the source beam is selected by the *wavelength selector* to reach the detector. In fact, the purpose of the monochromator is to isolate the measured line (resonance radiation) coming from the hollow cathode lamp and to reject as much emission as possible from the other emission lines of the cathode material, lamp filler gas and the flame or furnace materials.

Most AAS instruments use photomultiplier tubes as *detector*s, in which the cathode is coated with a photosensitive material, ejecting a photoelectron whenever struck by a light photon. The electrons are accelerated towards several additional electrodes, increasingly more positives than the cathode, originating an electron cascade that is collected at the anode with the resulting current being electronically amplified and measured.

When absorption by ground state atoms occurs, the intensity of the radiation beam leaving the analyte diminishes. In theory, the attenuation of the incident beam is proportional to the number of absorbing atoms that in turn is directly proportional to the concentration of the measured element in the sample, following Beer's law. In fact, departures from linearity are often observed but quantitative measurements are possible using external calibration where light attenuation is measured using standards of known concentration within the range of that expected in the sample. The analyte signal is converted in a concentration using the constructed calibration curve that is afterwards converted to a weight percent of the total sample material.

Overall, the AAS technique is subject to chemical, spectral and ionisation interferences that change the AAS signal while the analyte concentration remains unchanged. The general and specialised literature should be consulted to find ways to circumvent these interferences [96–102].

As already mentioned, the most common way of introducing the sample in the flame atomizer is in the form of an aqueous solution. However, many cultural heritage materials like ceramics, glasses, textiles, mortars, metals and pigments are either insoluble or only partially soluble in water. Common sample preparation methods include weighing, gridding, digestion, filtering and dilution of the sample previous to AAS analysis and the reagents used ought to be trace analysis grade to avoid sample contamination [96].

Wet chemical digestion procedures with several hot mineral acids (HCl, HNO_3, HF, H_2SO_4), alone or in mixtures, hydrogen peroxide and other reagents are used in a step process in open systems under atmospheric pressure, like the traditional "hot plate" or with recycling devices, or in high pressure closed vessels, like in a microwave apparatus [103–105].

Different methodologies are described in the literature for the study of the archaeological, ceramic and geological materials, since they contain insoluble inorganic fractions usually combined with an organic fraction [106–110]. The usage of acetic acid for digestion of carbonate material is also recorded [107]. Metal samples can be analysed by acidic digestion procedures as well [111]. Textiles are easily degradable materials and, in general, a lighter acidic attack is enough for the complete digestion [112, 113]. For main element analysis in glass samples, namely the analysis of Si, heating procedures that involve sample fusion with a flux (e.g. $LiBO_2$) followed by solubilisation in an acidic solution (e.g., HNO_3) are also described [114, 115].

ETAAS enables, in theory, the analysis of solid samples avoiding therefore the sample digestion process, simplifying the entire analytical procedure. Nevertheless, calibration is usually difficult and requires standards that match the sample in composition [96, 98].

4.5.3 AAS in Conservation Science

As it has now become common practice in the chemical characterisation of cultural heritage and archaeological materials, AAS has almost never been used as a standalone technique but rather as part of a multi-analytical methodological approach. A variety of materials have been studied with this technique: ceramics, metals, glass, stone, mortars, textile, paper and pigments [116, 117]. The following list of case studies is by no means intended to be exhaustive but rather to give just a "feeling" of the wide range of applications of AAS in Conservation Science.

The main scope for conducting detailed chemical analysis of archaeological pottery/ceramic materials is usually related to the identification of raw materials provenance (clay and aggregate) and production technologies [106–108, 118–120]. FAAS followed by multivariate statistical analysis has been used for example to determine the chemical composition (major and minor elements) of late Roman amphorae found in underwater environments from the European and African coasts across the Straits of Gibraltar, i.e. from Ceuta and Cadiz sites [119]. The relative increase in the content of specific elements, such as Cu, Cr and Ca, has allowed to distinguish amphorae produced in the two sites investigated supporting the archaeological hypothesis that some amphorae of South-Hispanic typology found in Ceuta did not represent imports but were produced locally. AAS has also been used to investigate the composition of the ceramic body of atypical mediaeval proto-majolica fragments of polychrome painted glazed pottery from Apulia, Southern Italy [106]. In this study, a weighed aliquote of about 30 mg from each ceramic sample was dissolved by acid attack and 13 elements were quantified by AAS: Na, K, Fe, Ca, Mg by air acetylene FAAS and Cr, Sr, Ba, Cu, Pb, Mn, Ni and Zn by graphite furnace ETAAS. The statistically treated (PCA) results permitted to identify the presence of several manufacturing locations of these atypical mediaeval pottery artefacts in the Apulia region.

Of particular interest in Conservation Science research is the application of AAS to determine decay effects induced by environmental agents in urban environments on various heritage materials such as ornamental stone [109, 121, 122], metals [123–125], ceramics [119, 126] and mortars [127]. AAS has been used, for instance, to investigate severely damaging acetate salt crystallisation on Dutch tiles of glazed ceramics stored in a Museum environment [126]. In this study, the combination of qualitative XRD with quantitative data on ionic species by AAS, AES and Ion Chromatography (IC) lead to the identification of the major crystalline species present in the efflorescences (together with their crystallisation pathways), enabling the planning of the correct preventive conservation measures to be applied to the salt contaminated ceramics under investigation. Particulate air pollutants deposited on soiled building facades in urban environments are known to act as catalysts (mainly through redox reactions) in soluble salt decay processes involving sulphate, nitrate and chlorides compounds [128–133]. Knowing the detailed chemical composition of the deposited particulate matter is therefore an important issue in Conservation Science. AAS has been used to investigate the elemental composition of atmospheric particulate on dust deposited in sheltered areas on buildings in Budapest [133]. Concentrations of water-soluble Fe, Mn and Zn reported in this study reached, respectively 126, 80 and 220 mg/Kg highlighting their mobility and potential to take part in surface oxidation reactions, especially under high relative humidity conditions.

Another interesting application of AAS that has important bearings in Conservation Science research is the evaluation of the chemical composition [134] and/or of the corrosion process [123–125, 135] of metal artefacts through field experimental studies in urban, polluted environments. The protective effects against environmental attack of wax-based antigraffiti coating on Cu patina composition and dissolution was tested, for example, by AAS by Goidanichand et al. [123]. The amount and rate of Cu release (reflecting the amount of patina dissolution) from artificially prepatinated and antigraffiti coated copper sheets was measured by AAS during a 4-year exposure period in the urban atmosphere. This was accomplished by analysing total copper concentration (limit of detection 6 μg/L) in the run-off water continuously collected over the exposure period. The artificial patina consisted of an inner layer of cuprite (Cu_2O) and a superficial layer of brochantite ($Cu_4SO_4(OH)_6$) to match naturally oxidised copper surfaces. FAAS has also been used to investigate the effect of leaching rain on the corrosion behaviour of bronze artefacts. Drop tests with a solution reproducing natural acid rain (which is known to affect the physicochemical stability of copper based patina in outdoor bronze surfaces) were performed and the leached water analysed by AAS with regards in particular to the content of alloying elements, Cu, Sn, Zn, Pb [125].

AAS has also been applied successfully to the characterisation of historical glasses and of their corrosion products [114, 115, 136, 137]. Major elemental analysis by AAS (Na and K) and by AES (Si, Al, Fe, Mg, Ca, Ti, P) combined with trace-element analyses by WDXRF, Pb isotopic data and archaeological evidence on Roman glass chunks from Sagalassos in Turkey, for instance, was used to confirm the existence of local secondary glass production centres in the region from imperial to Early Byzantine times [114]. AAS in combination with XPS was used to determine the chromophore elements content (Co, Cu, Fe, Mo in particular) in Bronze-age glasses from Northern Italy [137]: the analytical results revealed a remarkable level of know-how in the control of oxidation/reduction conditions during glass making attained by the proto-historic inhabitants of the region.

Another interesting application of AAS in Conservation Science is the analysis of metallic elements present in mordant and dyes in historical and modern textiles [112, 113]. FAAS, for instance, has been used in combination with ICP–OES to analyse the mordant metallic ions content in wool fibres dyed with madder, alum, copper and iron salts and artificially aged to assess the influence of mordant on madder chromophore photodegradation [113]. Mordant ion quantification by AAS and ICP–OES showed that the uptake of metal ion by the fibres was relatively small and that chromophore degradation was uneven and dependent on mordant composition which in turn influences the colour changes occurring after light exposure. Such experimental studies are of extreme importance in Conservation Science inasmuch as they provide much needed experimental data to support the selection of compatible materials and techniques to be applied to the conservation and restoration of historical textiles in museums and private collections.

4.5.4 Why AAS?

With AAS, it is possible to measure up to 65 elements with detection limits ranging from several ppm to a few ppb. The major strengths of AAS are the easiness of use, low detection limits, high specificity, low investment and minimal costs associated with it, well-documented interferences and methods for their elimination and access to a number of specialised techniques for the determination of non-metals and organic compounds. Flame and electrothermal atomization may be considered somewhat complementary techniques but while FAAS requires less operating skills, is faster, cheaper and less time-consuming, it also has poorer detection limits, especially for some refractory elements, cannot operate unattended and involves the use of larger sample volume.

These pros outlined above made AAS one of the mainly used analytical techniques for cultural heritage studies in the past. The major disadvantages of AAS are the hard task to do qualitative analysis and the time-consuming limitation of having to analyse one element at a time. As compared with AES where all atoms in a sample are excited and can be detected simultaneously, one evident drawback of AAS is that it is inherently a single-element analytical procedure (multielemental analysis, although possible, is rather complex, see below for more details). Despite this drawback, AAS, due to its selectivity, simplicity and ease of use, is one of the most user friendly and least expensive techniques available for elemental quantification.

More recent atomic techniques such as inductively coupled plasma atomic emission spectroscopy (ICP-AES), inductively coupled plasma mass spectrometry (ICP-MS), and/or X-Ray spectroscopy techniques (XRF and PIXE for example, see dedicated chapters in this volume) are nowadays routinely used in applications where AAS used to be the *gold standard.*

Analytical improvements can in fact be achieved with ICP techniques, but there are also some drawbacks which include the techniques implementation and running costs and the know-how needed to optimise the analytical methodologies. For example, considerable skill is required for ICP–MS technique development and the lower detection limits are only attainable under specific laboratory conditions, requiring sample treatment and analysis to be done in a clean room environment.

AAS is overall easier to set up and run, with simpler method development and vast bibliography data available. Whenever the researcher needs to choose the analytical methodology to apply to a set of samples, several considerations should be taken into account, namely, analytical performance (detection limits, precision, dynamic range, possible interferences and other analytical problems) and the costs associated. Compromises are often needed and, for cultural heritage studies, AAS techniques constitute still in many cases an adequate choice.

4.6 Thermal Analysis Techniques

K. Chrysafis

4.6.1 In Which Cases the Method Used is in Conservation Science?

The term thermal analysis (TA) denotes "a group of techniques in which a physical property of a substance and/or its reaction products is measured as a function of temperature whilst the substance is subjected to a controlled temperature program", as defined by the International Confederation for Thermal Analysis (ICTA). The most common TA techniques that are employed in conservation science are thermogravimerty (TG), differential thermal analysis (DTA) or differential scanning calorimetry (DSC), and thermo mechanical analysis (TMA).

Although TA methods are generally destructive, the relatively minor amount of samples required, which varies between a few μg up to a few mg, is the fact that lists them among the most commonly used techniques in art conservation science. Using only a minimum amount of sample each time is actually one of the most appealing features of TA techniques. These methods may provide the scientists, along with other analytical techniques, with useful information about the compositions of materials that belong to cultural heritage, confirming the complementary nature of TA. Thermoanalytical methods are not only related to the study of ceramic materials, which historically was among their most important applications, but also to other materials such as painting media, wood and synthetic polymer coatings. However, the fact that only a combination of different techniques and approaches will lead to a thorough investigation which would be considered as scientifically adequate, should not be overlooked in any case.

4.6.2 Principles and Concepts of the Instrument: Description of the Methodology

4.6.2.1 Differential Thermal Analysis

The difference in temperature between the sample and a reference material is recorded while both are subjected to the same heating programme. DTA is the oldest thermal analysis technique. In a DTA device, the sample holder assembly is placed in the centre of the furnace. One holder is filled with the sample and the other with an inert reference material. Thermocouples inserted in each holder measure the temperature difference between the sample and the reference while the temperature of the furnace is controlled by a temperature programmer. If the sample and reference pans are heated at the same rate, by placing them in the same furnace, the reference pan temperature will rise steadily, as the reference material is chosen to have no physical or chemical transitions, while the sample temperature will also rise steadily in the absence of any transitions. If a transition occurs in the sample, its temperature will lag behind the reference temperature as it absorbs or emits the heat energy necessary for that transition. The minimum temperature difference which can be measured by DTA is 0.01 K. A DTA curve plots the temperature difference as a function of temperature (scanning mode) or time (isothermal mode). When the transition is complete, steady heating is resumed once again. During a phase transition the programmed temperature ramp cannot be maintained owing to heat absorption or emission by the sample (Fig. 4.27).

4.6.2.2 Differential Scanning Calorimetry

The temperature and heat flow associated with the sample's transitions as a function of time and temperature are recorded. These measurements provide qualitative and quantitative information about physical and chemical changes that involve endothermic or exothermic processes, or changes in heat capacity, such as glass transitions, melting and crystallisation. Power compensated when an event occurs in the sample material, power (energy) is supplied to or removed from the sample furnace to compensate for the change in heat flow to or from the sample until the temperature difference is less than a threshold value. As a result, the sample and reference

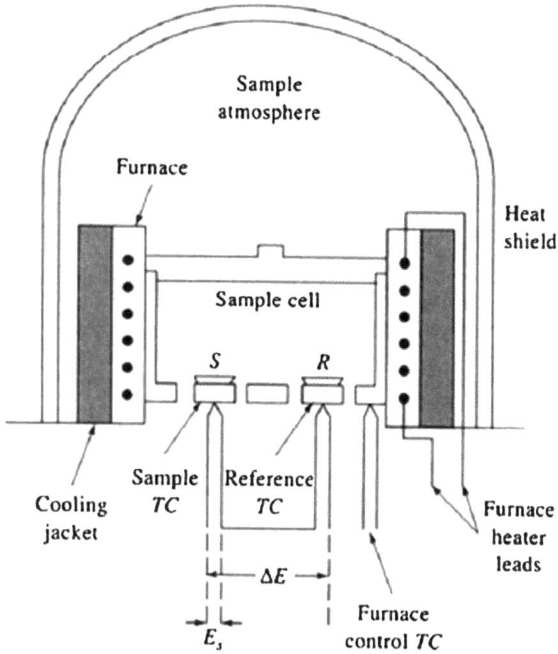

Fig. 4.27 Diagram of a DTA instrument

materials are maintained at the same temperature at all times. The amount of power required to maintain system equilibrium is directly proportional to the energy changes occurring in the sample. DSC directly measures energy flow to and from the sample. The temperature range of a power compensation DSC system is between 110 and 1000 K depending on the model of sample holder assembly chosen. Alternatively, in heat flux DSC, the sample and reference pans are heated by a single furnace. The difference in temperature between the two pans is recorded by means of thermocouples beneath the crucibles. Normally, there is a minimal difference in temperature of the pans but when a transition occurs in the sample this difference increases. A computer converts the recorded signal into heat flow (Fig. 4.28).

4.6.2.3 Thermogravimerty

The change of the mass of a sample in a controlled atmosphere is continuously recorded as a function of temperature and time. Not all thermal events bring about a change in the mass of the sample (for example melting, crystallisation or glass transition), but there are some very important exceptions which include desorption, absorption, sublimation, vaporisation, oxidation, reduction and decomposition. TG is used to characterise the decomposition and thermal stability of materials under a variety of conditions and to examine the kinetics of the physicochemical processes occurring in the sample. The mass change characteristics of a material are strongly dependent on the experimental conditions employed. Factors such as sample mass,

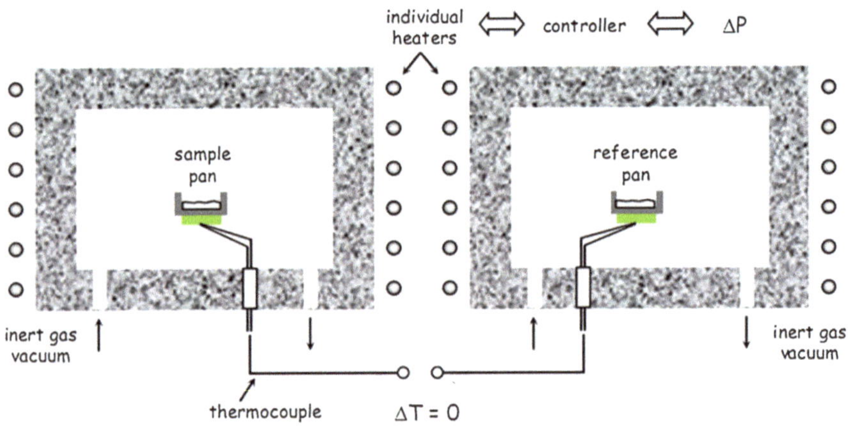

Fig. 4.28 Diagram of a power-compensated DSC Instrument

volume and physical form, the shape and nature of the sample holder, the nature and pressure of the atmosphere in the sample chamber and the scanning rate all have important influences on the characteristics of the recorded TG curve. TG curves are recorded using a thermobalance. The principal elements of a thermobalance are an electronic microbalance, a furnace, a temperature programmer and an instrument for simultaneously recording the outputs from these devices. A typical temperature operating range for the furnace is ambient temperature to 1000 °C, with heating rates up to 100 °C/min. The quality of the furnace atmosphere deserves careful attention, particularly the ability to establish an inert (oxygen-free) atmosphere, and it is useful to be able to quickly change the nature of the atmosphere. There is also available and widely used, TG-DSC as well as TG–DTA instruments that simultaneously perform the operations of both of the corresponding techniques (Fig. 4.29).

4.6.2.4 Thermomechanical Analysis

The change of a dimension or a mechanical property of a sample is recorded as a function of temperature and time. The sample is placed on a support within the furnace, where a probe is also present to sense changes in length, which are measured by a sensitive position transducer. The probe and support are made from a material of accurately known coefficient of thermal expansion, and also has low thermal conductivity, which helps to isolate the sensitive transducer from the changing temperatures in the furnace. A thermocouple near the sample senses its temperature. There is usually provision for establishing a flowing gas atmosphere through the instrument, to prevent oxidation for example, and also to assist in heat transfer to the sample. Helium is effective in this respect. TMA offers advantages, over the other thermoanalytical techniques, for certain types of studies samples.

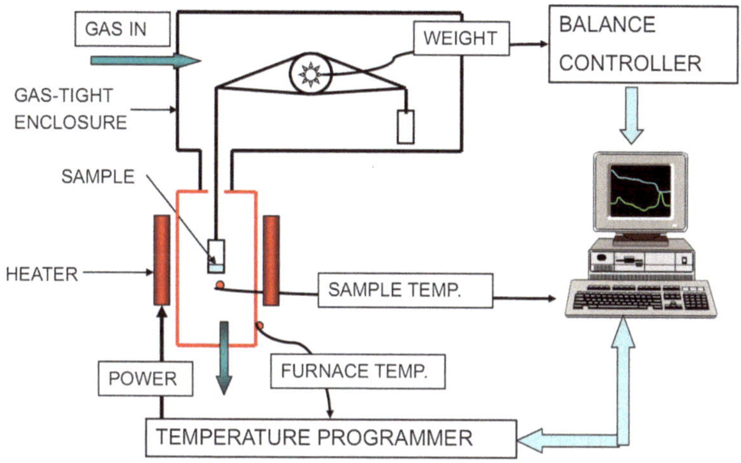

Fig. 4.29 Diagram of a TG instrument

Fig. 4.30 Diagram of a TMA instrument

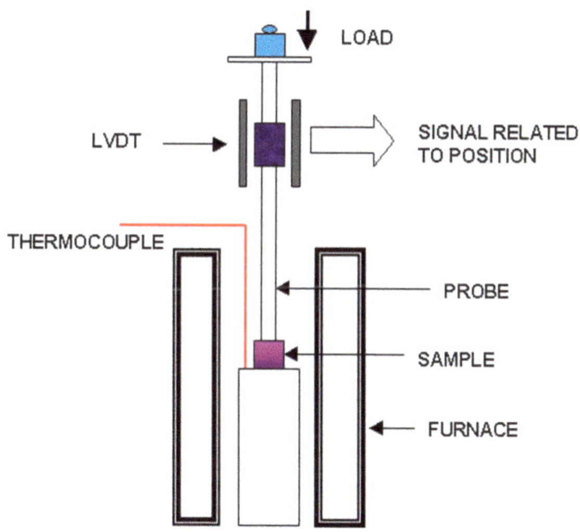

The indentation and penetration probes for instance, can follow transitions in very thin films, such as lacquer coatings on metals (Fig. 4.30).

4.6.3 Evaluation of the Method as used in Conservation Science

The characterisation of historical mortars is frequently difficult due to various aspects: (a) the variable chemical composition of materials of the same type, as a result of variable technology, (b) the chemical transformations that occur with time, (c) the double function (binder and aggregate) often carried out by the same substance and (d) the conservation and restoration actions. Only several and complementary studies, with information from different techniques, can give an appropriate insight on the composition of a historical mortar. The information so obtained can be used in the reconstruction of a given mortar, namely for conservation purposes.

The physical–chemical characterisation of historical mortars collected from the Arsenal of Venice [138] has provided historical and technological information as regards their manufacture. TG-DSC measurements indicated the effect of the presence of soluble salts, such as sodium chloride. When NaCl is present, instead of the expected one-step decomposition of $CaCO_3$, the TG-DSC curves showed a four-step decomposition process. The presence of NaCl resulted in a decomposition of the mortar at lower temperatures than those measured after removing the sodium chloride by washing the mortar with distilled water. The results from the chemical characterisation showed that the indoor masonry mortars are mostly air-hardening and the foundation ones are hydraulic. Thanks to this characterisation, it will be possible to prepare a restoration mixture having characteristics compatible with the masonry structure, through a 'reverse engineering' process. It allows advances to a possible consideration of the kinematics of decay, in order to estimate the level of danger for the conservation of the building, leading the planning of interventions as regards the real magnitude of the registered results, in order to select priorities of intervention and to quantify the conservation costs.

Different ancient mortar samples of Pamplona cathedral [139] have been analysed to characterise their binder and aggregate fractions. A lime binder with a silica aggregate has been established by thermal studies (simultaneous DTA and TG analysis). The approximate original compositions of the classes of mortars have been determined using the TG results. An incomplete carbonation in a sample has been discovered by the thermal analysis.

Thermal analysis has been used [140] to determine the weathering mechanisms of marble and sandstone taken from both the Captain Arthur Phillip Monument and St Mary's Cathedral in Sydney, Australia. The degradation of marble is largely due to the acidic environment created by the emission of SO_2 and CO_2 from motor vehicles and industry. The dissolution of $CaCO_3$ in marble by H_2SO_4 formed from SO_2 to produce $CaSO_4$ is a detrimental degradation phenomenon where the damage is totally irreversible. In addition, the formation of $CaSO_4$ encourages further attacks from environmental pollutants due to the increase in porosity of the stone surfaces. Sandstone degradation, on the other hand, resulted from the destruction of the crystal structure of the clay base binding material. Although

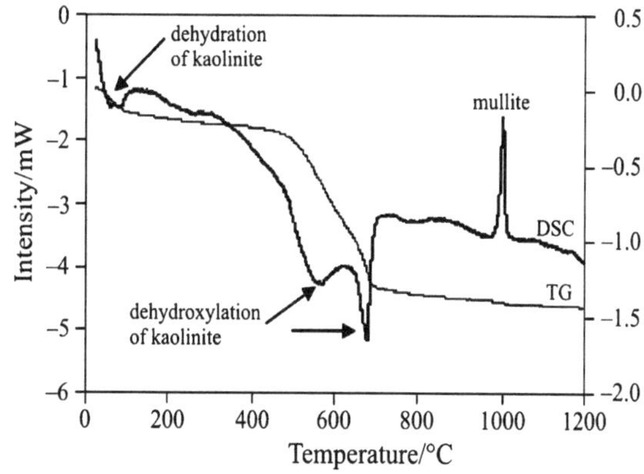

Fig. 4.31 DSC and TG curves of unweathered clay samples [140]

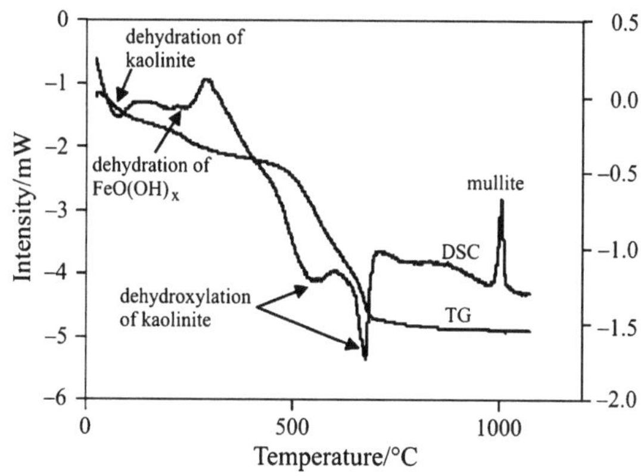

Fig. 4.32 DSC and TG curves of weathered clay samples [140]

the effects of the cation substitutions into the structure of clay may seem to be negligible, the destabilisation of the structure eventually destroys the binding ability of this cementing material. As a result, exfoliation and disintegration of the sandstone in our heritage buildings are sometimes observed. The decrease in intensity of the original kaolinite peak at 580 °C and the increase in intensity of the peak at 680 °C in the weathered sample suggest a greater substitution of Fe^{3+} in the clay structures upon weathering (Figs. 4.31 and 4.32).

Riontino Carlo et al. [141] developed an analytical methodology for the complete analysis of the black degradation crusts found on ancient buildings. The procedure permits the quantification of the main deterioration products of degradation and atmospheric deposition found on building materials, the measurement of carbonates by DTA–TGA, also in the presence of oxalate, and the identification and quantification of the non-carbonate, elemental and organic carbon. The data obtained consent a complete carbon balance in the damaged layers on monuments and historical buildings, which is of primary importance in the diagnosis of environmental damage and in planning correct conservation work.

Fig. 4.33 TG curves (mass loss) of three shards [142]

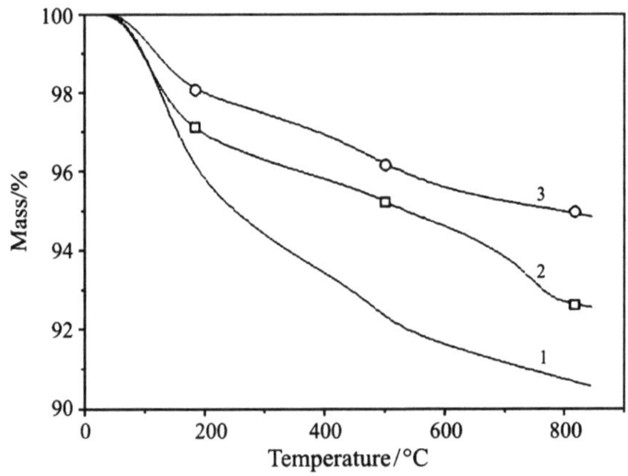

Ancient ceramic samples (single fragments and different parts of pots), dated from VIII to IV centuries BC, were investigated by thermal analysis together with other analytical techniques. In addition, in order to identify the clay sources for the ceramic manufacturing, samples of clays and soils found near archaeological digs and taken from the mineralogical museum were investigated. Drebushchak et al. [142] found out that the calcite content of ceramics is a very informative parameter for the identification of the clay source for the pottery manufactured at low technological level (low-temperature firing) (Fig. 4.33).

Thermal analysis has been used to analyse a number of terracotta materials [143] from different periods and yielded information not only about their composition but, at least in the first instance, also concerning the different levels of technical ability attained by the craftsmen. The terracotta material is characterised by significant differences between the upper and the lower parts, particularly as far as the calcite content is concerned, which is substantially different in the upper and lower parts of several of these statues. This indicates a substantial lack of homogeneity in these artefacts. The equivalent firing temperature is different, which would seem to indicate that the kiln operators experienced some difficulty in maintaining an even temperature throughout the kiln.

The results of thermoanalysis have been discussed in terms of the original assembly and firing of the archaeological finds [144] and their subsequent restoration. Several parts of the fictile statue were made in separate moulds and then assembled to form a single complex before firing. The firing temperature does not seem to have been completely uniform—in particular, the base seems to have reached a higher temperature than the upper parts of the statues (Fig. 4.34).

An integrated characterisation of ancient ceramics includes most often the description of their mineralogical, chemical and thermal properties. The interpretation of the analytical results in conjunction with the relevant archaeological information, may lead to significant conclusions in provenance and ancient technology studies. Ancient ceramic sherds (seventh century BC) from the archaeological excavation of Abdera, Greece and local raw clay bricks were characterised

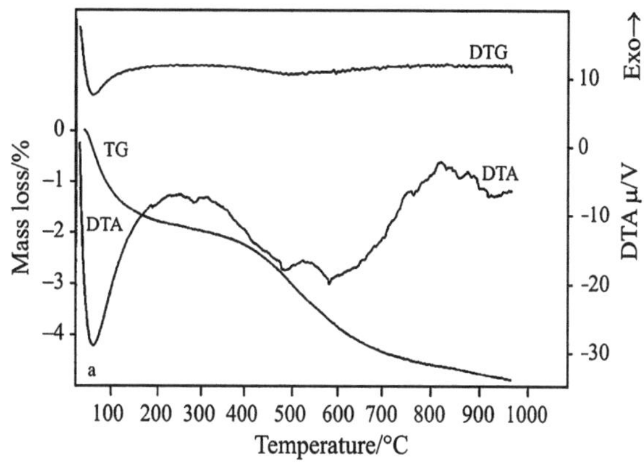

Fig. 4.34 TG/DTG-DTA curves of raw clay brick fired at 500 °C [145]

[145]. It has been found that the mineralogical composition of most of the studied shards is quartz, feldspars and micas, which is in agreement with the composition of the local bricks. Chlorite is also present in a few samples, while there is one completely different sherd, which belongs to the Ca-rich clays. From the TG/DTA data, they comment on the possible firing temperature and distinguish between samples of different origin. The existence of muscovite or illite in most of the samples denotes that the firing temperature was lower than 950 °C, while the existence of chlorite means that the firing process in these samples stopped before 700 °C. A very different thermogram gave the Ca-rich ceramic sherd, due to the existence of calcite, denoting that the firing temperature was about 700 °C.

Thermoanalytical techniques are used [146] to evaluate the mechanical properties and thermal stability of threads from historic tapestries in royal palaces and museums. To assist in this process, model tapestries have been prepared according to traditional techniques of weaving, mordanting and dyeing, and the threads from these tapestries have been characterised. The aim is to demonstrate that it is possible to quantify changes which occur in model tapestries after dyeing and mordanting and also accelerated light ageing, and to compare these with the observed changes in historical threads. The black dyes show different thermal stability (Fig. 4.35). This is probably due to the additional effect of the copper sulphate which is used together with iron sulphate and is not present in all the samples. Figure 4.36 shows that sample BXL1/12 [from Brussels tapestry BXL1 (1519–1524)] has the lowest enthalpy value and appears to be the most damaged. Woollen threads also show an exothermic peak in a similar temperature region as the silk samples but with a reduced intensity. Historical samples tested do not show values as low as those for silk. This approach may prove to be a useful rapid test for damage assessment of threads.

Parchment whether or not in the form of bookbindings or manuscripts is of considerable interest. dynamic mechanical analysis (DMA) and other techniques were used [147] to measure historical parchment samples. DMA was used in both thermal scan and creep modes. Thermal scans provided information on the

Fig. 4.35 TG and DTG curves for control (full line), BLKW2 (*dash*) and BLKW4 (*dot*) [146]

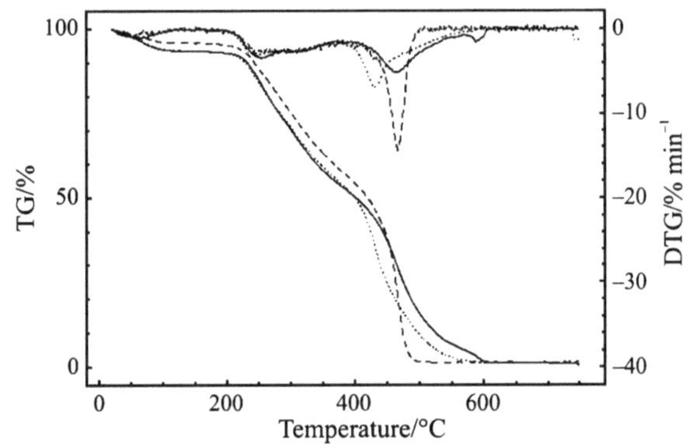

Fig. 4.36 DSC curves of model and historical tapestries, showing difference in peak between 210 and 420 °C. From *top* to *bottom* undyed unaged, light aged undyed for 400 h, BRU3/12, PNM2/11 and BXL1/12 [146]

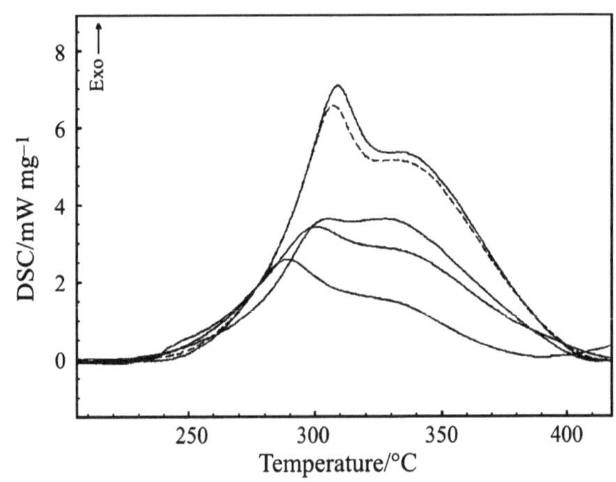

transitions associated with the collagen polymer. The behaviour of samples immersed in water and measured in DMA creep mode was used to measure the shrinkage behaviour of the parchment samples. The different but complementary techniques provided a means for characterising the physicochemical state of parchment samples (Fig. 4.37).

DSC, TG and other analytical techniques [148] have been used to validate the effects of a restoration method for fire injured ancient parchment codes. The results allowed understanding of modifications induced by the heat, namely collagen denaturation, and those by the restoring process on heat damaged ancient parchments. The study demonstrates that DSC is a suitable technique to detect the effects produced.

Environmental impact on parchment [149] was investigated by DSC. Parchments subjected to accelerated ageing and old parchments were compared to evaluate quality and extent of deterioration. DSC peak shape is an early indicator of the degree of deterioration. There is a relationship between deterioration and peak shape of old parchments (Fig. 4.38). The peaks for undamaged or slightly

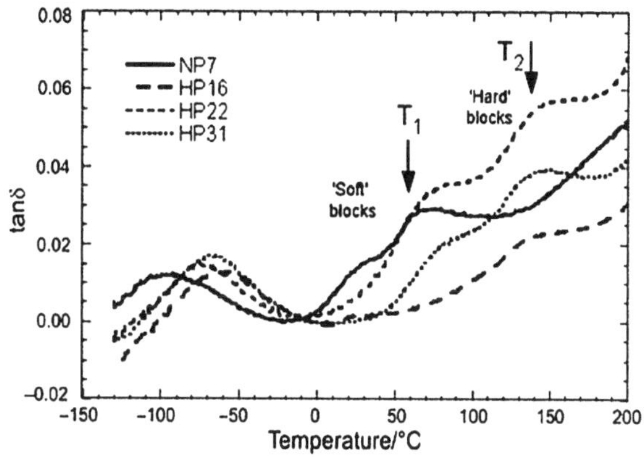

Fig. 4.37 DMA curves (tanδ) of unaged (NP7) and aged parchments (HP16, HP22 and HP31) as a function of temperature [147]

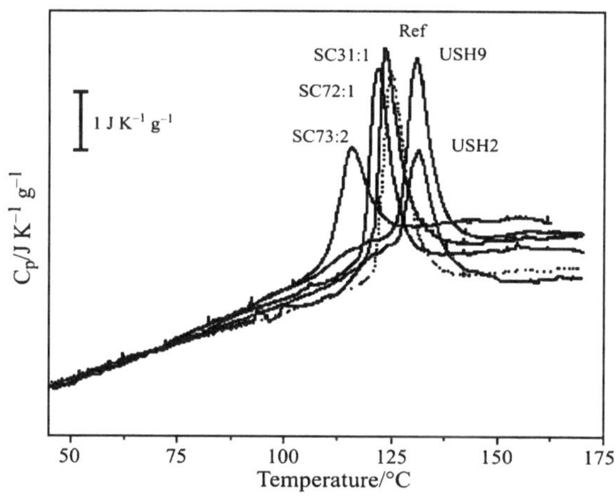

Fig. 4.38 DSC curves for some historical parchments showing thermal stability variation induced by environmental ageing: almost no changes (SC31:1); minor changes (SC72:2); major changes (USH9); heavy changes (USH2, SC73:2) [149]

damaged samples almost overlap the reference peak, sometimes at somewhat lower or higher temperatures.

The effect of oxidation on paper stability has been studied by means of DTA/TG. An application of this method was extended to the study of differently oxidised parts of a 1948 book. Aged paper shows discoloration and brittleness [150], which are caused by two main reactions: hydrolysis and oxidation. Each border sample showed a DTA peak temperature about 4.5 K lower than the samples drawn from the inner, less oxidised parts (Figs. 4.39, 4.40). This feature was in agreement with the results obtained on Whatman paper, a new one, although in the real case, degradation could be caused also by several other factors (e.g. acid-catalysed hydrolysis).

The applications of thermal analysis methods [Micro Hot Table, thermogravimetry/derivative thermogravimetry (TG/DTG), DSC, DTA and DMA] for the characterisation of the recently manufactured and old leathers are presented [151]. These methods can be used for the assessment of deterioration degree of leathers, and, therefore, the data obtained by these techniques could be useful for

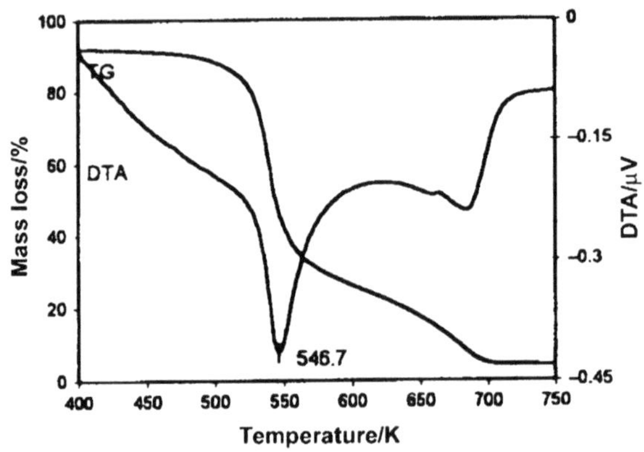

Fig. 4.39 Curves of sample from the inner part of a leave (a 1948 book) [150]

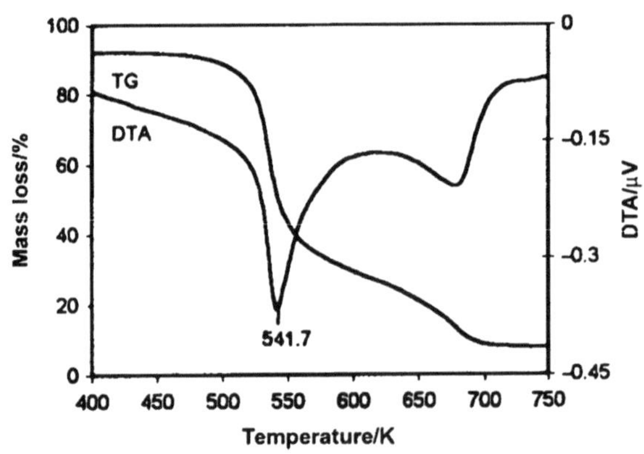

Fig. 4.40 Curves of the border of a leave (a 1948 book) [150]

achievement of the suitable preservation procedures as well as the effects of conservation treatments. In addition, it was pointed out that these methods are suitable for qualitative distinction between the recently manufactured leathers and heritage items.

According to Budrugeac et al. [152] in air atmosphere, if the rate of thermo-oxidation process is smaller than that corresponding to the recently manufactured leathers, then the analysed leather is patrimonial (historical). In water, the measurements of collagen-based materials show that each patrimonial leather exhibits 2–4 peaks. The number of peaks from DSC curve could be another qualitative criterion for distinction between recently manufactured leather and a patrimonial one. The two suggested criteria could be used for distinction between an original artefact and a bootleg (Figs. 4.41, 4.42).

In cultural historic terms the identification of the sources of functional materials used for decorating artefacts, i.e. of inorganic or organic pigments and colourants, is indicative of the technological skills, but also for the trade trajectories. From the

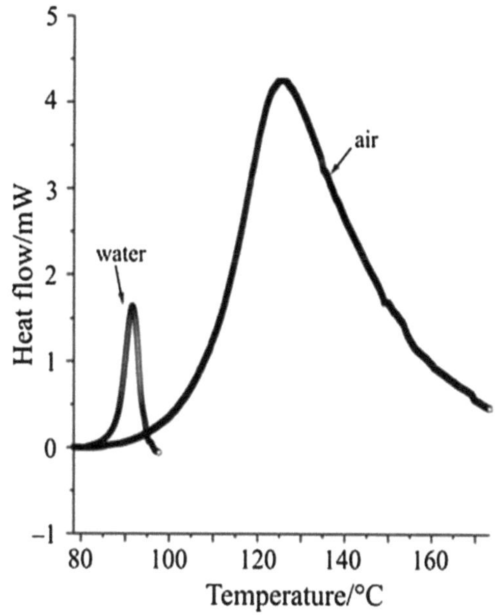

Fig. 4.41 DSC curves for denaturation in air and in water of recent manufactured leather [152]

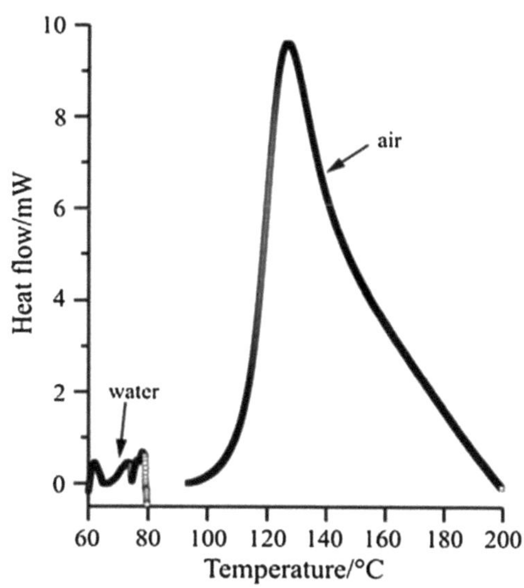

Fig. 4.42 DSC curves for denaturation in air and in water of a patrimonial leather [152]

comparison between the thermal degradation of the original wax/soot mixture collected of Nefertete's eye [153] with a phase mixture prepared from 3,000-year-old bee wax from the village of Deir-el-Medine (Ramesseum) and ancient soot from an oil lamp is shown that analogous phase mixtures are present, i.e. the black colourant has been prepared as mentioned.

References

1. Berger MJ, Hubbell JH (1978) XCOM: photon cross sections on a personal computer. US Department of Commerce, NBSIR, vol 87, pp 3597
2. Cesareo R, Gigante GE, Castellano A, Iwanczyk I (2000) Portable systems for energy dispersive X-ray fluorescence analysis: encyclopedia of chemistry. Wiley, Chichester
3. Cesareo R (1998) Photon induced X-ray emission. In: Cesareo R (ed) Nuclear analytical techniques in medicine. Elsevier, Amsterdam, pp 19–121
4. Van Grieken R, Markowicz A (ed) (2001) Handbook of X-ray spectrometry revised and expanded. CRC Press, Boca Raton
5. Cesareo R, Gigante GE (1998) Non destructive analysis of ancient metal alloys by in situ EDXRF transportable equipment. Radiat Phys Chem 51(4–6):689–700
6. Cesareo R, Brunetti A, Ridolfi S (2008) Pigment layers and precious metal sheets be EDXRF-analysis. X-ray Spectrom 37:309–316
7. Cesareo R et al (2010) Pre-columbian alloys from the royal tombs of Sipan analyzed with a portable EDXRF equipment. Appl Radiat Isot 68:525–528
8. Diana M, Gabrielli N, Ridolfi S (2007) Sulphur determination on stone monuments with a transportable EDXRF system. X-ray Spectrom 36(6):424–428
9. Kockelmann W, Pantos E, Kirfel A (2000) In: Creagh DC, Bradley DA (eds) Radiation in art and archeometry. Elsevier, Amsterdam, p 347
10. Kockelmann W, Kirfel A, Haehnel E (2001) Non-destructive phase analysis of archaeologiocal ceramics using TOF neutron diffraction. J Archaeol Sci 28:213
11. Siano S, Kockelmann W, Bafile U, Celli M, Iozzo M, Miccio M, Moze O, Pini R, Salimbeni R, Zoppi M (2002) Quantitative multiphase analysis of archaeological bronzes by neutron diffraction. Appl Phys A Mater Sci Process 74:1139
12. Siano S, Kockelmann W, Bafile U, Celli M, Iozzo M, Miccio M, Moze O, Pini R, Salimbeni R, Zoppi M (2002) Quantitative multiphase analysis of archaelogical bronzes by neutron diffraction. Appl Phys A 74:1139
13. Siano S, Bartoli L, Kockelmann W, Zoppi M, Miccio M (2004) Neutron metallography of archaeological bronzes. Phys B Condens Matter 350:123–126
14. Artioli G (2007) Small angle neutron scattering investigation of microporosity in marbles. Appl Phys A 74(Suppl 1):s1066–s1068. doi:10.1007/s003390201679
15. Barilaro D, Crupi V, Majolino D, Venuti V, Barone G, Kockelmann W (2007) Neutrons as probe of large volume specimen: the case of archaeological pottery findings. J Archaeol Sci 34:1148–1152. doi:10.1016/j.jas.2006.10.024
16. Bartoli L, Aliotta F, Imberti S, La Torre G, Ponterio R, Toscano Raffa A, Vasi C, Zoppi M (2009) Time-of-flight neutron diffraction characterization of ceramic findings from Southern and Western Sicily. Archaeometry 51:568–575. doi:10.1111/j.1475-4754.2008.00425.x
17. Siano S, Bartoli L, Santisteban JR, Kockelmann W, Daymond MR, Miccio M, De Marinis G (2006) Non-destructive investigation of bronze artefacts from the marches national museum of archaeology using neutron diffraction. Archaeometry 48:77–96. doi:10.1111/j.1475-4754.2006.00244.x
18. Dillmann Ph, Mazaudier F, Hoerlè S (2004) Advances in understanding atmospheric corrosion of iron. I. Rust characterisation of ancient ferrous artefacts exposed to indoor atmospheric corrosion. Corros Sci 46:1401–1429. doi:10.1016/j.corsci.2003.09.027
19. Fermo P, Cariati F, Cipriani C, Canetti M, Padeletti G, Brunetti B, Sgamellotti A (2002) The use of small angle X-ray scattering (SAXS) for the characterisation of lustre surfaces in Renaissance majolica. Appl Surf Sci 185:309–316. doi:10.1016/S0169-4332(01)00984-9
20. Kuczumow A, Pikus S, Un-Ro C, Sadowski P, Wajnberg P, Jurek M (2001) Structural investigations of a series of petrified woods of different origin. Spectrochimica Acta B 56:339–350
21. Kennedy CJ, Vest M, Cooper M, Wess TJ (2004) The use of small-angle X-ray scattering to study archaeological and experimentally altered bone. Appl Surf Sci 227:151–163. doi:10.1016/j.apsusc.2003.11.057

22. Hiller JC, Wess TJ (2006) The use of small-angle X-ray scattering to study archaeological and experimentally altered bone. J Arch Sci 33:560–572
23. Wess TJ, Drakopoulos M, Snigirev A, Wouters J, Paris O, Fratzl P, Collins M, Hiller J, Nielsen K (2001) The use of small angle X-ray diffraction studies for the analysis of structural features in archaeological samples. Archaeometry 43:117–129
24. Triolo R, Lo Celso F, Benfante V, Gorgoni C, Barker J, Butler P, Ruffo I (2007) Fingerprinting white marbles of archaeometric interest by means of combined SANS and USANS. Nuovo Cimento Soc Ital Fis C 30:129–138. doi:10.1393/ncc/i2006-10038-6
25. Anovitz LM, Lynn GW, Cole DR, Rother G, Allard LF, Hamilton WA, Porcar L, Kim M-H (2009) A new approach to quantification of metamorphism using ultra-small and small angle neutron scattering. Geochimica et Cosmochimica Acta 73:7303–7324. doi:10.1016/j.gca.2009.07.040
26. Guinier A (1939) La Diffraction des Rayons X aux Très Faibles Angles: Applications à l'Etude des Phénomènes Ultra-microscopiques. Ann Phys 12:161–236
27. Guinier A, Fournet G (1955) Small angle scattering of X-rays. Wiley, New York
28. Glatter O, Kratky O (1982) Small angle X-ray scattering. Academic Press, London
29. Fratzl P, Schreiber S, Klaushofer K (1996) Bone mineralization as studied by small-angle X-ray scattering. Connect Tissue Res 34:247–254O
30. Pèrez-Arantegui J, Molera J, Larrea A, Pradell T, Vendrell-Saz M, Borgia I, Brunetti B, Cariati F, Fermo P, Mellini M, Sgamellotti A, Viti C (2001) Luster pottery from the thirteenth century to the sixteenth century: a nanostructured thin metallic film. J Am Ceram Soc 84:442–446. doi:10.1111/j.1151-2916.2001.tb00674.x
31. Hosemann R, Bagchi SN (1962) Direct analysis of diffraction by matter. North-Holland Publishing Co., Amsterdam
32. Brunberger H (1965) Small angle X-ray scattering. Gordon and Breach, Science Publishers, New York
33. Chillura Martino D, Saladino ML, Nasillo G, Piga G, Enzo S, Caponetti E. J Appl Cryst (Submitted)
34. Axelos MAV, Tchoubar D, Bottero JY (1989) Small-angle X-ray scattering investigation of the silica/water interface: evolution of the structure with pH. Langmuir 5:1186–1190. doi:10.1021/la00089a010
35. Fratzl P, Groschner M, Vogl G, Plenk H Jr, Escheberger J, Fratzl-Zelman N, Koller K, Klaushofer K (1992) Mineral crystals in calcified tissues: a comparative study by SAXS. J Bone Miner Res 7:329–334
36. Fratzl P, Gupta HS, Paris O, Valenta A, Roschger P, Klaushofer K (2005) Diffracting ''stacks of cards''—some thoughts about small-angle scattering from bone. Prog Colloid Polym Sci 130:33–39. doi:10.1007/b107343
37. Hiller JC, Collins MJ, Chamberlain AT, Wess TJ (2004) Small-angle X-ray scattering: a high-throughput technique for investigating archaeological bone preservation. J Archaeol Sci 31:1349–1359
38. Entwistle KM, Terril NJ (2000) The measurements of the micro-fibril angle in soft wood. J Mater Sci 35:1675–1684. doi:10.1023/A:1004755913978
39. Beaucage G (1995) Approximations leading to a unified exponential/power-law approach to small-angle scattering. J Appl Crystallogr 28:717–728
40. Beaucage G (1996) Small-angle scattering from polymeric mass fractals of arbitrary mass-fractal dimension. J Appl Crystallogr 29:134–146
41. Botti A, Ricci MA, De Rossi G, Kockelmann W, Sodo A (2006) A Methodological aspects of SANS and TOF neutron diffraction measurements on pottery: the case of Miseno and Cuma. J Archaeol Sci 33:307–319. doi:10.1016/j.jas.2005.06.016
42. Barone G, Crupi V, Longo F, Majolino D, Mazzoleni P, Venuti V (2011) Characterisation of archaeological pottery: the case of ''Ionian Cups''. J Mol Struct 993:142–146. doi:10.1016/j.molstruc.2011.01.028
43. Teixeira J (1988) Small-angle scattering by fractal systems. J Appl Crystallogr 21:781–785. doi:10.1107/S0021889888000263

44. Ballirano P, Caracciolo G, Sadun C, Caminiti R (2008) The use of energy dispersive X-ray diffraction (EDXD) for the investigation of the structural and compositional features of old and modern papers. Microchem J 88:107–112. doi:10.1016/j.microc.2007.11.007
45. Chiari G, Giordano A, Menges G (1996) Non-destructive X-Ray diffraction analyses of non-prepared samples. Sci Technol Cult Herit 5(1):21–36
46. Vittiglio G, Bichlmeier S, Klinger P, Heckel J, Fuzhong W, Vincze L, Janssens K, Engström P, Rindby A, Dietrich K, Jembrih-Simbürger D, Schreiner M, Denis D, Lakdar A, Lamotte A (2004) A compact μ-XRF spectrometer for (in situ) analyses of cultural heritage and forensic materials. Nucl Instrum Method B 213:693–698. doi:10.1016/S0168-583X(03)01687-2
47. Johansson SAE, Campbell JL, Malmqvist KG (1995) Particle-induced X-ray emission spectrometry (PIXE). Wiley, New York
48. Jembrih D, Neelmeijer C, Schreiner M, Mäder M, Ebel M, Svagera R, Peev M (2001) Iridescent Art Nouveau glass—IBA and XPS for the characterization of thin iridescent layers. Nucl Instrum Method B 181:698–702. doi:10.1016/S0168-583X(01)00598-5
49. Wess TJ, Drakopoulos M, Snigirev A, Wouters J, Paris O, Fratzl P, Collins M, Hiller J, Nielsen K (2001) The use of small angle X-ray diffraction studies for the analysis of structural features in archaeological samples. Archaeometry 43:117–129
50. Giessen BC, Gordon GE (1968) X-ray diffraction: new high-speed technique based on X-ray spectrography. Science 159:973–975. doi:10.1126/science.159.3818.973-a
51. Ballirano P, Caminiti R (2001) Rietveld refinements on laboratory Energy dispersive X-ray diffraction (EDXD) data. J Appl Crystallogr 34:757–762
52. Caminiti R, Rossi AV (1999) The kinetics of phase transitions observed by energy-dispersive X-ray diffraction. Int Rev Phys Chem 18:263–299
53. Prober JM, Schultz J (1975) Liquid-structure analysis by energy-scanning X-ray diffraction: mercury. J Appl Crystallogr 8:405–414. doi:10.1107/S0021889875010898
54. Tokarski C, Cren-Olivè C, Rolando C, Martin E (2003) Protein studies in cultural heritage. In: Seiz-Jimenez C (eds) Molecular biology and cultural heritage. Taylor e Francis, London
55. Caponetti E, Caminiti R, Chillura Martino D, Saladino ML (2007) Energy dispersive X-ray diffraction potentiality in the field of Cultural Heritage: simultaneous structural and elemental analysis of various artefacts. Ann Chim-Sci Mat 97:473–490
56. Caminiti R, Portale G (2004) Analisi delle policromie in frammenti Pesaresi e Faentini mediante Fluorescenza X. In: Ciaroni A (ed) Maioliche del Quattrocento a Pesaro.Frammenti di storia dell'arte ceramica dalla bottega dei fedeli. Centro di Firenze– in italian, pp 255–258
57. Colombo C, Bracci S, Conti C, Greco M, Realini M (2011) Non-invasive approach in the study of polychrome terracotta sculptures: employment of the portable XRF to investigate complex stratigraphy. X-Ray Spectrom 40:273–279. doi:10.1002/xrs.1336
58. Rosi F, Miliani C, Borgia I, Brunetti B, Sgamellotti A (2004) Identification of nineteenth century blue and green pigments by in situ X-ray fluorescence and micro-Raman spectroscopy. J Raman Spectrosc 35:610–615. doi:10.1002/jrs.1180
59. Beckhoff B, Langhoff N, Kanngiefer B, Wedell R, Wolff H (2006) Handbook of practical X-ray fluorescence analysis. Springer, Berlin
60. http://www.artiscreation.com/Color_index_names.html
61. Artioli G (2010) Scientific methods in cultural heritage. Oxford University Press, Oxford
62. Verma HR (2007) Atomic and nuclear analytical methods. Springer, Berlin
63. Demortier G (2005) Ion beam techniques for the non-destructive analysis of archaeological materials. In: Uda M, Demortier G, Nakai I (eds) X-rays for archaeology. Springer, Berlin
64. Salomon J, Dran JC, Guillou T, Moignard B, Pichon L, Walter P, Mathis F (2008) Ion-beam analysis for cultural heritage on the AGLAE facility: impact of PIXE/RBS combination. Appl Phys A 92:43–50
65. Mahnke H, Denker A, Salomon J (2009) Accelerators and X-rays in cultural heritage investigations. Comptes Rendues Phys 10(7):660–675
66. Roumié M, Waksman SY, Nsouli B, Reynolds P, Lemaître S (2004) Use of PIXE analysis technique for the study of Beirut amphora production in the Roman period. Nucl Instrum Meth Phys Res B215:196–202

67. Schiavon N, Candeias A, Ferreira T, Conceiçao Lopes M, Carneiro A, Calligaro T, Mirao J (in press) A combined multi-analytical approach for the study of Roman Glass from Southwest Iberia: Synchrotron μ-XRF, external-PIXE/PIGE and VP-BSEM-EDS. Archaeometry
68. Guerra F (2005) Trace elements fingerprinting using accelerators and ICP-MS: circulation of gold from the 6th century BC to the 12th century AD. In: Van Grieken R, Janssens J (eds) Cultural heritage conservation and environmental impact assessment by non destructive testing and micro-analysis. Taylor and Francis Group, London
69. Spoto G, Torrisi A, Contino A (2000) Probing archaeological and artistic solid materials by spatially resolved analytical techniques. Chem Soc Rev 29:429–439
70. Zucchiatti A, Prati P, Bouquillon A, Giuntini L, Massi M, Migliori A, Cagnana A, Roascio S (2004) Characterization of early medieval frescoes by μ-PIXE, SEM and Raman spectroscopy. Nucl Instrum Meth Phys Res B 219:20–25
71. Constantinescu B, Bugoi R, Cojocaru V, Radtke M, Calligaro T, Salomon J, Pichon L, Röhrs S, Ceccato D, Eberländer-Tarnoveanu E (2008) Micro-SR-XRF and micro-PIXE studies for archaeological gold identification-the case of Carpathian (Transylvanian) gold and of Dacian bracelets. Nucl Instrum Meth Phys Res B 266:2325–2328
72. Ortega-Feliu I, Gómez-Tubio B, Respaldiza MA, Capel F (2011) Red layered medieval stained glass window characterization by means of micro-PIXE. Nucl Instrum Meth Phys Res. doi:10.1016/j.nimb.2011.02.023
73. Gersch HK, Robertson JD, Henderson AG, Pollack D, Munson CA (1998) PIXE analysis of prehistoric and proto-historic Caborn-Welborn phase copper artifacts from the lower Ohio River Valley. J Radioan Nucl Chem 234:85–90
74. Zhang B, Li YH, Li QH, Ma B, Gan FX, Zhang ZQ, Cheng HS, Yang FJ (2004) Non destructive analysis of early glass unearthed in south China by external beam PIXE. J Radioan Nucl Chem 261:387–392
75. Ruvalcada Sil JL (2005) PIXE analysis of pre-hispanic items from ancient America. In: Uda M, Demortier G, Nakai I (eds) X-rays for archaeology. Springer, Berlin
76. Sokaras D, Karydas AG, Oikonomou A, Zacharias N, Beltios K, Kantarelou V (2009) Combined elemental analysis of ancient glass beads by means of ion beam, portable XRF and EPMA techniques. Anal Bioanal Chem 395:2199–2209
77. Gómez-Tubio B, Ontalba Salamanca MA, Ortega-Feliu I, Respaldiza MA, Amores Carredano F, Gonzalez-Acuña D (2006) PIXE-PIGE analysis of late Roman glass fragments. Nucl Instrum Meth Phys Res B 249:616–621
78. Weber G, Strivay D, Martinot L, Garnir HP (2002) Use of PIXE-PIGE under variable incident angle for ancient glass corrosion measurements. Nucl Instrum Meth Phys Res B 189:350–357
79. Mäder M, Neelmeijer C (2004) Proton beam examination of glass-an analytical contribution for preventive glass conservation. Nucl Instrum Meth Phys Res B 226:110–118
80. Neelmeijer C, Mäder M (2005) Endangered glass objects identified by ion beam analysis. In: Van Grieken R, Janssens J (eds) Cultural heritage conservation and environmental impact assessment by non destructive testing and micro-analysis. Taylor and Francis Group, London
81. Climent-Font A, Muñoz-Martin A, Ynsa MD, Zucchiatti A (2008) Quantification of sodium in ancient Roman glasses with ion beam analysis. Nucl Instrum Meth Phys Res B 266:640–648
82. Ruvalcaba-Sil JL, Manzanilla L, Melgar M, Lozano Santa Cruz R (2008) PIXE and ionoluminescence for Mesoamerican jadeite characterization. X-Ray Spectrom 37:96–99
83. Calligaro T (2005) The origin of ancient gemstones unveiled by PIXE, PIGE and μ-Raman spectrometry. In: Uda M, Demortier G, Nakai I (eds) X-rays for archaeology. Springer, Berlin
84. Calligaro T, Dran JC, Poirot JP, Querré G, Salomon J, Zwaan JC (2000) PIXE-PIGE characterization of emeralds using an external micro-beam. Nucl Instrum Meth Phys Res B 161–163:769–774
85. Bourgarit D, Plateau J (2005) Quand l'aluminium valait de l'or: peut-on reconnaitre un aluminium "chimique" d'un aluminium "electrolytique"? Revue d'ArcheoSciences 29:95–105
86. Guerra MF (2004) Fingerprinting ancient gold with proton beams of different energy. Nucl Instrum Meth Phys Res B 226:185–198

87. Demortier G (2004) Precious metals and artifacts. In: Janssens K, Van Grieken R (eds) Non destructive micro-analysis of cultural heritage materials. Elsevier, Amsterdam
88. Sonck-Koota P, Lindroos A, Lill JO, Rajander J, Viitanen EM, Marra F, Pehkonen MH, Suksi J, Heselius SJ (2008) External-beam PIXE characterization of volcanic material used in ancient Roman mortars. Nucl Instrum Meth Phys Res B 266:2367–2370
89. Santos Silva A, Cruz T, Paiva MJ, Candeias A, Adriano P, Schiavon N, Mirão JAP (2011) Mineralogical and chemical characterization of historical mortars from military fortifications in Lisbon harbour (Portugal). Environ Earth Sci 63:1641–1650
90. Zucchiatti A, Pascual C, Ynsa MD, Castelli L, Recio P, Criado E, Valle FJ, Climent-Font A (2008) Compositional analysis of XVIII century glazed, polychrome layered porcelain by non destructive micro α-PIXE. J Eur Ceram Soc 28:757–762
91. Sha Y, Zhang PQ, Wang GG, Zhang XJ, Wang X, Liu J (2005) Compositional differences of blue and white porcelain analyzed external beam PIXE. In: Uda M, Demortier G, Nakai I (eds) X-rays for archaeology. Springer, Berlin
92. Calligaro T, Dran JC, Dubernet S, Poupeau G, Gendron F, Gonthier E, Meslay O, Tenorio D (2005) PIXE reveals that two Murillo's masterpieces were painted on Mexican obsidian slabs. Nucl Instrum Meth Phys Res B 240:576–582
93. Schiavon N, Zhou LP (1996) Magnetic, chemical and microscopic characterization of urban soiling on historical monuments. Environ Sci Technol 30:3624–3629
94. Nava S, Becherini F, Bernardi A, Bonazza A, Chiari M, Lucarelli F, Garcia-Orellana I, Ludwig N, Migliori A, Sabbioni C, Udisti R, Valli G, Vecchi R (2010) An integrated approach to assess air pollution threats to cultural heritage in a semi-confined environment: the case study of Michelozzo's court yard in Florence (Italy). Sci Total Environ 408:1403–1413
95. Carmona N, Ortega-Feliu I, Gomez-Tubio B, Villegas MA (2010) Advantages and disadvantages of PIXE/PIGE, XRF and EDX spectrometries applied to archaeometric characterization of glasses. Mater Charact 61:257–267
96. Skoog DA, Holler FJ, Nieman TA (1998) Principles of instrumental analysis, 5th edn. Harcourt Brace & Company, Philadelphia
97. Harris DC (2007) Quantitative chemical analyis, 7th edn. W.H. Freeman & Company, New York
98. Lajunen LHJ (1992) Spectrochemical analysis by atomic absorption and emission. The Royal Society of Chemistry, Cambridge
99. Cresser MS (1994) Flame spectrometry in environmental chemical analysis: a practical guide. The Royal Society of Chemistry, Cambridge
100. Sneddon J (1997) Atomic absorption spectroscopy. In: Settle FA (ed) Handbook of instrumental techniques for analytical chemistry. Prentice Hall, London
101. Ebdon L, Evans EH, Fisher AS, Hill SJ, Evans EH (1998) An introduction to analytical atomic spectrometry. Wiley, Chichester
102. Artioli G (2010) Scientific methods and cultural heritage, an introduction to the application of materials science to archaeometry and conservation science. Oxford University Press, Oxford
103. Hjulström B, Isaksson S (2009) Identification of activity area signatures in a reconstructed Iron Age house by combining element and lipid analyses of sediments. J Archaeol Sci 36:174–183
104. Boguckyj AB, Łanczont M, Łącka B, Madeyska T, Sytnyk O (2009) Age and the palaeoenvironment of the West Ukrainian palaeolithic: the case of VelykyiGlybochokmulti-cultural site. J Archaeol Sci 36:1376–1389
105. Rasmussen KL, Bjerregaard P, Gommesen PH, Jensen OL (2009) Arsenic in Danish and Swedish Mesolithic and Neolithic human bones—diet or diagenesis? J Archaeol Sci 36(2009):2826–2834
106. Mangone A, Giannossa LC, Laganara C, Laviano R, Traini A (2009) Manufacturing expedients in medieval ceramics in Apulia. J Cult Herit 10:134–143
107. Shingleton KL, Odell GH, Harris TM (1994) Atomic absorption spectrometry analysis of ceramics from a protohistoric site in Oklahoma. J Archaeol Sci 21:343–358

108. Mangone A, Giannossa LC, Ciancio A, Laviano R, Traini A (2008) Technological features of Apulian red figured pottery. J Archaeol Sci 35:1533–1541
109. Matović V, Erić S, Kremenović A, Colomban P, Srećković-Batoćanin D, Matović N (2011) The origin of syngenite in black crusts on the limestone monument King's Gate (Belgrade Fortress, Serbia)—the role of agriculture fertilizer. J Cult Herit
110. Manzano E, Bueno AG, Gonzalez-Casado A, Olmoa M (2000) Mortars, pigments and binding media of wall paintings in the 'Carrera del Darro' in Granada, Spain. J Cult Herit 1:19–28
111. Kleina S, Hauptmann A (1999) Iron age leaded Tin Bronzes from Khirbet Edh-Dharih, Jordan. J Archaeol Sci 26:1075–1082
112. Rezić I, Ćurković L, Ujević M (2010) Simplemethods for characterizationofmetals in historicaltextilethreads. Talanta 82:237–244
113. Manhita A, Ferreira V, Vargas H, Ribeiro I, Candeias A, Teixeira D, Ferreira T, Barrocas-Dias C (2011) Enlightening the influence of mordant, dyeing technique and photodegradation on the colour hue of textiles dyed with madder—a chromatographic and spectrometric approach. Microchem J 98:82–90
114. Degryse P, Schneider J, Poblome J, Waelkens M, Haack U, Muchez P (2005) A geochemical study of Roman to early Byzantine Glass from Sagalassos, South-west Turkey. J Archaeol Sci 32:287–299
115. Degryse P, Schneider J, Haack U, Lauwers V, Poblome J, Waelkens M, Muchez P (2006) Evidence for glass 'recycling' using Pb and Sr isotopic ratios and Sr-mixing lines: the case of early Byzantine Sagalassos. J Archaeol Sci 33:494–501
116. Pollard AM, Batt CM, Stern B, Young SMM (2007) Analytical chemistry in archaeology. Cambridge University Press, Cambridge
117. Stuart BH (2007) Analytical techniques in materials conservation. Wiley, Chichester
118. Vigil de la Villa R, Garcia Gimenez R, Petit Domınguez MD, Rucandio MI (2003) Physicochemical and Chemometric Characterisation of Late Roman Amphorae from Straits of Gibraltar. Microchim Acta 142:115–122
119. Sanchez-Ramos S, Bosch Reig F, Gimeno Adelantado JV, YusaMarco DJ, DomenechCarbo A (2002) Study and dating of medieval ceramic tiles by analysis of enamels with atomic absorption spectroscopy, X-ray fluorescence and electron probe microanalysis. Spectrochim Acta B 57:689–700
120. Kneisel EA, Ciszkowsky NA, Bowyer WJ, Walker FS, Huntsberger TG, Foust RD (1997) Identifying clay sources of prehistoric pottery using atomic spectroscopy. Microchem J 56:40–46
121. Sweevers H, Peeters A, Van Grieken R (1995) Weathering of Leinster granite under ambient atmospheric conditions. Sci Total Environ 167:73–85
122. Zamudio TJ, Garrido-Alfonseca A, Tenorio D, Jimenez-Reyes M (2003) Characterization of 16th and 18th century building materials from Veracruz City, Mexico. Microchem J 74:83–91
123. Goidanich S, Toniolo L, Jafarzadeh S, Odnevall Wallinder I (2010) Effects of wax-based anti-graffiti on copper patina composition and dissolution during four years of outdoor urban exposure. J Cult Herit 11:288–296
124. Goidanich S, Brunk J, Herting G, Arenas MA, Odnevall Wallinder I (2011) Atmospheric corrosion of brass in outdoor applications: patina evolution, metal release and aesthetic appearance at urban exposure conditions. Sci Total Environ (in press)
125. Bernardi E, Chiavari C, Lenza B, Martini C, Morselli L, Ospitali F, Robbiola L (2009) The atmospheric corrosion of quaternary bronzes: the leaching action of acid rain. Corros Sci 51:159–170
126. Limnow K, Halsberghe L, Steiger M (2007) Analysis of calcium acetate efflorescences formed on ceramictiles in a museum environment. J Cult Herit 8:44–52
127. Perry SH, Duffy AP (1997) The short term effects of mortar joints on salt movement in stone. Atmospheric Environ 31:1297–1305
128. Schiavon N (2003) Microfabrics of weathered granite in urban monuments. In: Thiel MJ (ed) Conservation of stone and other materials. E & FN Spon, London, pp 271–278

129. Schiavon N, Chiavari G, Schiavon G, Fabbri D (1995) Nature and decay effects of urban soiling on granitic building stones. Sci Total Environ 167:87–101
130. Schiavon N, Zhou L (1996) Magnetic, chemical and microscopical characterization of urban soiling on historical monuments. Environ Sci Technol 30(12):3624–3629
131. Brimblecombe P, Camuffo D (2003) Long-term damage to the built environment. In: Brimblecombe P (ed) The effects of air pollution on the built environment. Imperial College Press, London, pp 1–30
132. Saiz-Jimenez C (2004) Air pollution and cultural heritage. A.A. Balkema Publishers, Leiden
133. McAlister JJ, Smith BJ, Torok A (2008) Transition metals and water-soluble ions in deposits on a building and their potential catalysis of stone decay. Atmospheric Environ 42:7657–7668
134. Scott DA, Seeley NJ (1983) The examination of a pre-Hispanic gold chisel from Colombia. J Archaeol Sci 10:151–163
135. De Ryck I, VanBiezen E, Leyssens K, Adriaens A, S. Storme P, Adams F (2004) Study of tin corrosion: the influence of alloying elements. J Cult Herit 5:189–195
136. Schreiner MR, Prohaska I, Rendl J, Weigel C (1999) Leaching studies of potash-lime-soda glass with medieval glass composition. In: Tennent NH (ed) Conservation of glass and ceramic research, practice and training. James & James, London, pp 72–83
137. Angelini I, Artioli G, Bellintani P, Diella V, Gemmi M, Polla A, Rossi A (2004) Chemical analyses of Bronze Age glasses from Frattesina di Rovigo, Northern Italy. J Archaeol Sci 31:1175–1184
138. Biscontin G, Pellizon Birelli M, Zendri E (2002) Characterisation of binders employed in the manufacture of Venetian historical mortars. J Cult Herit 3:31–37
139. Alvarez JI, Navarro I, Garcia Casado PJ (2000) Thermal, mineralogical and chemical studies of the mortars used in the cathedral of Pamplona (Spain). Thermochim Acta 365:177–187
140. Friolo KH, Ray AS, Stuart BH, Thomas PS (2005) Thermal analysis of heritage stones. J Therm Anal Calorim 80:559–563
141. Riontino C, Sabbioni C, Ghedini N, Zappia G, Gobbi G, Favoni O (1998) Evaluation of atmospheric deposition on historic buildings by combined thermal analysis and combustion techniques. Thermochim Acta 321:215–222
142. Drebushchak VA, Mylnikova LN, Drebushchak TN, Boldyrev VV (2005) The investigation of ancient pottery. Application of thermal analysis. J Therm Anal Calorim 82:617–626
143. Campanella L, Flamini P, Grossi R, Tomassetti M (1998) Study and characterisation by thermoanalysis of statues and fictile finds of different historical and prehistoric ages. Thermochim Acta 321:167–174
144. Tomassetti M, Campanella L, Flamini P, Bandini G (1997) Thermal analysis of fictile votive statues of 3rd century BC. Thermochim Acta 291:117–130
145. Papadopoulou DN, Lalia-Kantouri M, Kantiranis N, Stratis JA (2006) Thermal and mineralogical contribution to the ancient ceramics and natural clays characterization. J Therm Anal Calorim 84:39–45
146. Odlyha M, Wang Q, Foster GM, Groot J, Horton M, Bozec L (2005) Thermal analysis of model and historic tapestries. J Therm Anal Calorim 82:627–636
147. Odlyha M, Cohen NS, Foster GM, Aliev A, Verdonck E, Grandy D (2003) Dynamic mechanical analysis (DMA), ^{13}C solid state NMR and micro-thermomechanical studies of historical parchment. J Therm Anal Calorim 71:939–950
148. Fessas D, Schiraldi A, Tenni R, Vitellaro ZL, Bairati A, Facchini A (2005) Calorimetric, biochemical and morphological investigations to validate a restoration method of fire injured ancient parchment. Thermochim Acta 348:129–137
149. Della GG, Badea E, Ceccarelli R, Usacheva T, Masic A, Coluccia S (2005) Assessment of damage in old parchments by DSC and SEM. J Therm Anal Calorim 85:637–649
150. Franceschi E, Palazzi D, Pedemonte E (2001) Thermoanalytical contribution to the study of paper degradation. Characterisation of oxidised paper. J Therm Anal Calorim 66:349–358

151. Budrugeac P, Cucos A, Miu L. The use of thermal analysis methods for authentication and conservation state determination of historical and/or cultural objects manufactured from leather. J Therm Anal Calorim. doi:10.1007/s10973-010-1183-0
152. Budrugeac P, Miu L, Popescu C, Wortmann F-J (2004) Identification of collagen-based materials that are supports of cultural and historical objects. J Therm Anal Calorim 77:975–985
153. Wiedemann H-G, Arpagaus E, Muller D, Marcolli C, Weigel S, Reller A (2002) Pigments of the bust of Nefertete compared with those of the Karnak Talatats. Thermochim Acta 382:239–247

Chapter 5
Applying the Techniques on Materials I

Turkan Yurdun, Emre Dolen, Recep Karadag, Carole Mathe, Andreas K. Tsakalof, Kyriaki A. Bairachtari, Evangelia A. Varella, A. Spinella, D. Capitani, S. Bastone, C. Di Stefano, E. Caponetti, Eleni Pavlidou, Maria Kyranoudi, Leopold Puchinger, Friedrich Sauter and Andreas Gössl

T. Yurdun (✉)
Department of Pharmaceutical Toxicology, Faculty of Pharmacy, Marmara University, 34668 Haydarpasa-Istanbul, Turkey
e-mail: tyurdun@marmara.edu.tr

E. Dolen
Department of Analytical Chemistry, Faculty of Pharmacy, Marmara University, 34668 Haydarpasa-Istanbul, Turkey
e-mail: emredolen@gmail.com

R. Karadag (✉)
Laboratory of Natural Dyes, Faculty of Fine Arts, Marmara University, Kadikoy, 34718 Istanbul, Turkey
e-mail: rkaradag@marmara.edu.tr

R. Karadag
Turkish Cultural Foundation, Natural Dyes Research and Development Laboratory, Istanbul, Turkey

C. Mathe (✉)
Laboratoire de chimie appliquée à l'Art et à l'Archéologie, Université d'Avignon, 33, rue Louis Pasteur, 84000 Avignon, France
e-mail: Carole.Mathe@univ-avignon.fr

A. K. Tsakalof (✉)
School of Medicine, University of Thessaly, Larissa, Greece
e-mail: atsakal@med.uth.gr

K. A. Bairachtari
National Center of Scientific Research "Demokritos", Institute of Nuclear Technology & Radiation Protection, Athens, Greece
e-mail: kyriaki@ipta.demokritos.gr

E. A. Varella (✉)
Department of Chemistry, Aristotle University of Thessaloniki, 54124 Thessaloniki, Greece
e-mail: varella@chem.auth.gr

E. A. Varella (ed.), *Conservation Science for the Cultural Heritage*,
Lecture Notes in Chemistry 79, DOI: 10.1007/978-3-642-30985-4_5,

A. Spinella (✉) · S. Bastone · E. Caponetti
Centro Grandi Apparecchiature (CGA) UniNetLab, Università degli Studi di Palermo, Via F. Marini, 14, 90128 Palermo, Italy
e-mail: aspin.cga@unipa.it

D. Capitani
Laboratorio di Risonanza Magnetica "Annalaura Segre", Istituto di Metodologie Chimiche CNR, Area della Ricerca di Roma, Via Salaria Km 29.300, 00015 Monterotondo, RM, Italy

C. Di Stefano
Centro Regionale per la Progettazione e il Restauro (CRPR), Via C. Colombo, 52, 90100 Palermo, Italy

E. Caponetti
Dipartimento di Chimica "S. Cannizzaro" Parco d'Orleans II, Università degli Studi di Palermo, Viale delle Scienze Pad.17, 90128 Palermo, Italy

E. Pavlidou (✉)
Department of Physics, School of Science, Aristotle University of Thessaloniki, 54124 Thessaloniki, Greece
e-mail: elpavlid@auth.gr

M. Kyranoudi
Conservation Department, Archaelogical Museum of Amphipolis, KH Ephorate of Prehistoric and Classical Antiquities, Hellenic Ministry of Culture, 62052 Amphipolis, Greece

L. Puchinger (✉) · A. Gössl
Institute of Chemical Engineering, Vienna University of Technology, Getreidemarkt 9, 1060 Vienna, Austria
e-mail: leopold.puchinger@tuwien.ac.at

F. Sauter
Institute of Applied Synthetic Chemistry, Vienna University of Technology, Getreidemarkt 9, 1060, Vienna, Austria
e-mail: fsauter@pop.tuwien.ac.at

5.1 Identification of Natural Dyes in Thirteenth Century Mevlana's Kaftans and Dervish's Dresses by HPLC–DAD Detection

Turkan Yurdun and Emre Dolen

Abstract The knowledge of the composition of the natural dye mixtures gives hints on appropriate procedures for restoration strategies. Identification of the dyes used, the establishment of their biological source, how they have been prepared, how these historical textiles were made, can provide important information. The dyestuffs used up to nineteenth century were either of plant (weld, madder, indigo, etc.) or animal origin (cochineal, shellfish, etc.) and belonged to various chemical types, such as flavonoids (yellow), anthraquinones (red) and indigoids (blue and violet). In this study microsamples (19) were examined from 11 Mevlana's kaftans, dervishs' dresses and dated from the thirteenth century. The objects belong to the collection of the Mevlana Museum in Konya, Turkey. The analysis of dyestuffs from historical object with microdestructive method as High Performance Liquid Chromatography (HPLC) with diode-array detector was performed. Historical samples (0.5–1 mg) were used as extracted with the HCl/methanol/water (2:1:1) extraction method. In this work, colour measurements of the cotton fabric samples dyed with indigotin and ellagic acid were performed by CIELab system.

5.1.1 Introduction

Dyestuffs used in dyeing of textiles are made from natural sources such as plants, insects and mineral sources. These dyes include mainly anthraquinones (red), flavonoids (yellow), indigotin (blue) and other compounds. The problems of identification of colouring materials in the historical textiles are directly linked to their chemical stability, fading, degradation, extensive use and other properties. Dyestuffs have been analysed, initially by thin layer chromatography [1] and more recently by high performance liquid chromatography (HPLC) with diode-array detection (DAD) [2–14] and mass spectrometric detection (MS). HPLC–DAD and HPLC–MS method was utilised for the identification of organic colouring ingredients in historic yarns, textiles fibres [15–21], historic maps [22], paintings [23] and icons [24–26].

The HPLC techniques provide qualitative and quantitative analyses of compounds of natural dyes. Before HPLC analysis the sample should be hydrolysis from the material of textile fibres. Acid hydrolysis has been suggested for mordant dyes. The acidic hydrolysis procedures are with HCl/methanol/water (widely used extraction method) [2–4, 7, 9, 14, 15, 20], HCl/ethanol [8, 19], formic acid/methanol [18], dimethyl sulfoxide (DMSO)/HCl [16, 17], oxalic acid [14, 21], or citric acid, trifluoroacetic acid (TFA) [14, 15], formic acid/ethylene diamine tetraacetic acid [14] extraction for textile samples. Many natural dyestuffs after hydrolysis are soluble mixture of water–organic solvent (e.g., methanol, acetonitrile), but indigotin is insoluble in methanol–water solvents therefore, soluble with *N,N*-dimethylformamide or DMSO.

Mevlana, a well-known poet in the world of mysticism and the pioneer of the Mevlevi path. In the present study, HPLC–DAD is employed for the identification of the dyestuffs in samples extracted from the Mevlana's kaftans and dervish's dresses. Identification is based on comparison of the dye absorption spectrum, retention time and reference standards.

5.1.2 Experimental Section

5.1.2.1 Chemicals

All reagents were analytical grade, unless stated otherwise. High purity water was obtained with a Milli-Q water purification system (Millipore, Bedford, MA, USA) and the HPLC mobile phase was prepared using Milli-Q water. Indigotin employed as reference standard material was purchased from Kremer Pigmente GmbH & Co. KG/Germany, ellagic acid was obtained as 95 % pure from Sigma E 2250-1G; TFA was from Riedel-de-Haën. All solvents used were HPLC grade.

5.1.2.2 HCl Extraction Method

Dyestuff extraction was performed according to previously described methods [2–4]. Historical textile samples (0.5–1.0 mg) were hydrolysed in 400 μl of H_2O: MeOH: 37 % HCl (1:1:2; v/v/v;) in conical glass tubes for precisely 8 min in a water bath at 100 °C to extract the organic dyes. After rapid cooling under running cold water, the solution was evaporated just to dryness in a water bath at 50–65 °C under a gentle stream of nitrogen. The dry residue was dissolved in 200 μl of a mixture of MeOH:H_2O (2:1; v/v) and was centrifuged to separate the particulate matter (2,500 rpm for 10 min.). Then, 25 and/or 50 μl of the supernatant were injected into the HPLC instrument. The residue was dissolved in 0.2 ml of *N,N*-dimethylformamide [indigotin is not soluble in MeOH:H_2O (2:1; v/v)] and was injected into the HPLC apparatus.

5.1.2.3 Chromatographic Analysis with Diode-Array Detection

Sample extracts and standards were analysed using an HPLC Agilent 1100 series system (Agilent Technologies) equipped with a delivery pump with a 50 μl loop and Rheodyne valve (7725i sample injector), a G1315A diode-array detector (chromatograms were obtained by scanning the sample from 191 to 799 nm with a resolution of 2 nm and chromatographic peaks were monitored at 255, 268, 276, 350 and 491 nm), a G1322A vacuum degasser and a G1316A thermostatted column compartment; the data station was an Agilent Chemstation. A Nova-Pak C_{18} analytical column (3.9 × 150 mm, 4 μm, Part No WAT 086344, Waters) protected by a guard column filled with the same material, was used for separation of the analytes. Analytical and guard columns were maintained at 30 °C. The HPLC gradient elution was performed using the method of Halpine [27] and Karapanagiotis et al. [9], 15]. The mobile phases used were the following: as phase A H_2O–0.1 % TFA and phase B CH_3CN–0.1 % TFA. The following elution programme was used: 0–1 min 5% B; 1–20 min, linear gradient to 30% B; 20–25 min, linear gradient to 60% B; 25–28 min, 60% B isocratic; 28–33 min, linear gradient to 95% B; 33–35 min 95% B isocratic. Flow rate was 0.5 ml/min. Injection volume was 25–50 μl. Retention time for indigotin was 28.977 min and ellagic acid was 17.735 min.

5.1.2.4 Colour Measurements

The *L**, *a** and *b** values for the Mevlana's kaftans and dervish's dresses were measured with a Konica Minolta CM 2003d spectrophotometer and calibrated against a standard white cap [28, 29].

5.1.3 Results and Discussion

The composition of natural dyes used in historical textile is important for conservation purposes and to determine the date of a textile. In this investigation, 11 Mevlana's kaftan samples which were provided by the Mevlana Museum in Konya-Turkey collection were studied. Dyestuffs were detected using HPLC with a diode-array detector. HPLC results collected for some selected fibres are presented in the Table 5.1.

In this study, indigotin was detected all historical blue samples and found the presence of ellagic acid, derived from hydrolysable tannins by the acid hydrolysis recovery procedure (yellow object inventor no. 690) (Fig. 5.1). These components of samples were identified based on the absorption spectra acquired with reference standard compounds. For identification of dyes on textiles the standards are needed and UV–Vis absorption spectra of these compounds were also acquired as depicted in Fig. 5.2. Natural dyestuffs not found in the cream cotton samples. Identified dyes of textile samples are discussed. Nowadays, the CIELab system is being used in a wide variety of applications such as textile, pharmaceutical and food

Table 5.1 The results of dyestuff analysis of investigated old historical samples

Inventory number	Sample (colour)	Detected compounds	Identified dye source (botanical name)
689	Blue	Indigotin	*Isatis tinctoria* or Indigofera species (e.g., *Indigofera tinctoria* L.)
690	Yellow Blue	Ellagic acid Indigotin	*Quercus infectoria* Oliv. or *Quercus ithaburensis* Decne *Isatis tinctoria* or Indigofera species (e.g., *Indigofera tinctoria* L.)
692	Dark blue	Indigotin	*Isatis tinctoria* or Indigofera species (e.g., *Indigofera tinctoria* L.)
693	Blue	Indigotin	*Isatis tinctoria* or Indigofera species (e.g., *Indigofera tinctoria* L.)
695	Blue	Indigotin	*Isatis tinctoria* or Indigofera species (e.g., *Indigofera tinctoria* L.)
696	Blue	Indigotin	*Isatis tinctoria* or Indigofera species (e.g., *Indigofera tinctoria* L.)
697	Blue	Indigotin	*Isatis tinctoria* or Indigofera species (e.g., *Indigofera tinctoria* L.)
698	Blue	Indigotin	*Isatis tinctoria* or Indigofera species (e.g., *Indigofera tinctoria* L.)
699	Blue	Indigotin	*Isatis tinctoria* or Indigofera species (e.g., *Indigofera tinctoria* L.)
700	Blue	Indigotin	*Isatis tinctoria* or Indigofera species (e.g., *Indigofera tinctoria* L.)
703	Blue and cream fibre	Indigotin	*Isatis tinctoria* or Indigofera species (e.g., *Indigofera tinctoria* L.)

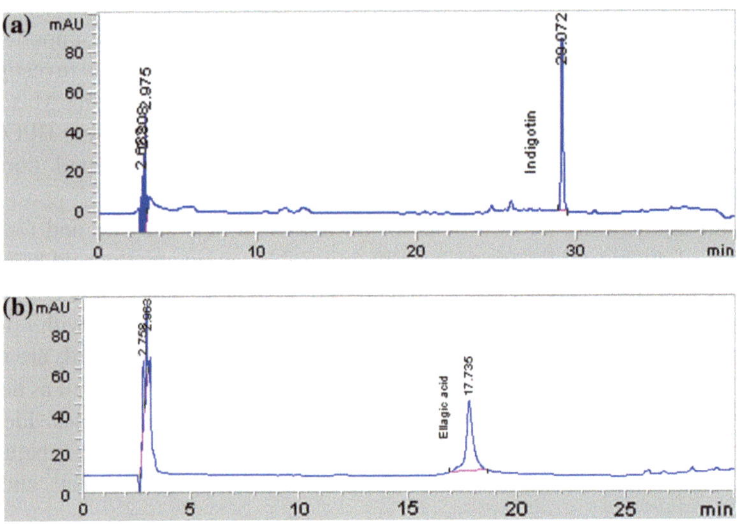

Fig. 5.1 HPLC-DAD Chromatograms at 276 nm of the *blue* sample (**a**) and at 255 nm of *yellow* sample (**b**)

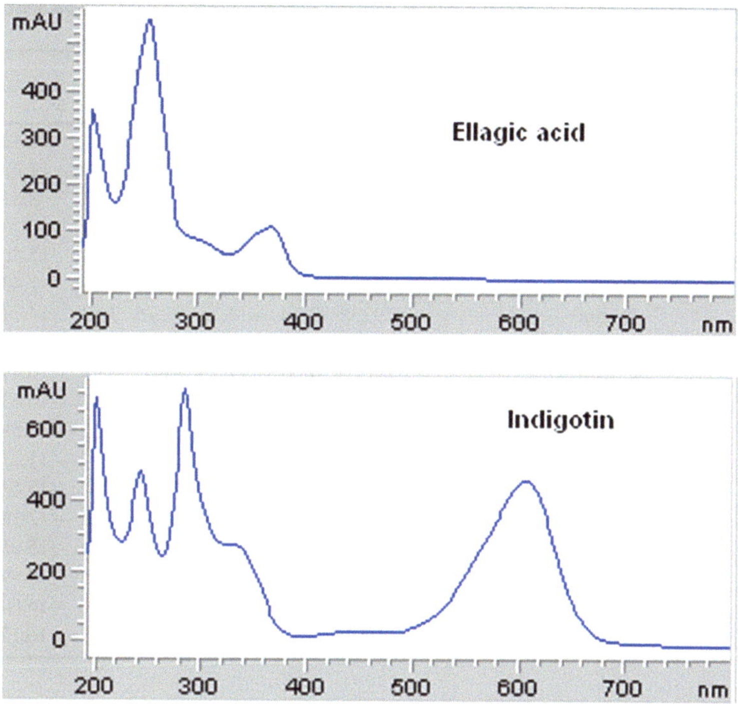

Fig. 5.2 Absorption spectra in UV–Vis regions of the dye standards

technology. The colour of the fibres of Mevlana's kaftans and dervish's dresses were measured by CIELab system. The ΔE represents the total colour difference. The colour difference between samples of the unwashed and washed with ether are not significant (Table 5.2). ΔE values were calculated as follows:

$$\Delta E = [(\Delta L*)^2 + (\Delta a*)^2 + (\Delta b*)^2]^{1/2}$$

Colour differences for Mevlana's kaftans and dervish's dresses, ΔE, are given in Tables 5.3 and 5.4. For colour differences ΔE values up to 5 is not significant, ΔE values from 5 to 20 showed important colour difference and also from 20 to the above showed extreme colour difference.

Indigo and woad (*Indigotin: blue dye*)

Indigoid dyes are the oldest natural dyes and have been known in India for about 4,000 years. The principle blue dye used was indigo the main sources of which were woad (*Isatis tinctoria* L., Brassicaceae) and the indigo plant (*Indigofera tinctoria L.*). Historically, woad (*Isatis tinctoria* L. also known as dyer's woad) has been used in Central Europe since antiquity as indigo dye. Indigotin, the main colouring matter in indigo, has been obtained from the leaves of *Isatis*

Table 5.2 L^*, a^* and b^* (CIELab) values for the cotton samples of Mevlana's kaftans and dervish's dresses

Inventory number	Colour	After washed			Unwashed			ΔE
		L^*	a^*	b^*	L^*	a^*	b^*	
689	Blue	89.92	3.44	−16.60	89.71	3.34	−16.42	0.294
	Cream	83.44	1.58	9.92	80.48	1.88	11.77	3.503
690	Blue and yellow plaid fabric	45.91	2.43	10.55	44.67	1.23	8.51	2.672
692	Blue	83.94	3.07	−16.33	83.65	2.73	−15.46	0.978
	Cream	81.18	3.94	−10.19	78.29	3.33	−7.43	4.042
693	Cream	92.96	3.78	−14.76	92.76	3.82	−14.71	0.210
695	Blue	82.02	1.60	−16.48	78.31	1.21	−15.86	3.782
	Cream	90.79	4.00	−12.73	89.03	3.91	−10.49	2.850
696	Blue	90.92	3.34	−16.71	90.60	3.27	−16.63	0.337
697	Blue	91.18	3.49	−16.76	89.61	3.22	−16.58	1.603
	Cream	90.10	3.91	−11.32	90.40	3.84	−11.13	0.362
698	Blue	77.92	2.42	−15.41	77.02	2.42	−15.64	0.929
	Cream	91.23	3.57	−11.73	91.23	3.72	−12.18	0.474
699	Blue	67.48	2.04	−14.91	64.85	1.89	−15.05	2.638
	Cream	82.24	2.31	14.01	81.86	2.39	12.99	1.091
700	Blue	87.07	3.35	−16.58	90.23	3.50	−16.60	3.164
703	Blue and cream fibre	90.54	2.81	−14.02	92.23	3.37	−16.12	2.753

Table 5.3 Colour difference for samples contain blue cotton fibres of Mevlana's kaftans and dervish's dresses

Inventory number	ΔE	Inventory number	ΔE
689–697	0.224	689–695	11.611
696–700	0.437	695–700	12.161
689–700	0.573	698–699	12.196
697–700	0.681	695–696	12.485
689–696	0.917	697–698	12.650
696–697	0.993	689–698	12.747
695–698	1.782	698–700	13.289
692–695	5.567	695–699	13.501
692–697	6.084	696–698	13.643
689–692	6.166	692–699	18.823
692–698	6.640	697–699	24.843
692–700	6.722	689–699	24.940
692–696	7.068	699–700	25.478
695–697	11.500	696–699	25.835

tinctoria L. by a complex process that involved fermentation (enzymatic hydrolysis) [30]. In some historical textiles dyed with natural indigo were detected to contain mostly indigotin and indirubin by HPLC [17, 31].

Table 5.4 Colour difference for samples contain cream cotton fibres of Mevlana's kaftans and dervish's dresses

Inventory number	ΔE	Inventory number	ΔE
697–698	1.344	689–692	19.379
695–697	1.514	692–699	20.751
689–699	1.911	689–695	23.932
695–698	2.781	695–699	24.597
693–698	2.958	689–697	25.033
693–695	5.633	697–699	25.628
693–697	4.288	689–693	29.253
692–695	11.182	689–698	26.316
692–697	12.673	698–699	26.891
692–698	13.789	693–699	29.802
692–693	16.206		

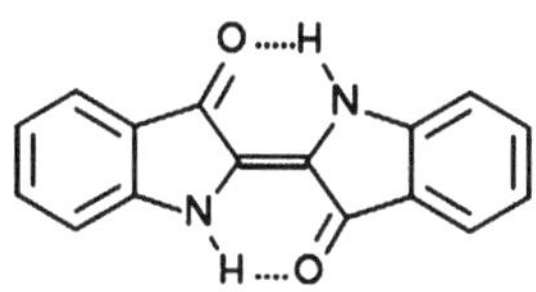

Indigotin

Ellagic acid is contained in several plants (*Quercus infectoria* Oliv. or *Quercus ithaburensis* Decne) and has been used in the natural dyeing of yellow, brown, grey and black as a mordant [30, 32]. Tannins are found in small quantities in plant tissues but are abundant in bark and damaged tissues such as gals and wounds [4, 32].

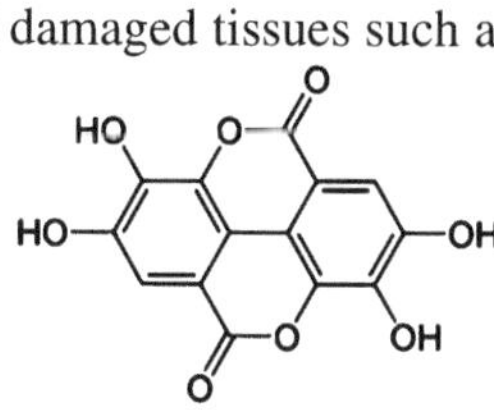

Ellagic acid

5.1.4 Conclusions

A chromatographic method for the separation and identification of 11 colour compounds presents in blue dyestuffs. Combination of instrumental analytical techniques was applied to analyse dye extracts of historical wool and cotton art objects (Mevlana's kaftans and dervish dresses) from the Mevlana Museum in Konya, Turkey. In this study, only fibres of yellow and blue colours were selected for analysis. HPLC with DAD has yielded good results in analyses of extracts from historical materials and the most dominant identification dyestuffs is indigotin. Natural dyes of plant origin—indigotin and ellagic acid—were found.

5.2 Identification of Natural Dyes by RP-HPLC–DAD and Natural Dyeing Procedures

Recep Karadag

Abstract Natura dyes are among the most significant markers of works of art and archaeological findings, historically. Therewithal, in the last few decades, new dyeing processes, analytical procedures and techniques have been developed and improved for dyeing quality and the identification of microsamples containing natural dyes. This review looks at the dyeing processes with natural dyes, the chemical composition of natural dye source materials used in the field of the cultural heritage and focuses on analytical method based on a chromatographic technique—reversed-phase HPLC with diode-array detection (RP-HPLC–DAD). RP-HPLC–DAD was utilised for the identification of dyes in different plants such as dyers weld, madder, dyer's oak, woad, indigo shrub, cochineal and lac. The extraction of dyes was carried out with an HCl/MeOH/H_2O (2:1:1; v/v/v) mixture. The most important dyestuffs detected were natural yellow, red, green, black and blue dyes which are found in historical textiles. In addition, RP-HPLC–DAD was employed in the identification of natural dyes in extracts of historical silk and wool art objects, from historical textiles and newly dyed silk and wool fibres.

5.2.1 Introduction

Reconstruction of dyeing receipts and identification of colouring materials in an art object related to cultural heritage has received significant attention because of their importance in the development of conservation and restoration works [33]. Natural dyes are derived from naturally occurring sources such as plants (e.g., weld, madder and indigo); insects (e.g., cochineal beetles and lac scale insects); animals (e.g., some species of mollusks or shellfish). Natural dyes are not only used for colouring of materials such as textiles, paper, ink, etc. but also in cosmetic, food and pharmaceutical industry as basis for drug products [34]. Alum ($KAl(SO_4)_2.12H_2O$) is widely used as mordant material for natural dyeing especially wool and silk fibres with polyphenolic dyes (flavonoid, anthraquinone, and indigoid) from natural sources (weld, madder, indigo, etc.) to obtain fine bright colours and a good wash-fastness [35]. Because of many natural dyes are water insoluble, before HPLC analysis the sample should be extracted with typically, strong acid, e.g., 37 % HCl/methanol/water, 2:1:1 (v/v/v), from the textile [36]. Historical textiles were dyed with natural dyes which were mainly of plant or insect origin. Dye identification of historical textiles is usually based on comparison with known references. In our case, dye sources used as references were as follows:

Weld (*Reseda luteola* L.): The dye plant is one of the most important natural yellow dyes used to produce fast yellows [37] owing to the presence of the flavones luteolin and apigenin as the major constituents [38]. Weld appears to have been used in Europe since prehistoric times. It is likely to have been widely used in the Mediterranean area in the Hellenistic period and in the Roman Empire era [37]. Weld was used in the first century Masada textiles, third century Palmyra textiles, thirteenth century Seljuk carpets and fifteen to twentieth century Ottoman textiles for yellow and green colours [39].

Madder (*Rubia tinctorium* L.): Madder is an herbaceous plant, light green in colour, with a perennial root system of long-branching roots in which are concentrated the numerous red colourants. In Egypt, the earliest examples of linen textiles dyed with madder date from much later periods than had previously been thought [37]. Madder was used in the first century Masada textiles, nineteenth century Daghestan textiles (for example suzani) and fifteen to twentieth century Ottoman textiles for red shades [39]. The anthraquinone contents of madder roots are complex and vary with the age of the plant. So far, some 36 different anthraquinones (alizarin, rubiadin, munjistin, purpurin, etc.) have been detected in *R. tinctorium* [40].

Walloon oak: The acorn caps of *Quercus ithaburensis* were used for black dyeing in the Ottoman textiles [39]. They contain mainly tannin compounds (gallic acid, ellagic acid, etc.) [41].

Woad (*Isatis tinctoria* L.): The plant appears to have been utilised since very early times for dyeing textiles, and possibly even earlier, for body paints. Egypt was one of the early centres of woad dyeing; woad was most probably one of the plant sources of blue dye mentioned in documents from the Pharaonic and Ptolemaic periods. Moreover, woad dyeing processes were well known to the Egyptians of the Hellenistic period [37]. Woad was also used in the first century Masada textiles, third century Palmyra textiles, thirteen century Seljuk carpets, and fifteen to twentieth century Ottoman textiles for blue and green colours [39]. There is a dispute, however, about whether dyeing with indigo was first developed in India or in Egypt. The name *indigo*, the most important dye of the ancient world, suggests India. For hundreds of years, perhaps even thousands of years, India was the greatest producer of indigo and the most important indigo plant is Indigo (*Indigofera tinctoria* L.) [37].

Cochineal (*Dactylopius coccus* Costa): Cochineal is one of the most important natural dyes used to produce fast reds. This dye was discovered by Mexican dyers in about 1000 BC. Eventually, the cultivation of Mexican cochineal spread to other countries with suitable climates and it began to play a very important part in various kind of dyeing. The cochineal insect was brought to Europe by Spaniards in the early sixteenth century after the discovery of America. Until Spanish conquest of Mexico the cochineal dye remained unknown by rest of the world, although it has been suggested that true cochineal was also known in the ancient world where it was produced from insects native to the Ararat valley in a mountain of Turkey [39, 42]. The major component of all cochineals (American cochineal [43], Polish cochineal, Armenian cochineal) is carminic acid [38].

Lac (*Kerria lacca* Kerr): Dyes containing lac was obtained from the female insects of the species. Major components are laccaic acid A (>70 %) and laccaic acid B (<20 %) [38]. This insect is found in India and Southeast Asian countries such as Thailand, Malaysia, Cambodia, Laos and Indonesia. Around 1500 BC, lac was mentioned in one of the oldest kingdom texts of India *Atharvaveda*. The insect was imported from India to the Near East more than 2,000 year ago. After the introduction of the American cochineal, the use of lac in dyeing was decreased [39, 42].

Case Studies

We will now comment on the key features of recent chromatographic and dyeing process studies on wool and silk fibres.

5.2.2 Materials and Methods

5.2.2.1 Chemicals

All reagents employed in this investigation were of analytical grade and were used as received, unless stated otherwise. High purity water was obtained by passing water though a Milli-Q treatment system (Millipore, Bedford, MA, USA) and the HPLC mobile phase was prepared using Milli-Q water. The following standard materials were obtained from commercial sources and used as references: Ellagic acid and gallic acid from Sigma, luteolin and apigenin, fisetin from Roth (Karlsruhe, Germany), natural indigotin (production from *Isatis tinctoria* L. or *Indigofera tinctoria* L.) from Kremer Pigmente GmbH & Co. KG. Silk and wool reference samples dyed with dyes weld, madder, dyer's oak, woad, indigo shrub (indigo production from *Indigofera tinctoria* L.), cochineal and lac were used as dyed standards. The standard coloured pieces of silk and wool were prepared in the Laboratory of Natural Dyes, Faculty of Fine Arts, Marmara University.

5.2.2.2 Mordanting Procedure

The mordanting and dyeing of wool and silk fibres were performed according to literature procedures as described below [39, 44–46]. In this study, mordanting was done prior to dyeing. Traditionally, a mordant was applied prior to dyeing (called premordanting) to assist the adsorption of the dye and to promote good bonding of the dye and the fibre. The mordant materials are metal salts which normally hydrolyse in hot aqueous solutions. The most commonly used mordants such as alum [potassium aluminium sulphate [$(KAl(SO_4)_2.12H_2O)$], iron [ferrous sulphate ($FeSO_4.\ 7H_2O$)] and tin [stannous chloride ($SnCl_2.\ 2H_2O$)] were chosen.

Fibres are immersed in the aqueous metal salt solution (mordant bath) and heated to 90 °C for 1 h and then, the mordanted fibres are removed from the bath and left to dry in the open air.

Steps of Mordanting

a. The mordant materials are dissolved in the mordant bath.
b. The silk or wool is added to the mordant bath.
c. The mordant bath is heated to 80 °C slowly.
d. At 80 °C the dye bath is hold for minimum 60 min.
e. The silk or wool is removed from the mordant bath and dried.

5.2.2.3 Dyeing Procedure

The dyeing procedures were performed in accordance with the historical dyeing methods [39]. In the dye bath, the ratio of fibres to dye extract is 1:25. For example, 25 g of dye extract is used for 1 g of wool. To prepare the dye bath, insect's or plant's dyes are extracted with water at 90 °C for 1 h and then the residues are filtered. Mordant fibres are immersed in the dye bath and the temperature is gradually raised to 90 °C and is kept at this value for about 30–60 min. The dye bath is left to cool down to around 30 °C and then the dyed fibres are removed from dye bath, squeezed, rinsed thoroughly with water and dried in the open air in the shade.

In this investigation, silk and wool were dyed with five different plant dyes and one insect dye (weld = *Reseda luteola* L., gall oak = *Quercus infectoria* Olivier, acorn caps = *Quercus ithaburensis* Decaisne, *dyer's sumac* = *Cotinus coggygria* Scop., cochineal = *Dactylopius coccus* Costa) and two indigotins, one from *Isatis tinctoria* L. and the other from *Indigofera tinctoria* L., as this was used in historical recipes.

Steps of Dyeing

a. Dye material: dried and powdered (plants or insects sources) dye materials are extracted with water.
b. The extract are added to the dye bath.
c. The earlier mordanted silk is added to the dye bath.
d. The dye bath is heated to 80 °C slowly.
e. At 80 °C, the dye bath is hold for 30–60 min.
f. The silk or wool is removed from the dye bath, washed, rinsed and dried.

5.2.2.4 Dye Extraction Procedure

The samples were prepared as follows: Historical art objects from Topkapi Palace Museum, dye extracts from dye insects and dye plants and newly dyed silk and wool samples [39, 44], were obtained using (weld = *Reseda luteola L.*; gall oak = *Quercus infectoria* Olivier; lac = Kerria lacca Kerr; cochineal = *Dactylopius coccus* Costa; Ararat kermes = *Prophyrophora hameli* Brand; kermes = *Kermes vermilio* Planchon; Anatolian buckthorn = *Rhamnus petiolaris* Boiss and indigotin (from *Isatis tinctoria* L. and from *Indigofera tinctoria* L.) according to historical recipes and were prepared in the Laboratory of Natural Dyes, Faculty of Fine Arts, Marmara University. Dyes were extracted directly from dye plants (weld, gallnut, cochineal, lac, Ararat kermes, kermes, Anatolian buckthorn, indigotin, etc.). Alternatively, mordant silk (undyed fibres) and wool were extracted and dyes were extracted from historical textile samples.

The dye extraction of all samples was performed according to previously described methods [47–49]. Samples (0.4–1.0 mg) were hydrolysed with a solution mixture of H_2O: MeOH: 37 % HCl (1:1:2; v/v/v; 400 μl) in conical glass tubes for precisely 8 min in a water bath at 100 °C to extract the organic dyes. After cooling under running cold tap water, the solution was evaporated just to dryness in a water bath at 65 °C under a gentle stream of nitrogen. The dry residue was dissolved in 200 μl of a mixture of MeOH:H_2O (2:1; v/v) and was centrifuged at 2,500 rpm for 10 min. Then 25 and/or 50 μl of the supernatant were injected into the HPLC instrument. The residue was dissolved in 0.2 ml of *N,N*-dimethylformamide [indigotin is not soluble in MeOH:H_2O (2:1; v/v)] and was injected into the HPLC apparatus.

5.2.3 Instrumentation

Chromatographic separations were performed using an Agilent 1100 series system (Agilent Technologies) equipped with a delivery pump with a 50 μl loop and Rheodyne valve (7725i sample injector), a G1315A diode-array detector (chromatograms were obtained by scanning the sample from 191 to 799 nm with a resolution of 2 nm and chromatographic peaks were monitored at 255, 268, 276, 350 and 491 nm), a G1322A vacuum degasser and a G1316A thermostatted column compartment; the data station was an Agilent Chemstation. A Nova-Pak C_{18} analytical column (3.9 × 150 mm, 4 μm, Part No WAT 086344, Waters) protected by a guard column filled with the same material, was used. Analytical and guard columns were maintained at 30 °C. The HPLC gradient elution was performed using the method of Halpine et al. [50] and Karapanagiotis et al. [51, 52]. Chromatographic separations of the hydrolysed samples were performed using a gradient elution programme that utilises two solvents: solvent A: H_2O–0.1 % TFA (trifluoroacetic acid) and solvent B: CH_3CN (acetonitrile)–0.1 % TFA. The flow rate was 0.5 ml/min and the elution programme was as described earlier [41–43, 53–55].

Table 5.5 Identified main colouring compounds by RP-HPLC–DAD and identified dye source

Inventory number or dyed sample or extract	Sample colour	Detected compound	Identified dye source
13/1900	Green	Luteolin Apigenin	Reseda luteola L.
		Indigotin	Isatis tinctoria L. or Indigofere tinctoria L.
	Yellow	Luteolin Apigenin	Reseda luteola L.
	Red	Carminic acid	Dactylopius coccus Costa
		Ellagic acid	Quercus infectoria Oliver
		Alizarin Purpurin	Rubia tinctorum L.
13/1671	Red	Carminic acid Kermesic acid	Dactylopius coccus Costa
		Ellagic acid	Quercus infectoria Oliver
		Alizarin Purpurin	Rubia tinctorum L.
	Light blue	Indigotin	Isatis tinctoria L. or Indigofere tinctoria L.
Extract	Red	Carminic acid	Dactylopius coccus Costa
Extract	Red	Carminic acid	Prophyrophora hameli Brand
Dyed silk	Red	Carminic acid	Prophyrophora hameli Brand
Extract	Red	Kermsic acid Flavokermesic acid	Kermes vermilio L.
Dyed wool	Red	Kermesic acid	Kermes vermilio L.
Extract	Red	Laccaic acid	Kerria lacca Kerr
Dyed wool	Red	Laccaic acid	Kerria lacca Kerr
Extract	Red	Alizarin Purpurin munjistin	Rubia tinctorum L.
Dyed silk	Red	Alizarin Purpurin munjistin	Rubia tinctorum L.
Extract	Yellow	Isorhamnetin Rhamnetin	Rhamnus petiolaris Boss.
Dyed silk	Yellow	Isorhamnetin Rhamnetin	Rhamnus petiolaris Boss.
Dyed wool	Yellow	Ellagic acid	Quercus infectoria Olivier
Extract	Yellow	Ellagic acid	Quercus infectoria Olivier
Extract	Yellow	Luteolin Apigenin	Reseda luteola L.
Dyed wool	Yellow	Luteolin Apigenin	Reseda luteola L.
Dyed silk	Yellow	Luteolin Apigenin	Reseda luteola L.

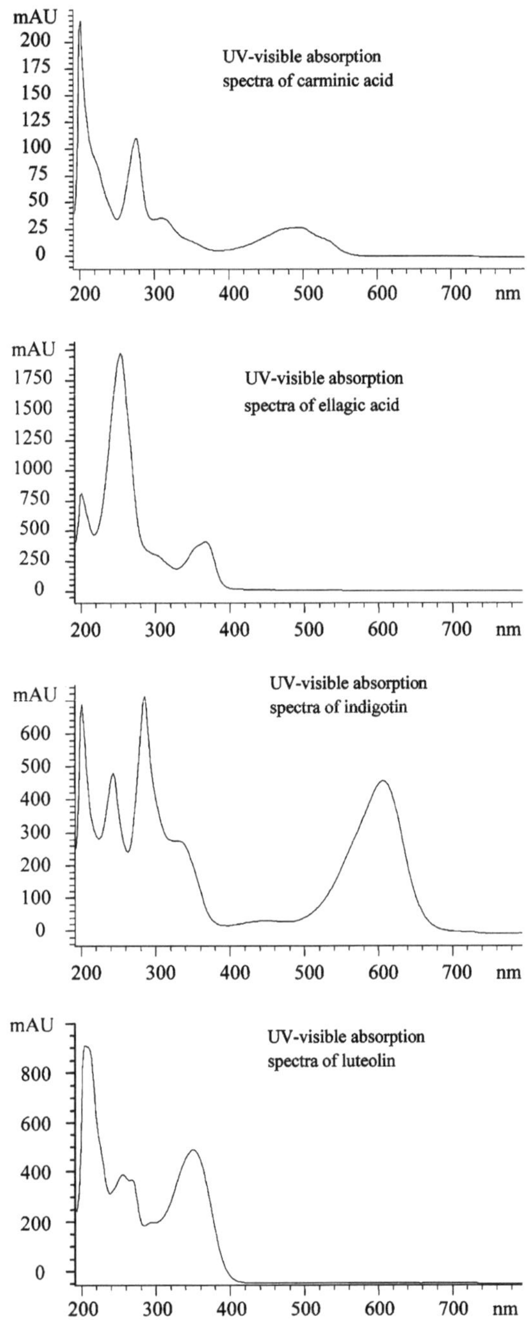

Fig. 5.3 Different spectras of natural dyes

Fig. 5.3 continued

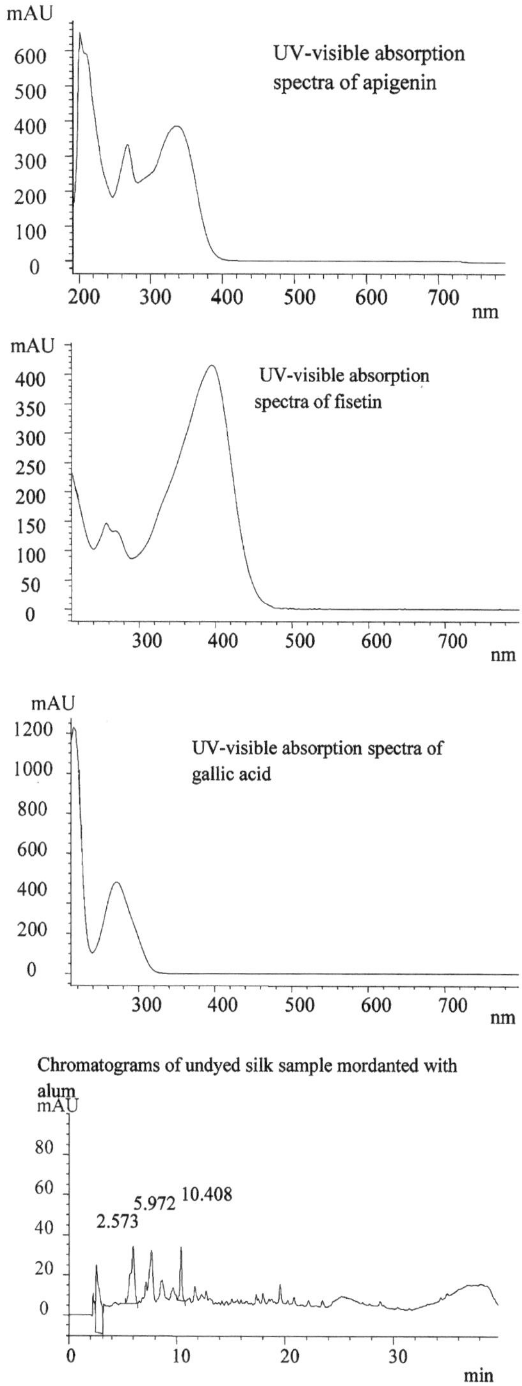

Fig. 5.3 continued

Chromatograms of undyed wool with alum mordant

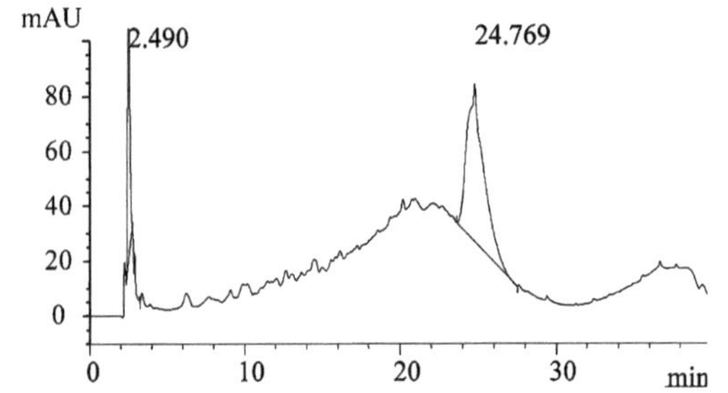

Chromatograms of *Dactylopius coccus* Costa extract

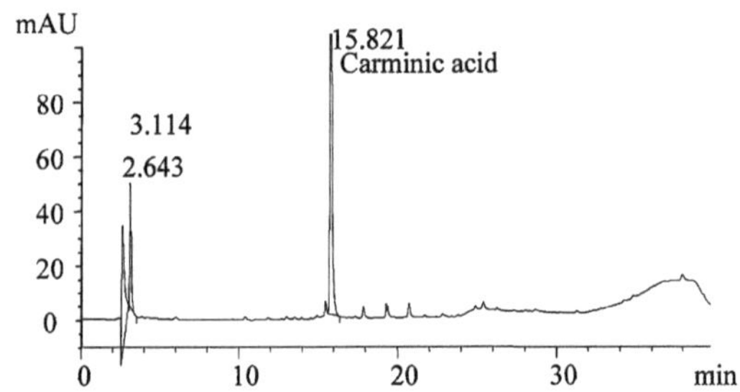

Chromatograms of dyed silk with Ararat kermes (*Porphyrophora hameli*)

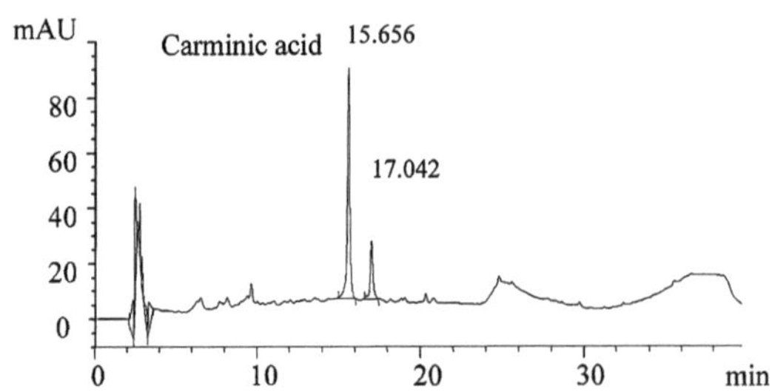

Chromatograms of *Kermes vermilio* extract

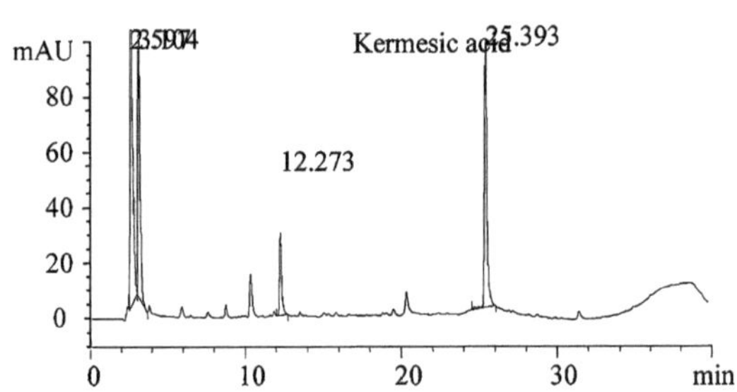

Fig. 5.3 continued

Chromatograms of dyed wool with *Kermes vermilio* (tin mordant)

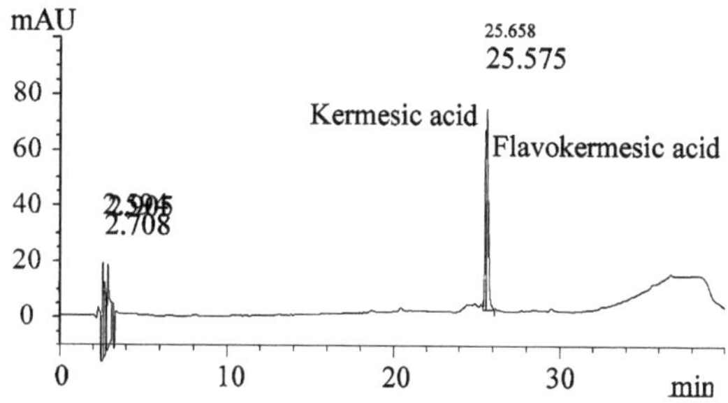

Chromatograms of *Kerria lacca* Kerr extract

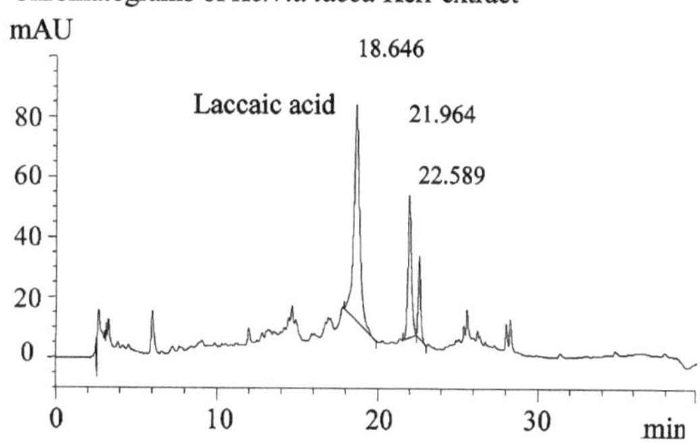

Chromatograms of dyed wool with *Kerria lacca* Kerr

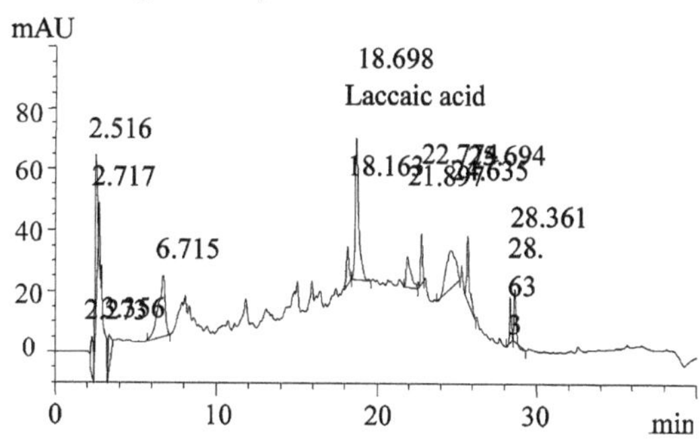

Chromatograms of *Rubia tinctorum* L. Extract

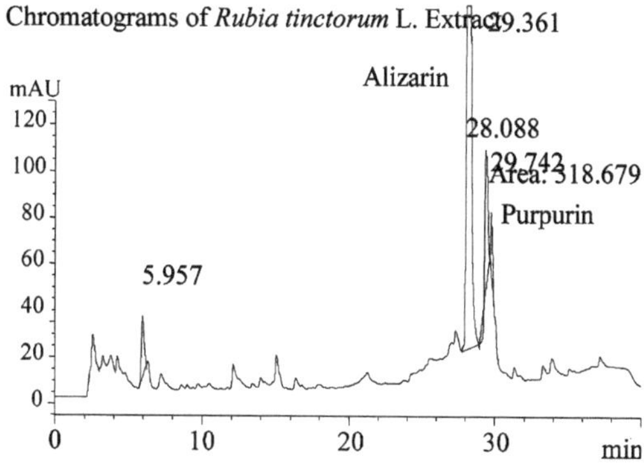

Fig. 5.3 continued

Chromatograms of dyed silk with madder (*Rubia tinctorum* L.)

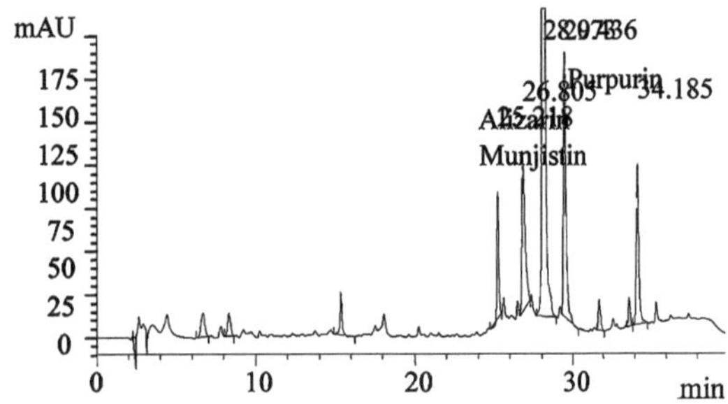

Chromatograms of *Rhamnus petiolaris* Boiss. Extract

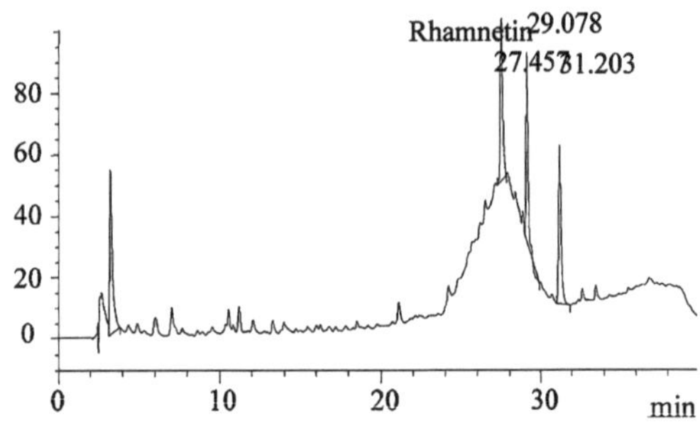

Chromatograms of dyed silk with *Rhamnus petiolaris* Boiss

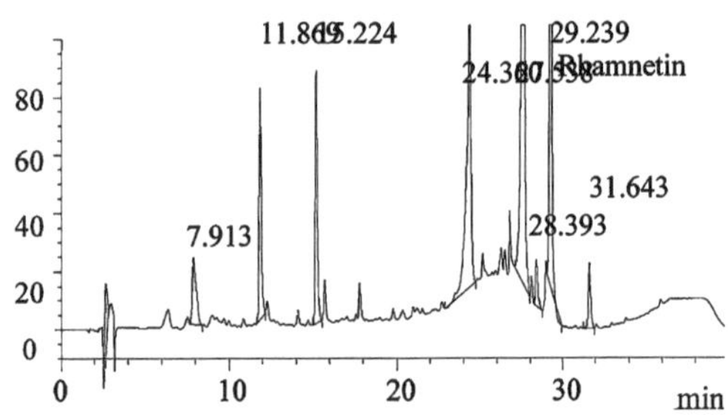

Chromatograms of Dyed wool with *Quercus infectoria* Olivier (iron mordant)

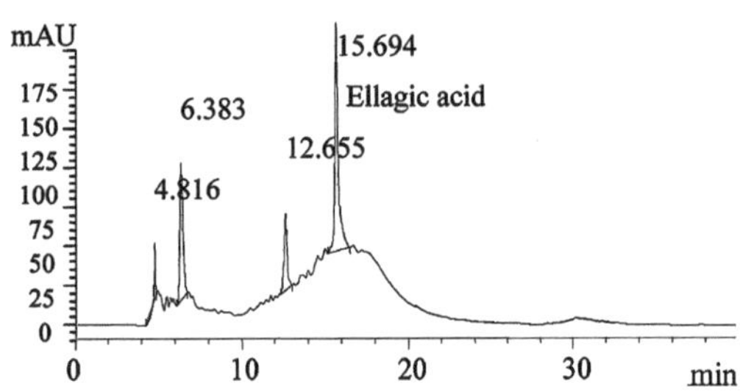

Fig. 5.3 continued

Chromatograms of dyed wool with *Rhamnus petiolaris* Boiss.

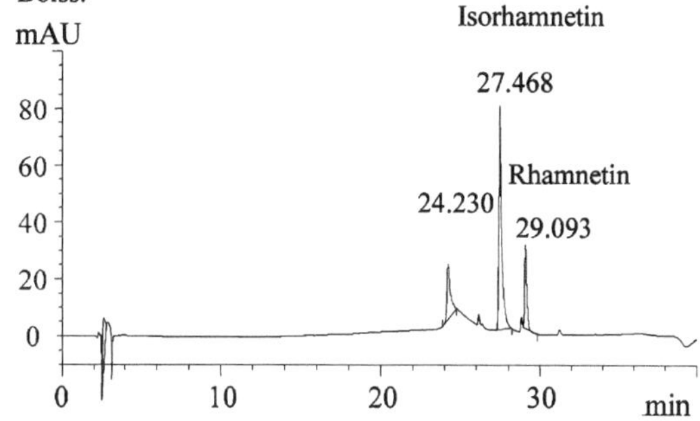

Chromatograms of *Quercus infectoria* Olivier extract

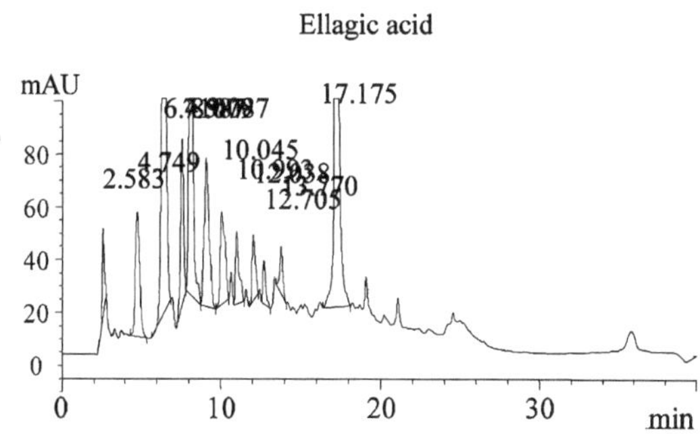

Chromatograms of weld *(Reseda luteola* L.*)* extract

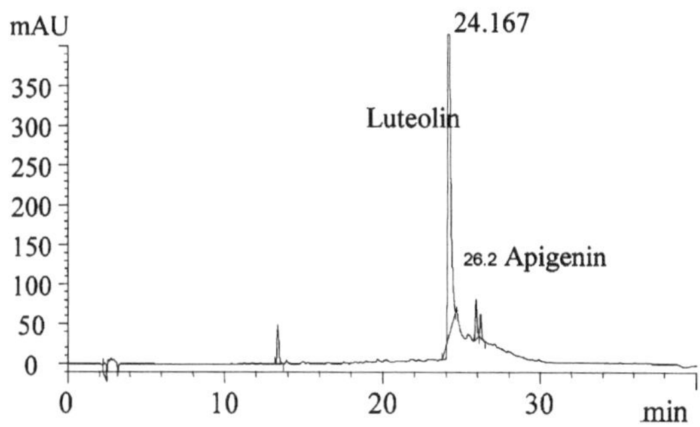

Chromatograms of dyed wool with weld (*Reseda luteola* L.)

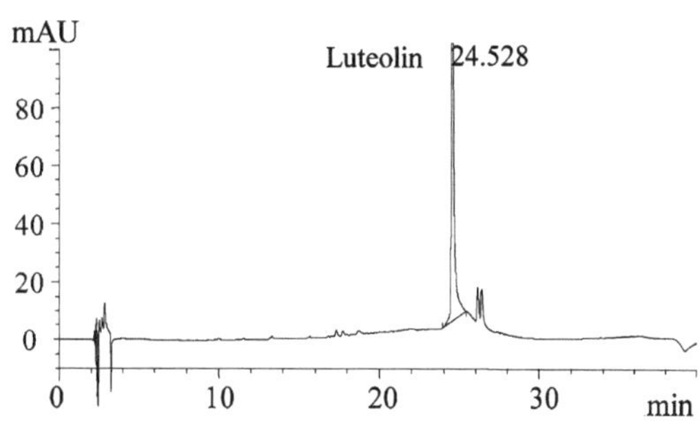

Fig. 5.3 continued

Chromatograms of dyed silk with weld

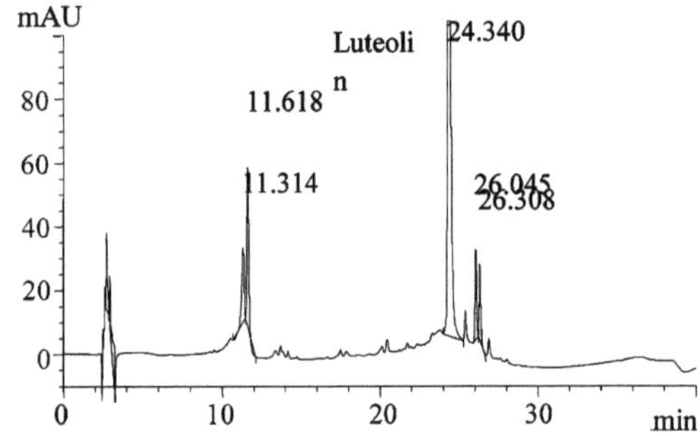

Chromatograms of dyed wool with *Quercus ithaburensis* Decaisne.

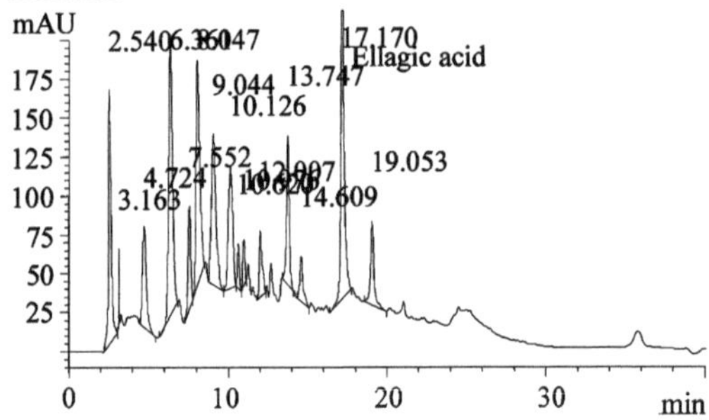

Chromatograms of red sample (inventory number 13/1671)

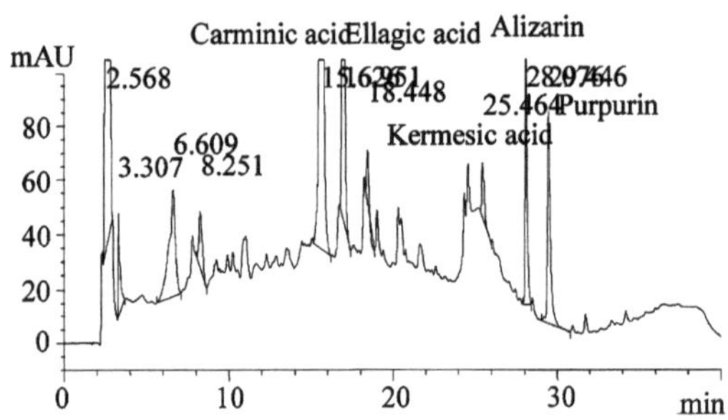

Chromatograms of red sample (inventory number 13/1900)

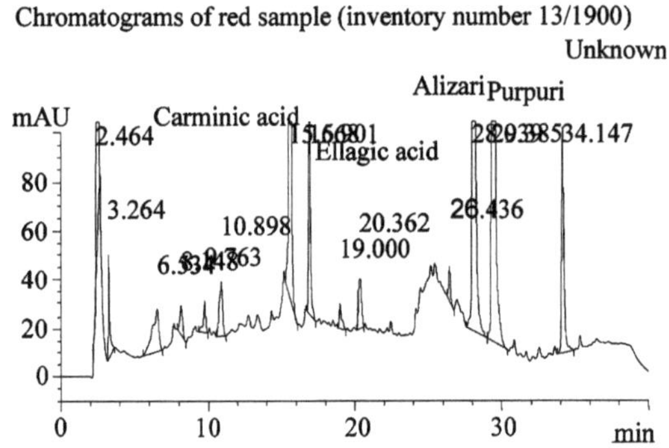

Fig. 5.3 continued

Chromatograms of yellow sample (inventory number 13/1900)

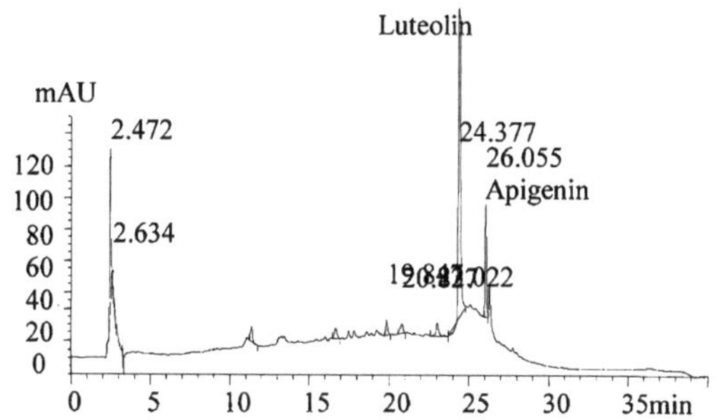

Chromatograms of green sample (inventory number 13/1900)

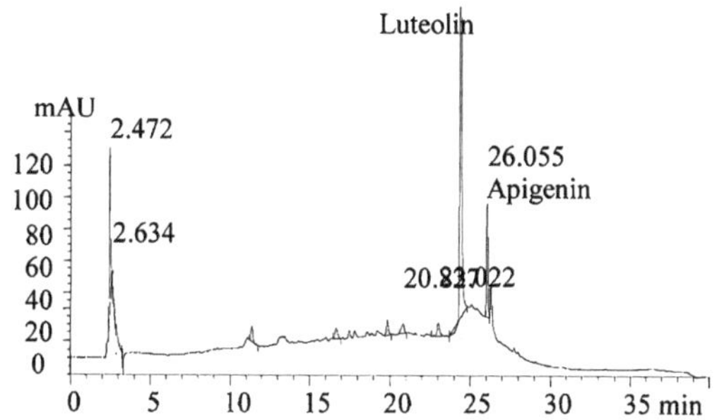

Chromatograms of green sample (inventory number 13/1900

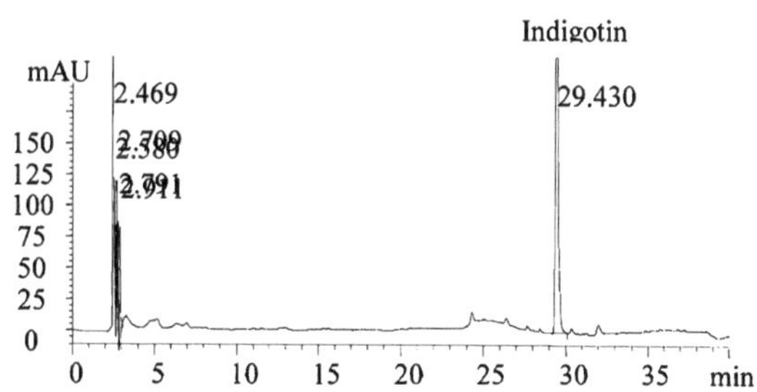

Chromatograms of blue sample (inventory number 13/1671)

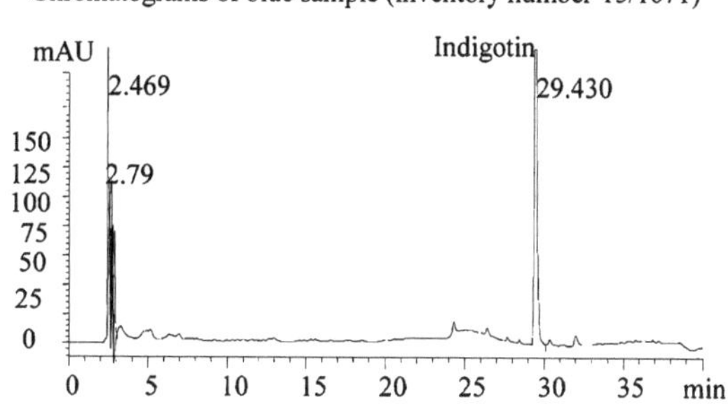

Parameters for HPLC gradient elution.

Time (min)	Flow rate (ml/min)	H_2O-0.1 %TFA (v/v)	CH_3CN-0.1 %TFA (v/v)
0	0.5	95	5
1	0.5	95	5
20	0.5	70	30
25	0.5	40	60
28	0.5	40	60
33	0.5	5	95
35	0.5	5	95
40	0.5	95	5

5.2.4 Conclusions

Combination of instrumental analytical techniques was applied to analyse dye extracts of historical art objects from the Topkapi Palace Museum collection. In this study, only fibres of yellow, blue, red and green colours were selected for analysis. HPLC with DAD has yielded good results in analyses of extracts from historical materials. Natural dyes of plant origin—ellagic acid, gallic acid indigotin, luteolin, apigenin, rhamnetin, isorhamnetin, alizarin, purpurin and tannin derivatives—and of insect origin—carminic acid, laccaic acid, flavokermesic acid and kermesic acid were found (Table 5.5 and Fig. 5.3).

5.3 Liquid Chromatographic Analysis of Ancient Egyptian Organic Material

Carole Mathe

Abstract An archaeological sample coming from Ancient Egypt was analysed by liquid chromatography coupled to a photodiode-array and fluorimetric detectors (LC-PDA-Fluorimetry). The obtained chromatograms traduced the simultaneous presence of two different resinous materials. The characterisation of terpenic compounds had permitted to clearly identify a mixture of Pinaceae resin and olibanum.

5.3.1 Introduction

Many resinous materials of vegetable origin were used during Ancient Egypt. The organic material characterisation of resinous origin in archaeological samples provides invaluable information for the ancestral use of these natural products. Among the substances used in the ancient world, the resins were frequently employed either directly, or in the development of real formulations [56, 57]. Their diversity results as well of their different botanical origins as from their use in the form of mixtures. They were employed during ritual ceremonies as incense or as adhesives, perfumes or cosmetics. The chemical stability of these materials or their evolution during time (anthropic alteration or natural ageing) makes them particularly interesting during their analysis [58]. The characteristic biochemical components composing these substances make it possible to define with a high degree of certainty their botanical and/or geographic origin [59, 60]. The analyses of such samples present a great interest for the scientist from the knowledge of the techniques employed [61, 62] but also of the inherent phenomena of ageing in these materials [57].

This part presents a case of an analytical study of an archaeological resinous substance. It is coming from the collection of the famous Victor Loret Egyptologist (University of Lyon II, France). This material was recovered from excavations at Dashour in 1894–1895 by the archaeologist J. de Morgan on the vestiges of pyramids of the North-western sector of Dashour [63]. The precise place of the sampling is in the pyramid of Amenemhat II, third sovereign of XIIth dynasty (1929–1892 BC). This studied sample comes from a flask of princess Sat-mer-hout's funeral endowment (XIIth dynasty). As written, it may be a ritual fragrance or, at least, an artificial composition.

This sample coming from Ancient Egypt was analysed by liquid chromatography coupled to a double detection: UV and fluorimetry (LC/PDA/fluorimetry). Fluorimetry is generally more sensitive than other detection systems, such as classical UV absorption, and less expensive than mass spectrometry (MS) detection. Additionally, fluorescence detectors are very selective, overcoming matrix interference, so it is possible to decrease the detection threshold of the organic compounds. According to the literature, this analytical tool consisting of a technique of separation was employed to characterise organic matter used in cultural heritage [64–66]. Peaks in the chromatogram were identified on the basis of the retention times and UV–Vis absorption spectra of the references molecules injected in the same conditions. In our analytical conditions, the archaeological sample was solubilized in methanol and submitted to ultrasounds for 10 min. The solution was then filtered and directly injected in the chromatographic system. One of the most advantages of the liquid chromatographic technique that no sample derivatization is necessary to analyse the substance. The chromatographic conditions of this analysis were described by Mathe et al. [59].

The obtained results show the presence of a mixture of two different resinous materials including terpenic compounds. The first resin was identified to a diterpenic resin because of the presence of chemical biomarkers of Coniferales class belonging to the Pinaceae family [67, 68]. The second one is a triterpenic resin named olibanum or frankincense (*Boswellia* spp. Burseraceae family). The terpenic derivatives correspond to a very important group of natural substances in the vegetable kingdom. It can be subdivided as various families of molecules classified according to their carbon atoms number: mono (C10), sesqui- (C15), di- (C20), triterpenes (C30) and carotenoids (C40).

According to the literature, the identified diterpenic resin has compounds with an abietane or a pimarane chemical skeleton [69, 70]. These molecules contain almost exclusively a carboxylic acid group and they are very abundant in this botanical family. Among these compounds a native diterpene is predominant named abietic acid ($C_{20}H_{30}O_2$) resulting from the rapid acid-catalysed isomerization of the double bonds of the resinous acids precursors: acids levopimaric, palustric and neoabietic [71]. This reaction pathway gives the retene (ultimate product of alteration) [57, 72] (Fig. 5.4).

A preliminary study of commercial diterpenic standard compounds was realised in order to establish our analytical conditions. Most of these diterpenes can be detected with a photodiode-array detection (PDA) and fluorimetry. Indeed they have characteristic UV spectra and because of their chemical structures, they have the particularity to give a response in fluorimetry. The fluorimetric detection permits to characterise compounds with sometimes a weak UV absorption. In some case of studies, the fluorimetry can give complementary information permitting to distinguish two different molecules having same UV spectra. The Table 5.6 resumes all of the spectroscopic and chromatographic data of these diterpenes. The chromatogram obtained by liquid chromatography shows the presence of two abietane compounds, respectively eluted at 17.8 and 32.4 min (Fig. 5.5).

The first one corresponds to 7-oxodehydroabietic acid (1) and was identified according to its UV spectrum. No corresponding peak was observed in fluorimetric detection because this molecule does not have this spectroscopic property. The second characterised diterpene is the retene (2). Despite of its small peak area, it was unambiguously identified because of its characteristic UV spectrum and its

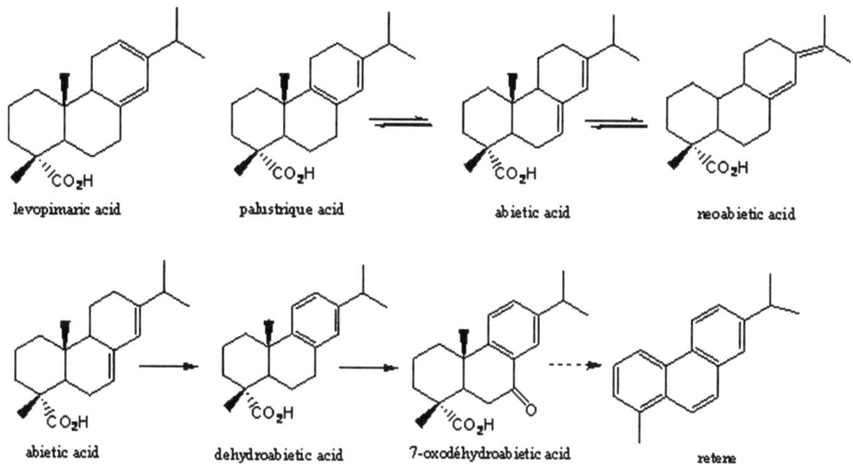

Fig. 5.4 Alteration pathway of abietanes compounds occurring in the Pinaceae family

Table 5.6 Chromatographic data of diterpenic standards compounds

Compounds	Retention time (min)	Absorption maxima (nm)	Fluorimetric conditions λ excitation/ λ emission (nm/nm)
7-oxodehydroabietic acid	17.8	209; 252; 299	–
Dehydroabietic acid	20.6	202; 274	271/576
Abietic acid	25.3	240	271/576
Retene	32.4	190; 213; 257; 299	240/368

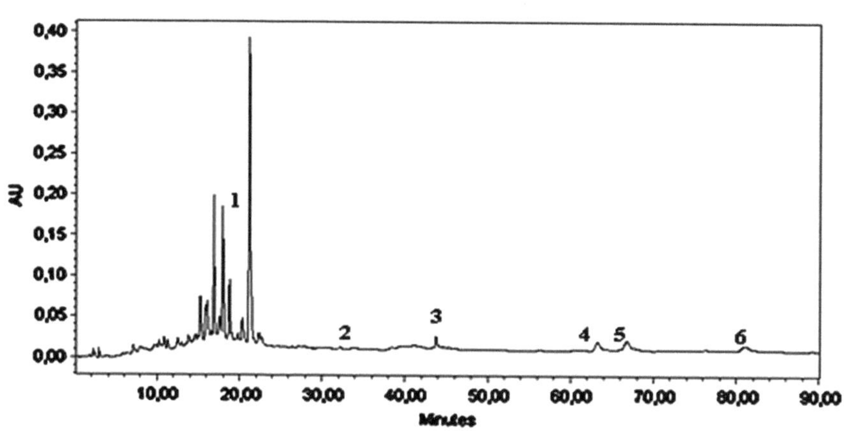

Fig. 5.5 Chromatogram at 210 nm of the archaeological sample

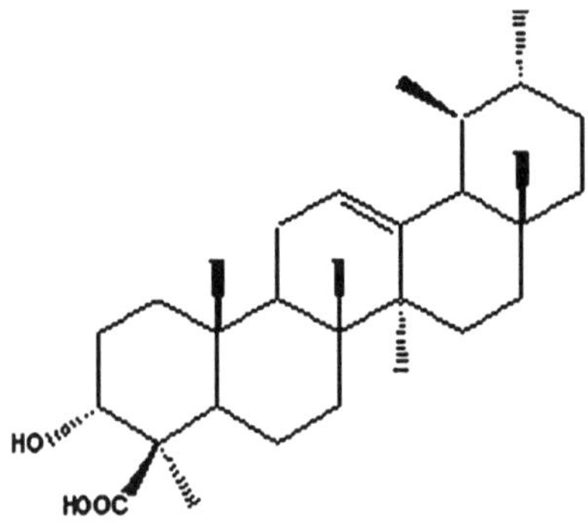

β-boswellic acid (3)

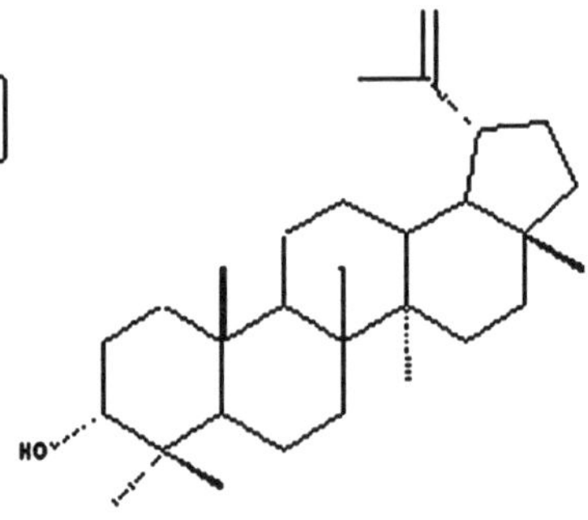

3-*epi*-lupeol (4)

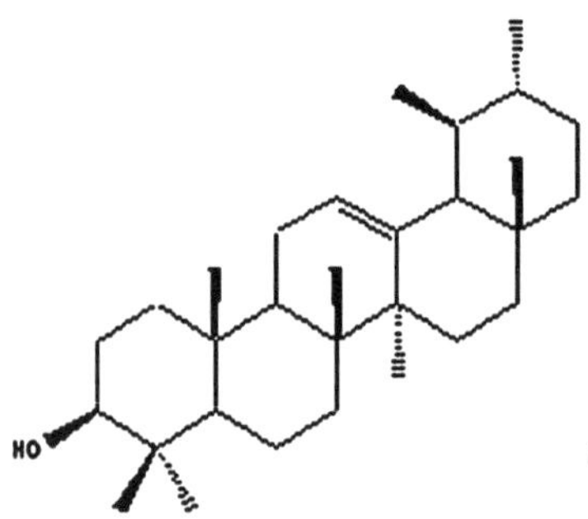

lupeol (5)

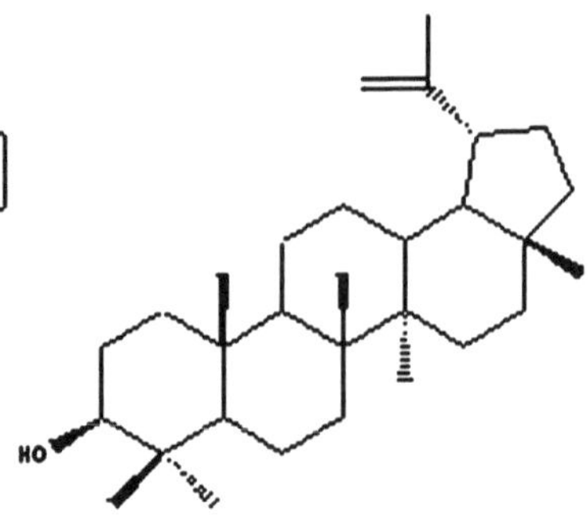

α-amyrine (6)

Fig. 5.6 Chemical structures of identified triterpenoids

response in fluorimetry. It can be noted that abietic and dehydroabietic acids are not detected or at least in significant proportion. The occurence of oxidised abietic acid derivatives and the absence of abietic acid are in agreement with the great age of the sample and/or could confirm that the resin was heated during the preparation of this archaeological mixture.

The liquid chromatography analysis carried out on this archaeological sample shows also the presence of a triterpenic fraction corresponds to pentacyclic compounds with ursane and lupane skeletons. Indeed, the study of the second part of the obtained chromatogram reveals triterpenoids and more particulary the presence of β-boswellic acid (3) which is one of the chemical markers of olibanum. This triterpenic resin has others specific compounds like α-boswellic, lupeolic acids and their respective *O*-acetyl derivatives which are no detected by LC-PDA-fluorimetry. However, three other triterpenic compounds were also detected; they were eluted at 65.5, 67.7 and 84.4 min, and respectively correspond to 3-*epi*-lupeol (4), lupeol (5) and α-amyrine (6) (Fig. 5.6).

Table 5.7 Chromatographic data of identified triterpenes

Compounds	Retention time (min)	Absorption maxima (nm)
β-boswellic acid	44.4	206
3-epi-lupeol	65.5	204
Lupeol	67.7	205
α-amyrin	84.4	203

These last compounds were described to belong to olibanum resin but they are not specific to the *Boswellia* genus. Indeed they occur largely in the Burseraceae family and in the vegetal kingdom in general [73, 74]. They generally occur in the chemical composition of the main vegetal resins attested in archaeology [56]. Most of the terpenoid compounds in frankincense show only absorption at 210 nm because of the lacking specific chromophores. So, it was difficult to identify these triterpenic compounds only by their single UV spectrum, because these molecules showed very similar maximal absorption wavelength. So, most of them have been isolated to furnish standard compounds and in order to give a sure identification. Moreover, they do not have fluorimetric properties (Table 5.7).

To conclude, the study of archaeological organic samples is an analytical challenge because of the wide range of classes of molecules which are present and because of the damage which has affected the products through the action of time. The result of this archaeological substance shows the presence of mixture of diterpenic and triterpenic resinous materials corresponding to a real formulation composed by a pine resin and frankincense. So, in accordance with the inscriptions on the flask, this ancient material is an artificial composition.

5.4 Binding Media Identification in Art Objects by Gas Chromatography–Mass Spectrometry

Andreas K. Tsakalof and Kyriaki A. Bairachtari

5.4.1 Introduction

The precise knowledge of the materials employed for creation of art object provides essential and some times crucial information for art historians and conservators. This knowledge can contribute to the revelation of the painting technique used, certification of authenticity and determination of the time and place of the object creation. It is also crucial for the realisation of appropriate restoration and conservation of the art object which will not alter the initial characteristics of the object. In particular, the knowledge of materials initially used by the creator can guide the proper selection of materials for the conservation and/or restoration of a particular work of art.

Binding media are among the materials used for the creation of painting object and their identification is a part of the art object investigation. Binding media can be defined as fluid substances which in time solidify; and thus serve initially to disperse and eventually to bind the colourants to the ground. But apart from this pure mechanical role, binding media properties influence the surface gloss, depth and richness of colours obtained by artists in their paintings. Over centuries, the main binding media have embraced gums (such as arabic gum, cherry gum), beeswax (encaustic painting), proteinaceous binding media (egg, casein, animal glue), drying oils (linseed oil, walnut oil, poppy seed oil) and also mixtures of the above (tempera grassa mix of egg/oil).

The identification of these materials is nowadays achieved by different instrumental methods. Benchmark in the introduction of instrumental analysis to the investigation of artworks is considered the study of Mills [75], who first applied gas chromatography with flame ionising detector (GC-FID) for the identification of drying oils. He suggested the use of the palmitic to steatic acid ratio (P/S) for the identification of drying oils and this ratio is still a basis for the drying oils identification [75]. The GC–FID technique was later also applied for the identification of proteinaceous binding media and varnishes.

During the 1990s, the gas chromatography combined with mass spectrometry (GC–MS) was introduced as result of systematic studies by M. Colombini and other research groups and has found wide application in the chemical characterisation of binding media [76–78]. The main advantage of GC–MS is high sensitivity—very small amount of sample is required for the analysis.

5.4.2 What is GC–MS?

For non-chemists involved in the investigation of art objects we describe in few words the GC–MS method. Gas Chromatography–Mass Spectrometry belongs to so-called hyphenated analytical methods, methods which combine two different techniques with different physicochemical basis and different role in the analysis.

Gas Chromatography is a separation technique destined to separate a mixture of volatile compounds for further individual analysis of each component by MS. The separation is conducted in chromatographic column—a capillary tube internally coated with special material called stationary phase. A gas, usually helium, flowing through the column is called mobile phase or carrier gas. A mixture of compounds is introduced to a column and carried through to the column exit by carrier gas. The separation is based on compounds' differences in volatility and affinity to a stationary phase. The compounds with low volatility and high affinity to a stationary phase are hardly transferred to the air phase, and thus carried later to the column exit in comparison to more volatile and low affinity compounds. The eluted from the column compounds are detected by the detector and detection is recorded as chromatographic peaks.

With the help of GC not only intrinsically volatile compounds can be separated. A number of non-volatile organic compounds, e.g. amino acids and fatty acids, can be chemically transformed to volatile derivatives and analysed by GC in this derivatized form.

Mass Spectrometry is an identification technique capable to detect and identify compounds eluted from chromatography column. As a standalone technique MS is used for the elucidation of chemical compounds structure. In combination with GC, MS plays the role of very versatile and most sensitive detector. In mass spectrometer, the compounds are ionised and fully or partially fragmented after their collision with high energy electrons in the ion source of mass spectrometer. The resultant ions are analysed in another part of the mass spectrometer called mass analyzer and registered in turn as ion current. The fragmentation pattern is recoded as mass spectrum—a plot of relative abundances of ions resultant as a function of their m/z values, where m is mass and z charge of the ion. The mass spectrum is unique for each compound and used for compound unequivocal identification and/or clarification of its structure. This is very brief description of GC–MS. More detailed and in depth information can be found elsewhere [76].

Gas chromatography–mass spectrometry is the most often method used nowadays for binding media identification due to its sensitivity and specificity. This method needs very small amount of sample and identify unequivocally the compounds of interest, in particular the binders used by the creator of the art object.

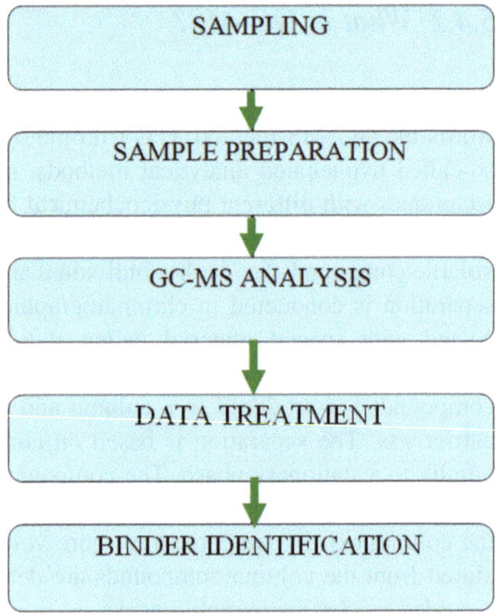

Fig. 5.7 Main steps of the binding media identification in art object by GC–MS

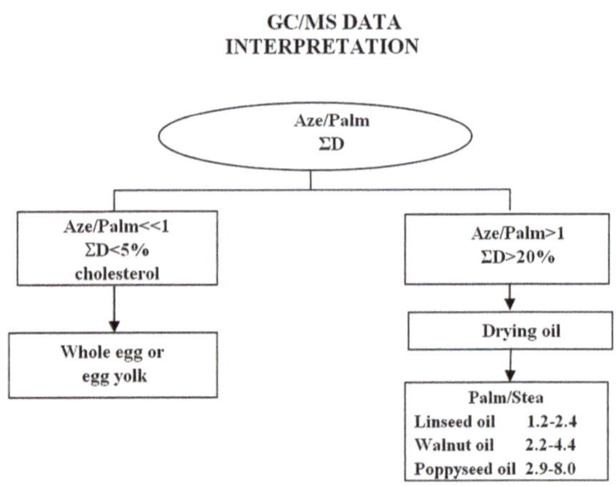

Fig. 5.8 Logical flow chart for the identification of lipid binders on the basis of characteristic fatty acids ratios and sum of dicarboxylic acids percentages

5.4.3 Main Steps in Binding Media Identification

The identification of binding media comprises several consecutive steps presented in Fig. 5.7.

The first step-sampling, e.g. acquisition of a representative sample from the art object, is usually performed by the conservator. The sample is submitted to an analytical laboratory for the analysis. The analytical investigation starts from the proper sample preparation for the subsequent instrumental analysis. In case of

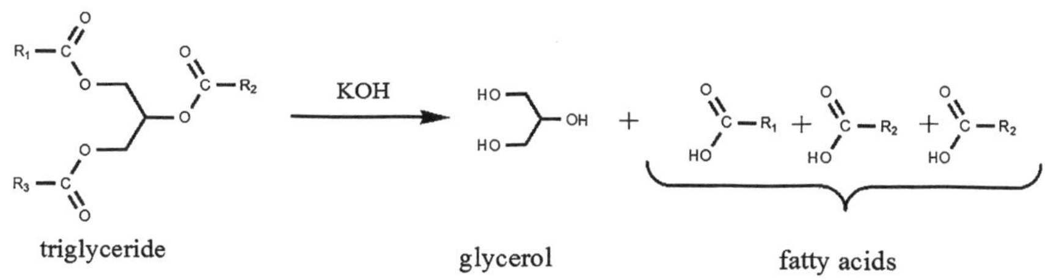

Fig. 5.9 Release of fatty acids from triglyceride by basic hydrolysis

organic binding media binders are initially separated from inorganic pigments and other compounds which can interfere with the analysis. Afterwards, they are hydrolysed to release representative chemical compounds which are amino acids for proteinaceous and fatty acids for lipid binding media. Amino or fatty acids are not volatile compounds so that they are chemically transformed to their volatile derivatives and submitted for GC–MS analysis. The GC–MS analysis quantifies the amino acid or fatty acid percentage content in the sample. These data are treated (data treatment) for the determination of the characteristic amino acids or fatty acid ratios on the basis of which the binder can be identified [77, 79] (binder identification, Fig. 5.8).

5.4.4 Case Study: Identification of Drying Oils in Paintings of N. Gyzis

Nikolaos Gyzis (1 March 1842–4 January 1901) is considered one of Greece's most important nineteenth century painters. He is the major representative of the so-called "Munich School", the major nineteenth century Greek art movement. Drying oils as a binding media were in his brushwork [80].

Drying oils are main components of oil paints. By far, the largest proportion of the oil constituents is triacylglycerols (TAG)—triesters of glycerol of saturated and unsaturated fatty acids (Fig. 5.9). The chain lengths of fatty acids occurring in TAG vary, but usually contain 16, 18 or 20 carbon atoms. The carbon–carbon bonds in saturated fatty acids are all single bonds, while unsaturated fatty acids have one or more carbon–carbon double bonds in their chains. Interaction with the atmospheric oxygen of the unsaturated double bonds leads to cross-linking between neighbouring fatty acid chains results in hardening, or, more properly, curing of oils and formation of solid film [75, 81–83]. During film degradation dicarboxylic acids are formed by chain scission of unsaturated linolenic (18:3), linoleic (18:2) and oleic (18:1) acids. Aged drying oil paint films generally contain substantial amounts of dicarboxylic acids such as suberic, azelaic and sebacic acid, with azelaic acid being the most abundant [75, 78, 81, 84–86].

Fig. 5.10 N. Gyzis, "Three sisters" 1882–1885

5.4.4.1 Sampling

Samples from the Gyzi's paintings were acquired by the conservator submitted for the analysis (Figs. 5.10 and 5.11).

Fig. 5.11 N. Gyzis, Penelope Gyzis, 1879

5.4.4.2 Sample Preparation for the Identification of the Lipid Binders

Drying oils consist of glycerol triesters of fatty acids. The fatty acids are released from triglycerides by their hydrolysis under basic conditions (reaction of saponification, Fig. 5.9). In this process the triglyceride react with a strong base such as sodium or potassium hydroxide to produce glycerol and fatty acid salts. After acidification of the reaction mixture free fatty acids are extracted from the reaction

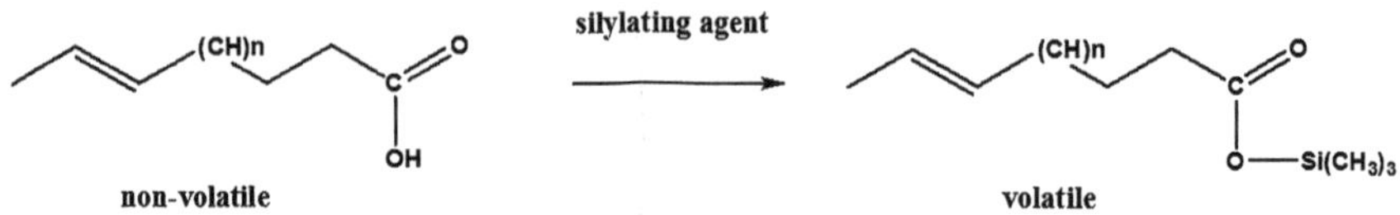

Fig. 5.12 Derivatization of fatty acid by silylation

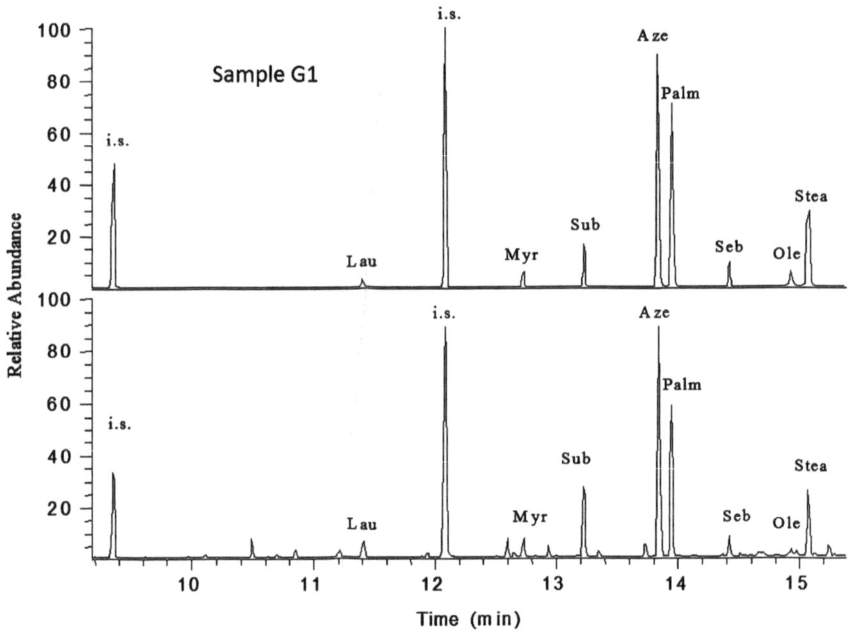

Fig. 5.13 Representative GC–MS chromatogram of the fatty acid derivatives in sample from N. Gyzi's painting "Penelope Gyzis" acquired in Selected Ion Monitoring (*top*) and Total Ion Current (*bottom*) mode

mixture by liquid–liquid extraction with diethyl ether. After ether evaporation fatty acids are derivatized to their volatile silylated analogues (Fig. 5.12) that can be analysed by the GC. Generally by derivatization reaction polar functional groups (e.g. hydroxyl, carboxylic, amine, thiol, phosphate) are chemically transformed to less polar, and thus the whole compound becomes less polar and more volatile.

5.4.4.3 GC–MS Analysis

GC–MS analysis of derivatized fatty acids includes their separation by GC and separate detection and identification by MS. Identification is based on the unique mass spectrum of each fatty acid. In the acquired chromatogram (Fig. 5.13), each peak corresponds to specific fatty acid and the area of the peak is proportional to

Table 5.8 Relative fatty acid percentage in samples from paintings of N. Gyzis

Paintings/sample		Relative percentages							
		Lau	Myr	Sub	Aze	Palm	Seb	Ole	Stea
Penelope Gyzis	G1	1.9	2.8	10.6	38.4	26.2	3.1	3.6	13.4
Still Life	G4	1.8	4.4	11.7	37.2	29.6	2.6	0.6	12.1
	G5	1.0	2.9	10.6	34.5	29.2	2.3	5.5	13.9
The New Century	G6	1.9	2.3	7.6	37.6	26.4	2.1	5.5	16.5
The Three Sisters	G7	0.3	1.7	17.7	45.0	20.2	3.9	0.5	10.7

Table 5.9 Mean value of the palmitic to stearic acid (P/S), azelaic to palmitic acid (A/P) and sum of dicarboxylic azelaic, suberic and sebacic percentage (ΣD) in analysed samples from Gyzis paintings

Paintings	Code	Layers	A/P	P/S	$\%\Sigma D$	Drying oil
Penelope Gyzis	G1	Preparation	1.5	2.0	52.1	Linseed oil
Still Life	G4	Preparation	1.3	2.4	51.5	Linseed oil
	G5	Preparation + painting	1.2	2.1	47.4	Linseed oil
The New Century	G6	Preparation	1.4	1.,6	47.3	Linseed oil
The Three Sisters	G7	Preparation	2.2	1.9	66.5	Linseed oil

the amount of the fatty acid in the sample. Internal standards tridecanoic acid and hexadecane were used as internal standards to control the efficiency of derivatization and injection steps, respectively.

The procedure permits the determination of eight fatty acids. The determination of the percentage content of five monocarboxylic fatty acids (lauric, miristic, palmitic, oleic and stearic acids) and of three dicarboxylic acids (suberic, azelaic and sebacic acids) allows us to identify the lipid binder used (Fig. 5.8, Tables 5.8 and 5.9).

5.4.4.4 Data Treatment

Binding material identification was based on content ratios of palmitic to stearic acid (Palm/Stea), azelaic to palmitic acid (Aze/Palm) and total content of dicarboxylic acids ΣD = azelaic acid + sebacic acid + suberic acid. The identification of cholesterol and of its oxidation products in a paint sample was a further confirmation of the presence of egg. If the ratio of azelaic over palmitic acid is lower than 1 (Aze/Pal $\ll$ 1) and overall content of dicarboxylic acid (ΣD) is lower than 5 % it can be concluded that whole egg or egg yolk were used in the binding media and are the source of the fatty acid detected. On the contrary, if Aze/Palm > 1 and $\Sigma D > 20$ % drying oils were used as binding media. Analysing the ratio Palm/Stea the conclusion of what concrete drying oil was used can be made.

Table 5.10 Characteristic fatty acid ratios and dicarboxylic acids percentage content as function of sample dilution in case of linear and nonlinear instrument response. Analysis of the sample from N. Gyzis' painting

Sample preparation		Nonlinear response			Linear response		
Aliquot of derivatization solution (μl)	Isooctane added (μl)	P/S	A/P	ΣD	P/S	A/P	ΣD
30	320	1.7	1.6	50.2	1.6	1.5	45.6
20	330	1.9	1.8	54.3	1.6	1.5	47.7
10	340	2.4	2.9	65.3	1.7	1.4	48.7

5.4.4.5 Binder Identification

In accordance with acquired data (Tables 5.8 and 5.9) the ratio Aze/Palm > 1 and $\Sigma D > 20\ \%$ for all submitted samples. These results indicate that drying oil was used in all submitted samples. From Palm/Stea ratios (Table 5.9), it can be concluded that is was linseed oil.

5.4.5 Concluding Remarks: Pitfalls in Drying Oils Identification in Art Objects

In spite that described procedure of drying oils identification seems relatively simple and straightforward, there are some precautions that should be taken to receive correct result. It has been demonstrated that if linear response of the GC–MS is not assured by multilevel calibration the ambiguous results could be acquired and erroneous conclusions could be made [87]. In particular, in the case of nonlinear instrument response the ratios of palmitic to stearic (P/S), distinctive for each oil type and used for drying oil identification, depends on sample dilution prior to GC–MS analysis so that different dilutions of the same sample can give different P/S ratios. Specifically, under nonlinear instrument response the analysis of samples from Gyzis' paintings gave ambiguous results as soon as the crucial ratios P/S and A/P were strongly dependent on the level of sample dilution and ranged from 1.7 to 2.4 and from 1.6 to 2.9, correspondingly (Table 5.10). While, in correspondence with the available literature data, P/S = 1.7 is indicative for linseed oil, P/S = 2.4 is more characteristic for walnut oil. This variability of P/S ratio was attributed to incomplete sample evaporation in the split/splitless injector and consecutive high boilers (in this case stearic acid) discrimination. The linear instrument response was restored by improving sample evaporation. This was achieved by packing of injector liner with deactivated glass wool. Under linear instrument response, the P/S and A/P characteristic ratios become stable and independent of sample dilution (Table 5.10). The oil used by the painter was unambiguously identified as linseed oil.

5.5 FT Infrared Spectroscopy as an Identification Tool in Polychrome Artefact Studies

Evangelia A. Varella

5.5.1 General Framework

Works of art, exhibited or stored, are susceptible to various external influences, and alterations occur with the passage of time. To achieve a better and more effective preservation, the behaviour of all materials involved should be clarified. Painting palettes represent an optimal operational model for examining the influence of environmental factors and historical processes to the chromatic profile originally sought by the artist; and for confronting preservation problems created by the coexistence of traditional and modern synthetic colourants or binding media.

Artificial ageing of experimental plates combined to a systematic comparative review of colourimetric and FT infrared spectroscopic data permits evaluating organic painting materials as to compatibility and stability toward extrinsic factors; and is proposing degradation routes at a molecular level, with the intention of contributing to the physicochemical elucidation and appropriate preservation of genuine polychrome artefacts and pictorial works.

In the following case studies, all experimental tables are subjected to the ageing tests in a Voetsch VC0018 climatic chamber. They are exposed to Philips Cleo 20 W fluorescence tubes, which emit highly concentrated ultraviolet radiation in the 300–400 nm range, peaking at 350 nm. The samples are placed at a distance of 2 cm from the radiation source in the case of paper ground, and at a distance of 25 cm in the case of glass ground. The average intensity measured is 1.7 mW/cm^2. A Miniscan XE Plus spectrophotometer (HunterLab) is used for colour measurements during the accelerated ageing. The surface is studied under an Olympus Bx60 optical microscope with a JVC TK-C1381 camera and Leica MW Software. The total colour difference ΔE^* between the sample prior to light exposure and at each measurement during the ageing is calculated according to the equation:

$$\Delta E^* = \left\{(\Delta L^*)^2 + (\Delta a^*)^2 + (\Delta b^*)^2\right\}^{1/2}.$$

5.5.2 Watercolour and Gouache Layers on Paper Ground

Particularly susceptible to injuries are works of art on paper, since cellulose undergoes natural ageing and decomposes, and can further interact with the actual paint layers, causing denaturation components. In this case, physicochemical research is proceeding by colourimetric and spectroscopic analysis of samples deriving from artificially aged experimental tables, prepared as watercolour and gouache layers on paper ground devoid of preparation. Respective binding media are gum Arabic or gum Arabic and chalk. Carmine lake, Brazil wood, alizarin and fuchsine are selected as widely used colourants bearing extensively studied degradation processes, and having structures easily correlated to further chemically related pigments. Environmental factors taken into account encompass humidity, temperature and ultraviolet radiation.

The experimental tables are subjected for a total time of 3 months to the influence of moist heat (90 °C, 60 % relative humidity), and the influence of UV radiation (30 °C, 50 % relative humidity). Colour measurements are performed during the accelerated ageing, and changes expressed using the colour space CIE 1976 ($L^*a^*b^*$). The surface of both untreated and aged paint layers is microscopically observed as well. In order to determine the degree, in which chemical and molecular alterations are related to colour changes, FT infrared spectra of paint layers before and after UV exposure are recorded.

Carmine Lake. Colour alterations observed in ultraviolet radiated carmine layers are due to partial loss of the lake complex structure, as well as to crack formation on the surface. In the IR spectra, a destabilisation of the complex, in which the alum salt is linked through the carbonyl group, could be detected (Fig. 5.14).

Brazil Wood. Under UV radiation, main parameter contributing to the total colour difference is L^*, confirming that the fugitive behaviour of the dye is playing the predominant role in colour change. The IR spectra are implying that even a 350 nm radiation is sufficient for inducing oxidation of phenolic brazilin to aromatic carboxylic brazilein. A small rise in brilliance under the effect of moist heat is pointing at a dye loss on the paper substrate.

Alizarin. Colour fading, observed in alizarin layers subjected to moist heat ageing, is due to alterations in the intramolecular hydrogen bonding between the carbonyl and the hydroxyl group, as well as to high crack formation on the surface (Fig. 5.15).

Fuchsine. As a solid, fuchsine forms greenish-yellow crystals, turning to purple when dissolved in water. Exposure in ultraviolet light causes an immediate substantial decrease of b values in both watercolour and gouache layers. Infrared spectra of ultraviolet radiated layers recorded amino group shifts due to the intramolecular charge transfer caused by ultraviolet excitation. An increase in brilliance, observed after 40 days of ageing, is justified on the basis of the more fugitive dye monomer. On the paint layer surface, the paper was at occasions totally revealed. The increase in brilliance and yellowness is related to the formation of bright yellow areas, probably due to aggregation of dye crystallites and the improvement of their reflecting surface. Further ageing induced intense crack formation, and subsequent redarkening of the surface.

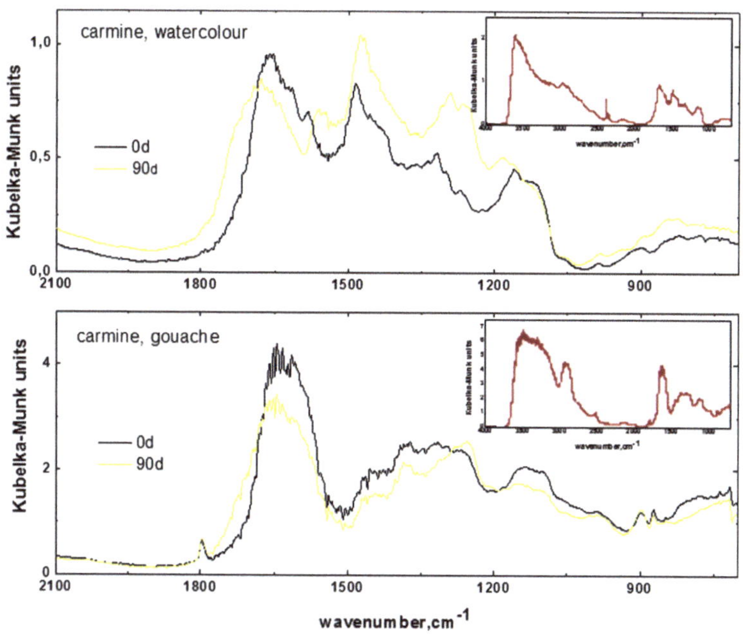

Fig. 5.14 FT-IR reflectance spectra of Carmine Lake layers before and after ageing (Ultraviolet radiation, 30 °C, 50 % relative humidity)

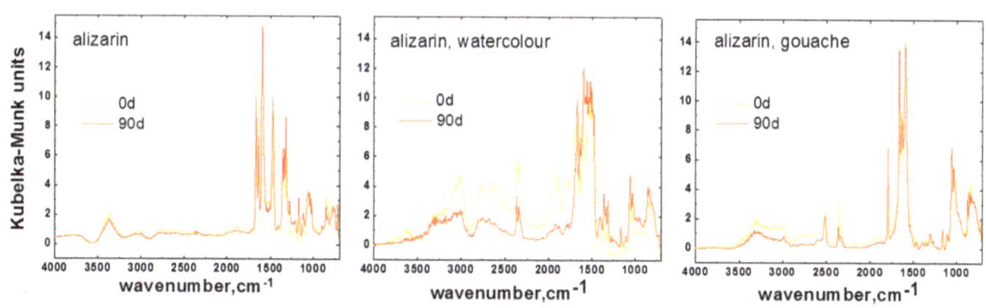

Fig. 5.15 FT-IR reflectance spectra of Alizarin layers before and after ageing (90 °C, 60 % relative humidity)

In general, the abovementioned colouring agents—with the exception of alizarin—were stronger influenced by UV light than by high temperature combined to humidity. Carmine lake and Brazil wood proved more sensitive; while alizarin and fuchsine exhibited a better stability.

5.5.3 Layers of Styrene–Acrylic Copolymer Dispersions on Glass Ground

A further approach is addressing contemporary pictorial works, in which synthetic polymeric binding media are combined with titanium dioxide. The physico-chemical research proceeds by colourimetric and FT infrared spectroscopic analysis of samples deriving from artificially aged experimental tables, prepared as homogenous layers of styrene–acrylic copolymer water-based dispersions on inert glass ground devoid of preparation.

Hansa yellow PY3, Hansa yellow PY74, and naphthol AS-D PR112 are selected as widely used colourants bearing extensively studied degradation processes, and having structures easily correlated to further chemically related pigments. Environmental factors taken into account encompass humidity, temperature and UV radiation. Subsequently, the pigments are mixed with various proportions of titanium dioxide, a material broadly used in modern and contemporary pictorial art, and acting both as a pure white pigment or a moderator of hue and saturation. The efficiency of the photo-catalytic oxidation on titanium dioxide is strongly dependent on the experimental conditions—presence of oxygen, temperature, relevant concentrations, light intensity, presence of electron donors and acceptors.

To reproduce a sequence of typical indoor environments, the distortion is measured on experimental tables at various proportions of colourant versus commercial rutile, with reference to samples not containing titanium dioxide. Paint layers contain 1.50 g of pigment, and 5.40 ml of styrene–acrylic copolymer dispersed in 3.80 ml of water. Whenever added, titanium dioxide constitutes one or three quarters of the initial weight.

In order to determine the degree, in which chemical and molecular alterations are related to colour changes, FT infrared spectra of paint layers before and after exposure are recorded.

Hansa yellow PY3—pure or mixed with titanium dioxide—experimental tables are subjected to the influence of environmental conditions under exposure to UV radiation for a total period of 3 months. At approximately mid-time, moist heat (90 °C, 60 % relative humidity) is applied for a total period of 3 days. Colour measurements are performed during the ageing procedure, and changes expressed using the colour space CIE 1976 ($L^*a^*b^*$). The surface of both untreated and aged paint layers is as well microscopically observed.

Hansa yellow PY3—pure or mixed with titanium dioxide—experimental tables are subjected to the influence of environmental conditions under exposure to ultraviolet radiation for a total period of 3 months. At approximately mid-time, moist heat (90 °C, 60 % relative humidity) is applied for a total period of 3 days. Colour measurements are performed during the ageing procedure, and changes expressed using the colour space CIE 1976 ($L^*a^*b^*$). The surface of both untreated and aged paint layers is as well microscopically observed.

During exposure to UV radiation under environmental conditions, paint layers of Hansa yellow PY3 containing titanium dioxide remain rather stable and solely

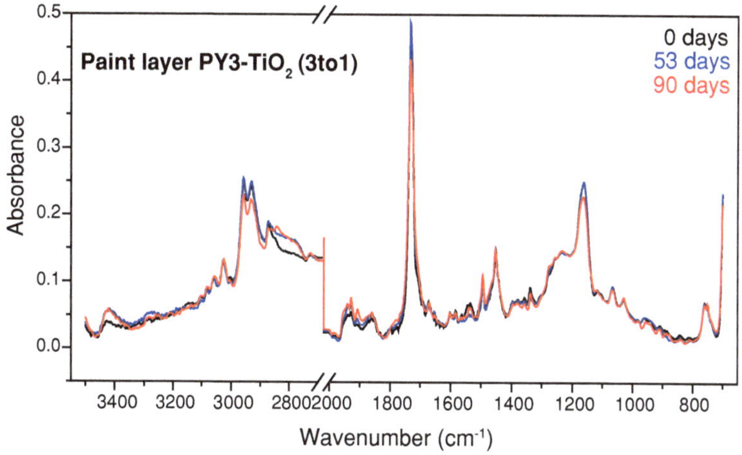

Fig. 5.16 FT Infrared Spectra of untreated and artificially aged paint layers—Hansa *Yellow* PY3 versus Rutile 3:1

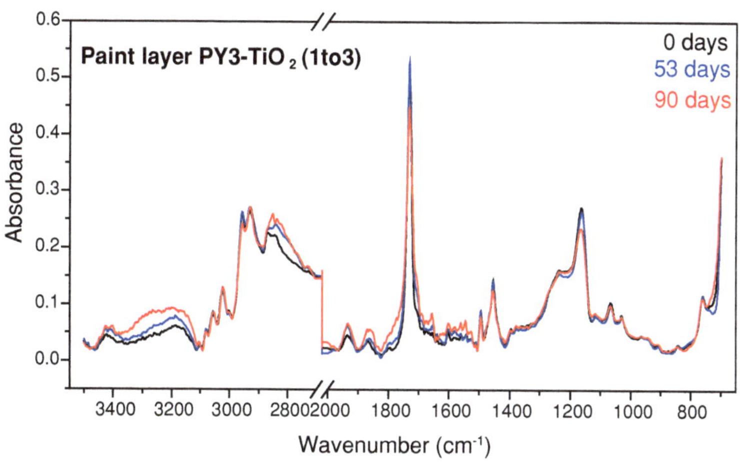

Fig. 5.17 FT Infrared Spectra of untreated and artificially aged paint layers—Hansa *Yellow* PY3 versus Rutile 1:3

moist heat yields a substantial colour change. On the other hand, in absence of rutile a decrease of reflectance values in the green region is noticed already during the first ageing phase, crediting the pure paint layer with the most important alterations. Hence, rutile is clearly promoting stability of paint layers (Figs. 5.16 and 5.17).

Alteration patterns in the IR spectra of untreated and aged Hansa yellow PY3 paint layers are essentially observed at the end of the ageing procedure. In the pure paint layers, the carbonyl stretching absorbance is broadening, but further pigment peaks—such as 1,605, 1,495, 1,456 cm^{-1}—are essentially stable. On the contrary, in paint layers containing rutile the carbonyl bands are broadening and loosing in intensity, and the hydroxyl bands are increasing in number; while binding medium peaks—such as 1150 cm^{-1}—are anew decreasing. Thus, titanium dioxide does not appear to create substantial interventions in the rate of molecular modifications. No qualitative changes, which would correspond to molecular structure modification of the pigments as a result of accelerated ageing, are observed in the IR spectra of untreated and aged Hansa yellow PY3 paint layers.

The overall results permit the assumption that the rutile form of titanium dioxide is protecting paint layers from degradation, in comparison to the ageing process observed in the pure pigment layers.

Hansa yellow PY74—pure or mixed with titanium dioxide—experimental tables are subjected to the influence of environmental conditions under exposure to ultraviolet radiation for a total period of 3 months. At approximately mid-time, moist heat (90 °C, 60 % relative humidity) is applied for a total period of 3 days. Colour measurements are performed during the ageing procedure, and changes expressed using the colour space CIE 1976 ($L^*a^*b^*$). The surface of both untreated and aged paint layers is as well microscopically observed.

During exposure of Hansa yellow PY74 tables to UV radiation under environmental conditions, the quantity of titanium dioxide is clearly affecting colour parameters, in what concerns alteration toward yellow, and this is further the case under moist heat. On the other hand, in absence of rutile the first ageing phase yields no alterations, and the second slight ones, proving that titanium dioxide is in fact promoting instability of paint layers.

Alteration patterns in the IR spectra of untreated and aged Hansa yellow PY74 paint layers are essentially observed at the end of the ageing procedure. In the pure paint layers, the carbonyl stretching absorbance is broadening, and several pigment peaks—such as 1,557, 1,522, 1,340, 1,255 cm^{-1}—are appearing increased in intensity, a fact probably due to a gradual binding medium degradation; while in paint layers containing rutile the carbonyl bands are broadening and loosing in intensity, and binding medium peaks—such as 1,150 cm^{-1}—are anew decreasing. Thus, titanium dioxide does not appear to create substantial interventions in the rate of molecular modifications. No qualitative changes, which would correspond to molecular structure modification of the pigments as a result of accelerated ageing, are observed in the IR spectra of untreated and aged Hansa yellow PY74 paint layers. The overall results permit the assumption that the rutile form of titanium dioxide is accelerating paint layers' degradation, in comparison to the ageing process observed in the pure pigment layers.

Naphthol AS-D PR112—pure or mixed with titanium dioxide—experimental tables are subjected to the influence of environmental conditions under exposure to UV radiation for a total period of 3 months. At approximately mid-time, moist heat (90 °C, 60 % relative humidity) is applied for a total period of 3 days. Colour measurements are performed during the ageing procedure, and changes expressed using the colour space CIE 1976 ($L^*a^*b^*$). The surface of both untreated and aged paint layers is as well microscopically observed.

During exposure of naphthol AS-D PR112 tables to UV radiation under environmental conditions, the quantity of titanium dioxide is not substantially affecting colour parameters, since the alteration toward yellow is almost identical in both cases. Under moist heat, however, ΔE^* and Δa^* are increasing in analogy to the amount of rutile present. On the other hand, in absence of rutile the first ageing phase yields no alterations, and the second slight ones, proving that titanium dioxide is in fact promoting instability of paint layers.

Alteration patterns in the infrared spectra of untreated and aged naphthol AS-D PR112 paint layers are confirming the fact that titanium dioxide is promoting instability. Thus, after day 53 exposure to UV radiation and moist heat is causing the aliphatic C–H and the carbonyl stretching bands to broaden and loose in intensity, and the hydroxyl bands to increase in number; while pure paint layers would exhibit the same behaviour solely after the total ageing procedure. The results permit the assumption that the rutile form of titanium dioxide is accelerating paint layers' degradation, in comparison to the ageing process observed in the pure pigment layers.

During exposure to UV irradiation under environmental conditions, molecular alterations—as observed in the IR spectra—are actually insignificant. After the intervention of moist heat, changes include loss of intensity for all aliphatic C–H and carbonyl bands, as well as for the ethyl acrylate peak at 1,150 cm^{-1}. The comparative review of all data permits regenerating certain features of the chromatic profile, as originally created by the artist; is evaluating the colourants as to compatibility and stability toward extrinsic factors; and is proposing degradation routes at a molecular level.

5.6 Solid State NMR to Characterise Paper and Parchment Samples from a Sixteenth Century Book

A. Spinella, D. Capitani, S. Bastone, C. Di Stefano and E. Caponetti

Abstract In this chapter, a solid state Nuclear Magnetic Resonance (ss-NMR) study of a sixteenth century book is reported. Cross-Polarization Magic-Angle Spinning Nuclear Magnetic Resonance ($^{13}C\{^{1}H\}$ CP MAS NMR) spectra of paper samples collected from the book, allowed us to obtain information on its state of conservation. The physicochemical characterisation of two parchment samples collected from the cover book was performed through the evaluation of the changes in line width and intensity of signals in the spectra.

5.6.1 Introduction

Solid state NMR (ss-NMR) is a powerful tool for chemical characterisation of many types of materials and natural polymers based materials such as paper [88] and parchment [89].

The application of ss-NMR spectroscopy can provide valuable information to conservation scientists, conservators and archaeologists.

In this chapter, the characterisation and determination of the extent of degradation of a sixteenth century book is presented in order to show the potentiality of ss-NMR technique in this field.

The book under investigation consists of three different volumes bound together. The three volumes were printed in Paris in different years of the sixteenth century (1534, 1578, 1597) and bound together at the Monastery of San Martino delle Scale (near Palermo, Sicily), which was one of the most illustrious monastic libraries of Sicily and also a writing workshop producing good quality manuscripts not only for its own library but also for other Benedictine monasteries. The parchment cover was processed in the same monastery, and it shows typical incisions of Martinian binding [90]. For this reason, the book binding was dated between the late sixteenth and early seventeenth century as in the period after that time the monastery commissioned the bindings outside. An example of this kind of binding is shown in Fig. 5.18.

Paper is one of the most common, oldest and sometimes precious man-made materials. The main source of cellulose fibres for the paper production in Mediaeval Europe was rags made by cotton or flax. Modern paper is mostly made from wood cellulose that is difficult to purify. The chemical treatments involved in the purification process can cause significant damage to the cellulose fibres. An undamaged cellulose with long fibres is necessary in order to obtain strong papers with a good durability. For this reason, traditional Chinese and Japanese papers

Fig. 5.18 An example of Martinian binding

were made from cellulose fibres derived from local plants which had these characteristics.

The first paper was made in China in 105 AD from a suspension of cellulose fibres dispersed in water. The suspension was drained through a fine mesh so that the water drained away and the retained fibres formed a mat on the mesh. After drying the fibres stacked together forming a sheet of paper [91].

High quality paper is mainly a two-component material constituted by cellulose and by an almost equimolar amount of bound water, plus a variable amount of organic and inorganic additives and/or dia- and paramagnetic impurities. [92]. The paper structure can be schematically described as amorphous cellulose domains surrounding water pools, and the amorphous domains are surrounded by crystalline cellulose domains [93]. Some water molecules also permeate the cellulose fibres. A comprehensive microscopic view of paper could describe it as an interconnected structure of water, amorphous and crystalline cellulose domains on the nanometre scale [94].

On the basis of $^{13}C\ \{^{1}H\}$ CP MAS NMR spectra, Vanderhart and Atalla [95] found out that all native celluloses are a mixture of two crystalline modifications, called $I\alpha$ and $I\beta$, in different proportions. The two allomorphs are shown in Fig. 5.19.

The corresponding crystallographic units are characterised by one-chain triclinic and two-chain monoclinic unit cells, respectively. In fact, these crystalline forms have the same microfibril but in the monoclinic form, the cellobiose units stagger with a shift of a quarter of the crystallographic *c*-axis period, whereas the triclinic form exhibits a diagonal shift of the same amount. The ratio of the two phases depends on the origin of the cellulose. The $I\alpha$ form is dominant in cellulose produced by primitive organisms, such as the bacterium Acetobacter xylinum and the alga Valonia macrophysa, whereas the $I\beta$ form dominates in cellulose produced by higher plants.

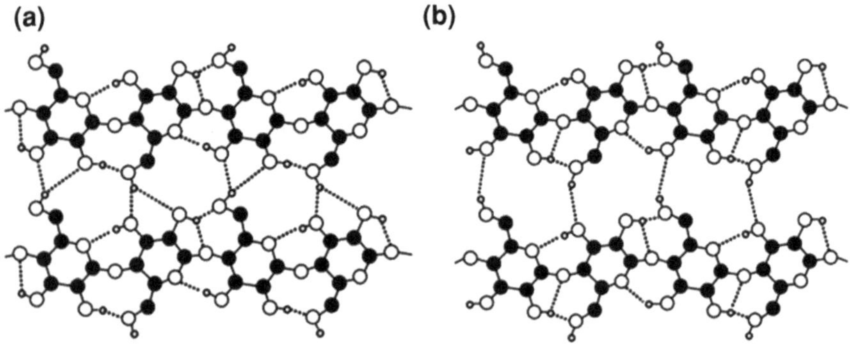

Fig. 5.19 Hydrogen bonding patterns in cellulose Iα (**a**) and Iβ (**b**) (Sturcova et al., Copyright 2006 American Chemical Society)

Cellulose in a well-preserved paper has a high degree of polymerization and the amorphous domains surround small water pools. A measurable loss of bound water and an increase of the amorphous fraction accompany the destruction of the texture of the paper. Also paramagnetic impurities play an important role in the degradation of the paper, as they may act as catalysts and initiators of hydrolytic and/or oxidation reactions. Furthermore, paramagnetic centres undergo profound changes upon thermal or photochemical ageing of paper [96, 97].

Parchment is a material made from calfskin, sheepskin or goatskin. Parchment used for manuscripts, scrolls, charters, book covers and substrata for artworks makes a major contribution to the most valuable objects of European cultural heritage [98]. Large collections exist in varying degrees of preservation in public and private libraries, archives, museums and in different religious foundations. At the end of the Middle Age, parchment was joined by paper; however, it has continued to be used for special purposes, such as official documents and book bindings. It is manufactured from animal hide after strong alkaline removal of the epidermis and subcutaneous tissue collagen that is a fibrous protein constituting the main structural component of parchment. The structure of collagen is well described in terms of three individual protein strands in the α-helix conformation [99]. These strands are rigidly held by strong hydrogen bond between the hydroxyl group of hydroxyproline and the amino function of adjacent glycine units, forming a rod like crystalline phase embedded into an amorphous matrix.

The degradation of parchment involves structure loss of the intact fibre through different stages of alteration to a terminal stage when the fibre structure is close to a complete hydrolysis. When in contact with water or stored in wet conditions, the fibres transform into a gelatinous substance. To improve preservation and conservation treatments of patrimonial skin objects can be, therefore, useful to study the chemical composition and the physical–chemical state of aged collagen. Many collagen-based objects are valuable treasures due to the history they represent, and their preservation is still a challenge for museums and private collectors as well.

Table 5.11 Paper samples and position of sampling

Sample	Position
1	First page of the first volume
2	Last page of the first volume
3	First page of the second volume
4	Centre of the second volume
5	Last page of the second volume
6	First page of the third volume
7	Last page of the third volume

The development of specific analytical techniques may improve the procedures to authenticate patrimonial objects made from collagen-based materials as well as methods to study the impact of environmental factors.

5.6.2 Experimental

The sampling was performed in such a way to minimally damage the product and to obtain a representative number of samples. The paper sampling was carried out by taking two small (10 mg) stripes from each volume, one from the first and one from the last page. In addition, another sample was collected from the centre of the second volume. The two parchment samples were taken from the back of the cover.

^{13}C {^{1}H} CP MAS NMR spectra were obtained at room temperature with a Bruker Avance II 400 MHz (9.4 T) spectrometer operating at 100.63 MHz for the ^{13}C nucleus, the spinning rate was 13 kHz, A total of 1,024 scans were collected, a contact time of 1.5 min and a repetition delay of 2 s were used. The Hartmann-Hahn condition [100] was optimised on adamantane which was also used as a chemical shift external reference. Samples were placed in 4 mm zirconia rotors and sealed with KEL-F caps using silica as a filler to avoid inhomogeneities inside the rotor.

Spectral deconvolution was performed using the DMFit software [101].

5.6.3 Analysis of NMR Spectra of Paper Samples

^{13}C {^{1}H} CP MAS NMR spectra were acquired on the seven paper samples reported in Table 5.11:

As an example, the spectrum of sample 1 is shown in Fig. 5.20. The peak assignment of carbons of cellulose repetitive unit is also reported.

The weak signal observed at 172 ppm is ascribed to carbonyl carbons. In the range between 110 and 100 ppm, assigned to C1, three intense signals are observed; two external signals are due to C1 of $I\beta$ polymorphous form, whereas the inner signal is due to C1 of $I\alpha$ polymorphous form. A shoulder due to anomeric carbons of oligomers is also present. The signal at 88.7 ppm is due to C4 of crystalline cellulose and the signal at 83.8 ppm is due to C4 of amorphous

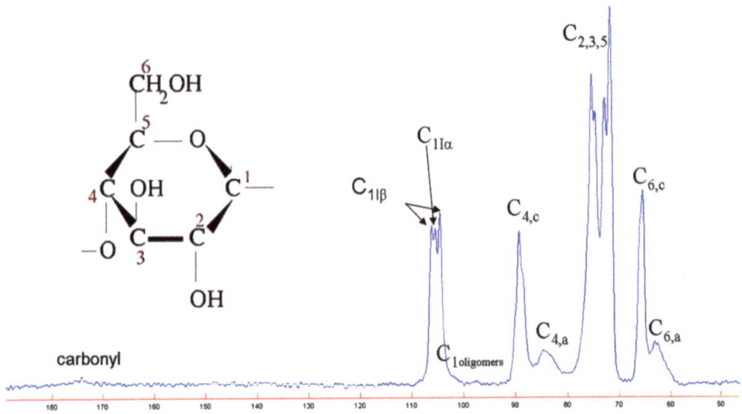

Fig. 5.20 $^{13}C\{^1H\}$ CP MAS NMR spectrum of sample 1

cellulose. Peaks between 80 and 70 ppm are assigned to C2, C3 and C5 carbons. The signal at 67 ppm is due to C6 of crystalline cellulose, whereas the one at 63 ppm is due to C6 of amorphous cellulose. The spectral deconvolution of the 110–55 ppm region is shown in Fig. 5.21.

Peak areas obtained from the spectral deconvolution of NMR spectra of all samples are reported in Table 5.12.

In the Table, the relative composition of $I\alpha$ and $I\beta$ polymorphous forms obtained from the deconvolution is also reported. This value is obtained from the ratio between the integral of Ix ($I\alpha$ or $I\beta$) and the total integral of C1.

The deconvolution of C4 signal allowed us to determine the degree of crystallinity (DCry) of the cellulose which is strictly correlated to the state of degradation of the paper. The latter parameter is calculated from the ratio between the integral of C4,c signal and the total integral of C4 [102].

All samples show a rather good degree of crystallinity between 0.5 and 0.6. Samples 1, 3, 4, 6 and 7 also show the presence of a certain amount of cellulose oligomers possibly due to the hydrolysis of the glucosidic bonds. The lowest amount of oligomers is observed in samples 2 and 5. In samples 1, 2, 3 and 5, $I\beta$ is the predominant polymorphous form; whereas, in samples 4, 6 and 7, $I\alpha$ polymorphous form prevails. Sample 1 is the only one showing a carbonyl carbon signal due to cellulose oxidation.

It is worth noting that the ss-NMR results obtained by Horii et al. [103] for pure cellulose samples were in accordance with the ones obtained by X-ray Diffraction. The advantage of using NMR is that, to perform the analysis, a small quantity of sample without any previous preparation is needed; on the contrary, to measure the degree of crystallinity by XRD technique the sample must be milled. As it has been noted, the grinding often causes a decrease of the cellulose degree of crystallinity; nevertheless the milling is necessary to obtain homogeneous samples and reproducible measurements.

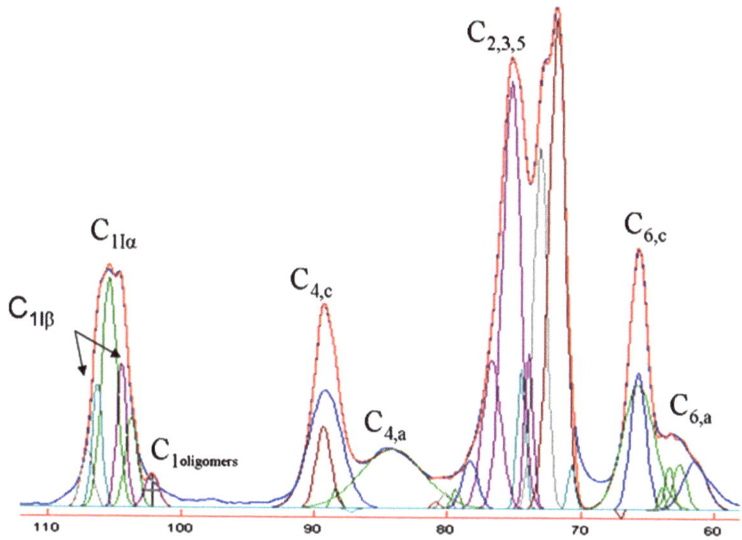

Fig. 5.21 Deconvolution of $^{13}C\{^1H\}$ CP MAS NMR spectrum of sample 1

Table 5.12 Peak areas of $C1_{I\alpha}$, $C1_{I\beta}$, oligomers, C4,*c* and C4,*a* signals obtained from the spectral deconvolution

Sample	Carbonyl 175 ppm	$C1_{I\alpha}$ 105 ppm	$C1_{I\beta}$ 105 ppm	I_α/I_β	$C1_{oligomers}$ 103 ppm	C4,*c* 89 ppm	C4,*a* 84 ppm	DCry
1	3.49	6.42 (44 %)	8.16 (56 %)	0.8	0.8	9.66	6.80	0.6
2	–	6.41 (48 %)	6.86 (52 %)	0.9	0.1	10.21	6.61	0.6
3	–	5.11 (35 %)	9.50 (65 %)	0.5	0.7	8.71	9.29	0.5
4	–	7.58 (52 %)	7.01 (48 %)	1.1	0.9	8.62	7.72	0.5
5	–	4.78 (36 %)	8.49 (64 %)	0.6	0.3	10.05	5.90	0.6
6	–	5.50 (53 %)	4.88 (47 %)	1.1	0.6	8.33	5.24	0.6
7	–	7.23 (57 %)	5.49 (43 %)	1.3	0.9	8.58	7.53	0.5

The relative composition of α and β polymorphous forms and the degree of crystallinity DCry are also reported

5.6.4 Analysis of NMR Spectra of Parchment Samples

$^{13}C\{^1H\}$ CP MAS NMR spectra were acquired on two parchment samples, A24 and A35, and on a sample of new parchment used as a reference. The spectrum of sample A24 is shown in Fig. 5.22.

The assignment of signals was carried out according to the literature [104].

The region from 185 to 160 ppm contains the carbonyl and carboxylic carbons of amino acid residues; in particular the most intense peak at 174 ppm is due to the

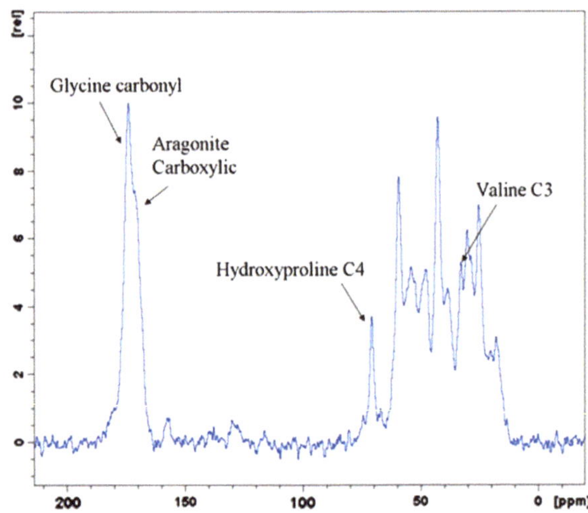

Fig. 5.22 $^{13}C\ \{^1H\}$ CP MAS NMR spectrum of parchment sample A 24

glycine carbonyl carbon. The strong peak at 171 ppm is ascribed to aragonite. The weak peak at 169 ppm is ascribed to calcite. Aragonite and calcite, two common polymorphs of $CaCO_3$ present in most of the historical parchments are probably due to the reaction of the atmospheric CO_2 with $Ca(OH)_2$ which was used to eliminate the fat from the skin. The weak peaks observed in the 127–138 ppm range are assigned to aromatic carbons of aromatic amino acids. The signals due to amino acids are observed in the aliphatic region (0–85 ppm), in particular the signal at 71 ppm is assigned to hydroxyproline C4, and the signal at 33 ppm is assigned to valine C3.

The spectrum of sample A35 reported in Fig. 5.23 shows the same features as those of sample A24; however, in the spectrum of sample A35, the peak at 33 ppm is more intense than expected. This peak is assigned to the methylene carbons $(CH_2)_n$ of hydrocarbon chains of fatty acids. Besides, a rather sharp signal overlapped with the signal of C3 of alanine, is also observed at 18 ppm. This signal was ascribed to CH_3 groups of fatty acids. Furthermore, a peak at 185 ppm due to free fatty acids is present. It is not possible to determine the type of lipid present. The expanded aliphatic carbons region of both spectra is reported in Fig. 5.24.

A semiquantitative comparison of the lipid content in the two samples can be obtained from the ratio between the intensity of the resonance at 33 ppm (lipid), I_{33}, and the intensity of the resonance at 43 ppm I_{43} (collagen).

The I_{33}/I_{43} ratio for samples A24 and A35 was found to be 0.4 and 0.9, respectively. In the case of a new parchment sample collected as a reference, the I_{33}/I_{43} ratio was found to be 0.4, i.e. the same value obtained for sample A24. Ghioni suggested a relationship between the collagen degradation and the increase of lipid content [105]. He showed that new parchments contain small quantities of lipids usually acquired through microbial attack, besides it is possible that the lipid

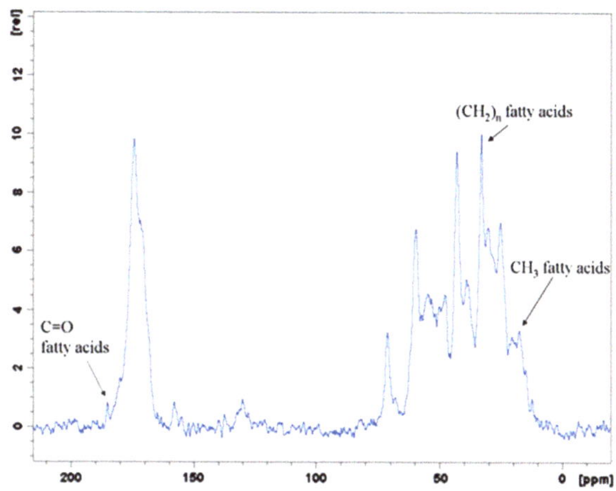

Fig. 5.23 $^{13}C\ \{^1H\}$ CP MAS NMR spectrum of parchment sample A 35

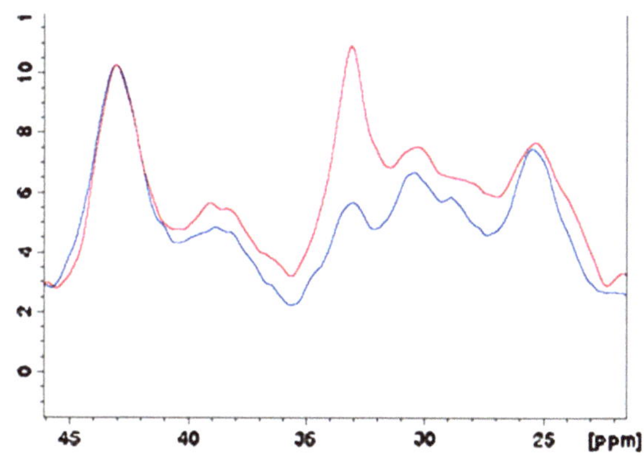

Fig. 5.24 Region between 20 and 45 ppm of the spectrum of sample A 24 (*bottom*) and sample A 35 (*top*)

fraction of parchment causes with its peroxidation a generation of reactive oxygen species that can damage the collagen structure of parchment [105].

It has been shown that, as gelatinization evolves through hydrolysis of polypeptidic chains of collagen, a loss of structural order occurs causing a significant broadening of the lines in the solid state ^{13}C NMR spectra [104]. In this regard, a method which makes possible to investigate the state of conservation of parchment was introduced. This method consists in determining a parameter R obtained by measuring the line width of the peak at 71 ppm, due to C4-hydroxyproline. The spectral region between 66 and 76 ppm of samples A24 and A35 is reported in Fig. 5.25.

The parameter is calculated using the following equation [104]:

$$R = \left(\Delta\nu_{1/2} - \Delta\nu_{1/2}^{c}\right) / \left(\Delta\nu_{1/2}^{g} - \Delta\nu_{1/2}^{c}\right)$$

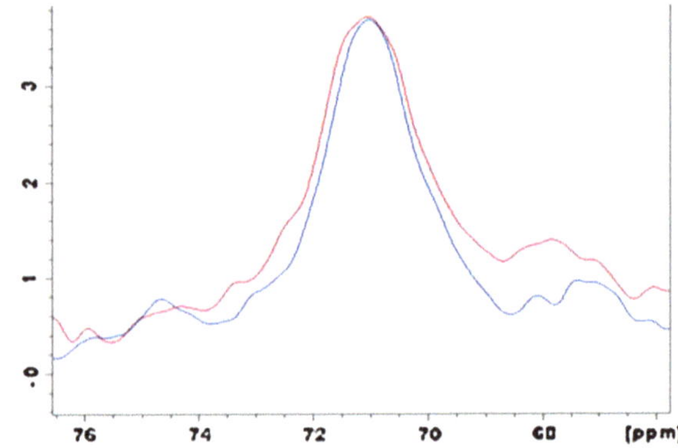

Fig. 5.25 Spectral region between 66 and 76 ppm of samples A24 (*bottom*) and A35 (*top*)

where $\Delta\nu_{1/2}$ is the line width of the signal at 71 ppm of the parchment under investigation, $\Delta\nu^c_{1/2}$ and $\Delta\nu^g_{1/2}$ are the line widths of the same signal measured in a sample of new parchment used as a reference and in gelatin, respectively.

By definition $R = 0$ in the case of new parchment, $0 \leq R \leq 1$ in the case of historical parchments in their pregelation state, with higher R values corresponding to a higher level of degradation. Compared to gelatin, hydrolyzed collagen is a mixture of smaller molecular weight polypeptides, as a consequence it can be considered as degraded gelatin, with $R > 1$ reflecting the extent of the disintegration relative to gelatin.

In samples A24 and A35, R was found to be 0.34 ± 0.03 and 0.47 ± 0.03, respectively. As a consequence, it can be inferred that sample A24 is less degraded than sample A35.

This result is further confirmed by the presence of lipids that is evident in the NMR spectrum of the A35 sample. Lipids are not generally present in parchment; however, their presence may be a consequence of an incomplete cleaning or a bacterial degradation. This type of degradation can cause the breakdown of protein structures and the formation of protein–lipid aggregates. Furthermore, the contact with the skin of those who handled them, especially at the edges and the surface can be the cause of the observed lipids. Therefore, the source can be intrinsic, extrinsic or both.

The presence of lipids increases the speed of degradation of collagen because the lipid peroxidation in the presence of atmospheric SO_2 causes the hydrolysis of collagen. Therefore, this observation must be carefully taken into account especially in the consolidation phase of the material.

5.6.5 Conclusions

NMR technique is a powerful investigation tool in the field of Cultural Heritage artworks. Parameters which may be obtained with this technique are useful to get information on the state of conservation of various items, as shown in the case of ancient paper and parchment. Analyses are performed on a small quantity (10 mg or less) of sample which can be utilised for further investigations as the method is not destructive. Besides, no treatment of samples is required.

The careful analysis of ^{13}C {^{1}H} CP MAS NMR spectra produced information on the state of conservation of the sixteenth century book investigated.

It was demonstrated that the paper is in a good state of conservation, with only a small amount of oligomers due to the hydrolysis of cellulose. Only the first sample, taken from the first page of the book, shows an oxidative degradation. All samples show a good index of crystallinity.

Ss-NMR also provided information on the state of degradation of the two ancient parchment samples analysed. Parchment A35 is more degraded than parchment A24. The degradation observed in A35 is related to the presence of lipids clearly observable in the NMR spectrum.

5.7 Case Studies on Textile Studies with Scanning Electron Microscopy

Eleni Pavlidou

Abstract Microscopy in general, has been employed to help art conservation science a lot in the past decades. Numerous case studies could be mentioned toward that direction. In the following, two very interesting case studies of scanning electron microscopy (SEM) and optical microscopy (OP) use in textiles preservation will be presented, illustrating several of the many aspects microscopy has to offer in that sector. Not only morphological and texture characteristics can be studied with OP and SEM, but also elemental analysis and thus identification of the samples is possible. Through observations using the above techniques, the state of degradation as well as technological aspects, such as the manufacturing techniques, can be derived. Finally, the well-established microscopic techniques can provide valuable historical information concerning the samples studied, such as their origin, the economical and cultural trading relationships between countries, or even the technological knowledge of the people at that time as well as the identity of the people using the artefacts.

5.7.1 Study of the Textiles Found in the Cave of Letters

In a recent study, Müller et al. [106] studied fibres from textile fragments almost 2,000 years old that were excavated from the Cave of Letters in the Dead Sea region. In comparison to modern reference samples, most of the fibres were identified as wool and some as plant bast fibres (flax). The molecular and supermolecular structure of both keratin (wool) and cellulose (flax) were found completely intact. In many fibres, mineral crystals were intimately connected with them. The analysis of the dyed wool textiles suggests the possible use of metal-containing mordants for the fixation of organic dyes. In this Cave, wool, linen and all kinds of garments were found, sometimes even near-complete tunics. From the archaeological point of view, the study of these textiles is without any doubt of great importance for determining what the Romans and the Jews wore in the second century AD in the eastern Mediterranean and where they got their materials from.

An indication of the type of textiles was provided by inspection under a stereoscopic microscope; this proved that the studied samples were S-spun (left-handed), which is normally typical for bast fibre handedness, whereas wool is usually Z-spun (right-handed). The samples were untreated and thus contained soil particles. Individual fibres of a few millimetres length were carefully extracted from the samples using tweezers under a stereo microscope. Handling of the archaeological fibres was more difficult than for modern fibres, since the material was much more brittle, indicating a certain degree of degradation. Fibre diameter was typically between 10

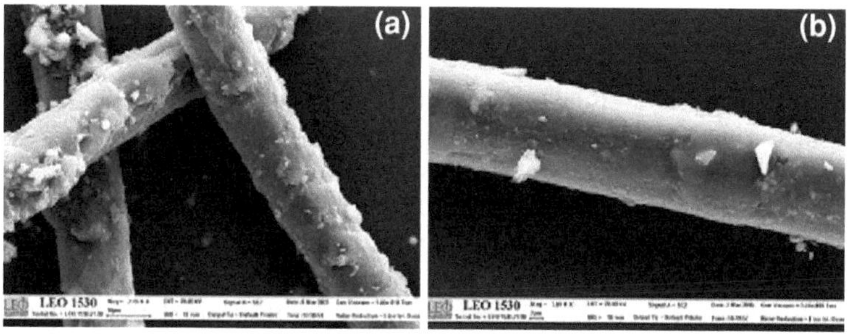

Fig. 5.26 SEM images of archaeological fibres. **a** Fibre type identified as wool; **b** Fibre diameter was found as 16.6 μm, preliminary identified as bast fibre

and 20 μm. In order to prepare the samples for scanning electron microscopy (SEM), a thin (10–20 nm) film of gold was deposited onto each one of them.

The textile samples from the Cave of Letters showed almost no signs of degradation in the optical microscope. Practically, all individual fibres in a single thread remained intact after nearly 2,000 years in the sediment of the caves. However, the fibres were found to be rather brittle, indicating some internal degradation of the fibre material. The colours of the textiles were still very bright and intense and thus probably still close to their initial state when the garments were worn. Some of the fibres were easily identified as wool by their very regular outer shape and the high transparency of the homogenous fibre material. The uncoloured fibres were probably plant fibres. However, not every sample was easily handled and identified; a bright red sample is an example of a far more difficult identification that required further inspection with a higher resolution SEM.

As revealed from the SEM images (Fig. 5.26a and b), the plant fibre had a smooth surface and showed the hint of a typical dislocation "knee defect" as found in bast fibres. More pronounced "knees" were seen in SEM images of other fibres of the same sample. From these images, the diameter of the fibre was found to be about 16.6 μm, which is a very typical value for flax. Other fibres exhibited a much more rugged surface while having a very similar diameter. This surface morphology could mean a further state of degradation, or a completely different fibre type like wool. Some of the SEM images of fibres already identified as wool with the optical microscope clearly showed the scale-like periodic structure of the wool fibre cuticula.

The SEM images revealed another very important feature of the ancient samples. Not only was the fibre surface covered with microscopic and submicroscopic soil particles as expected from finding the textile fragments in the cave sediments, but also there seemed to be aggregates of non-fibrous, probably inorganic, material intimately connected with the fibre surface (Fig. 5.26a). For fibres in these conditions, further cleaning procedures, e.g. washing in distilled water or mild solvent, were thus considered inappropriate.

Wool mainly consists of the fibrous protein α-keratin whose architecture is based on α-helices oriented parallel to the fibre axis. Through this investigation, Müller et al. proved that the molecular and supermolecular structure of wool was preserved during the two millennia that these textiles had been lying in these dements of the cave. Similarly, well-preserved structures of keratin have also been found in the hair of ancient Egyptian mummies of about the same age. Nevertheless, in this work too there were samples that could not be precisely identified with the use of optical microscopy (OP) and SEM images alone. Other techniques were employed and carefully chosen, in order for the complete characterisation and conservation to be performed. The complementarity of the analytical techniques is clearly demonstrated through this work. Finally, with the use of SEM, OP, diffraction patterns and fluorescence spectra that were collected from the European Synchrotron Radiation Facility, the textiles that were found in the Cave of Letters were proved to have been made from wool (dyed in intense colours) and only a few from linen (native undyed flax fibres). This finding is very interesting in the archaeological context and is in contrast to the results on textiles from Qumran (mostly undyed linen) indicating a significantly different cultural background of the people using the respective textiles.

5.7.2 *Metal Threads in Byzantine Textiles*

Precious metals, which have been used for the decoration of textiles since ancient times, have traditionally been associated with the use of silk. Gold threads, like silk, are luxury materials engaged in the manufacture of the finest and most expensive fabrics. Recently, Karatzani et al. [107] employed three analytical techniques, namely OP, scanning electron microscopy coupled with energy-dispersive X-ray spectrometry (SEM–EDS) and electron probe microanalysis, in order to study the nature of Byzantine-Greek metal threads. The aim was to identify the morphological and technological characteristics of the metal threads used for ecclesiastical textiles between the thirteenth and the nineteenth centuries. The study also aimed to determine the significance of these objects in comparison with the value of the materials used. The results helped to explore possible stylistic and technological relationships among the various embroidery workshops operating through that period and at the same time to examine exchanges of technological knowledge among the Greek, European and Ottoman craftsmen during this period. So the historical aspects of this study were really overwhelming.

Six main types of metal threads have been recorded so far: thin sheets of gold or silver cut into narrow strips; gold or silver wire either drawn or twisted; thin sheets of gold or silver (lamellae) spun around a silk or fine linen thread; gold or silver wire wound around a fibrous core of vegetable or animal origin (tirtir); gilt organic strips and combined metal threads. The metals mainly used were gold, silver and copper, either alone or combined. The organic supporting material of the gilt strips can be cellulose-based (paper) or protein-based (leather, parchment, and animal gut). The core around which the strips are spun can be protein-based fibre such as silk or wool or cellulose-based such as linen or hemp.

The metal threads in this study represent 300 samples obtained from 137 objects belonging to 16 different Greek and European institutions covering the period between the thirteenth and the nineteenth centuries. Samples of all the different metal threads used in each object were taken from damaged areas of the textiles, when possible, normally consisting of bits of loose threads measuring 5–7 mm in length. These cover all the known metal thread types; samples also were taken of other metallic decorations such as sequins (palettes) and combined threads.

The samples were studied with a stereo binocular microscope first to determine their morphological characteristics, such as the type of thread, the twisting of the metal strip, and the colour of the organic core thread. Then, a part of each sample was studied using the SEM to measure the strip width and to acquire more detailed surface information, such as the number of coils per millimetre of the thread sample and the state of preservation of the samples. The SEM study also helped to determine the manufacturing techniques, since details such as parallel lines characteristic of a drawn wire become visible as well as the marks of a cutting tool at the edges of a strip. Finally, the EDS technique was used for the surface elemental analysis of the samples.

The quite interesting results from this work indicate that the vast majority of the samples were found to be wound-strip type, while the strip and the wire types represented only 11 and 10 %, respectively; all the other types were less than 4 %. The EDS surface analysis showed that about half of the samples were gilt silver, followed by ones made of pure silver. The number of samples containing gold, silver and copper and those containing silver and copper was almost the same. The vast majority of the threads were made of pure silver or silver with copper. A total of 40 of the studied samples were made of pure silver. Also, 20 were silver and copper, while 13 samples consisted of pure copper. Only one sample of a gold–silver–copper alloy was identified and one of pure gold. The identification of the materials used in these threads was rather helpful to the historians and archaeologists that study the corresponding centuries.

The variations on the morphological and technological characteristics of the metal threads were also demonstrated through this work. It confirmed that the simple optical examination of the textile cannot provide accurate information about the materials and the manufacturing techniques used for the production of metal threads. It becomes clear that only the combination of surface analysis with the study of polished cross-sections can complete this investigation. Moreover, the results showed that most of the known metal thread types were used for the decoration of the Greek Orthodox ecclesiastical textiles. Furthermore, the diversity of the types used, especially in the same object, suggests that all these materials were available to the Greek craftsmen throughout that period. The choice of materials used is closely related to the final appearance of the object. The identification of such a large number of gilt threads, on the other hand, indicates that the church maintained its power and wealth during the period examined, making possible the production of such fabrics.

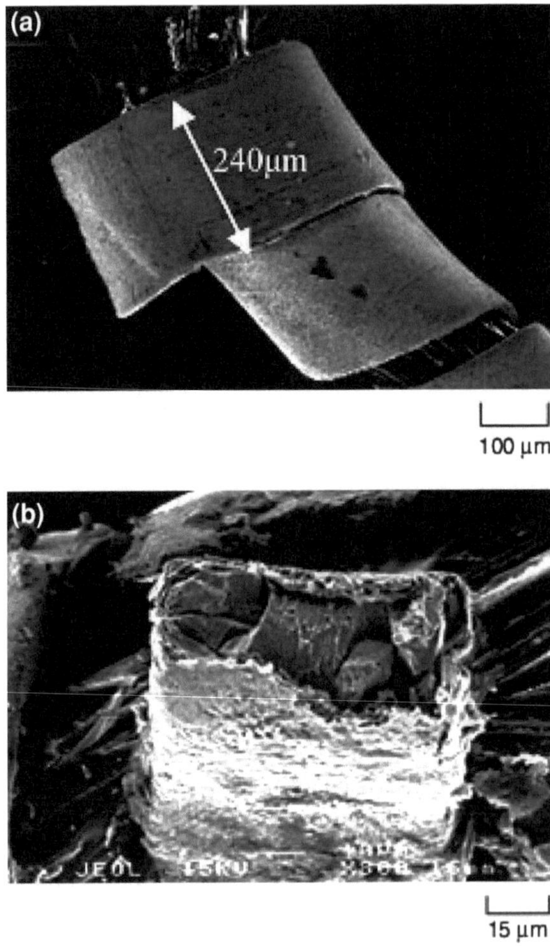

Fig. 5.27 SEM micrographs: **a** measurement of the width and **b** condition of the metal strip

Examples of different types of threads are also shown in Figs. 5.27 and 5.28. In Fig. 5.29 are shown SEM images of wool textile and metal thread from nineteenth century findings in a Greek area renown for its traditional costumes. In Fig. 5.30, images of cotton and wool threads from the same costumes are shown.

As shown by the above studies, the morphology and texture of the threads, as well as their preservation condition can be studied through SEM and OM investigations. Also, with the use of EDS such samples can be identified, and thus important aspects such as their conservation methods and details about their historical meaning and background can be derived.

In conclusion, the importance of complementary analytical techniques being used in textile and art preservation and particularly the use of optical and SEM have been clearly demonstrated through the above studies. Also, the help that these techniques can provide toward a better understanding and clarification of historical aspects was also discussed.

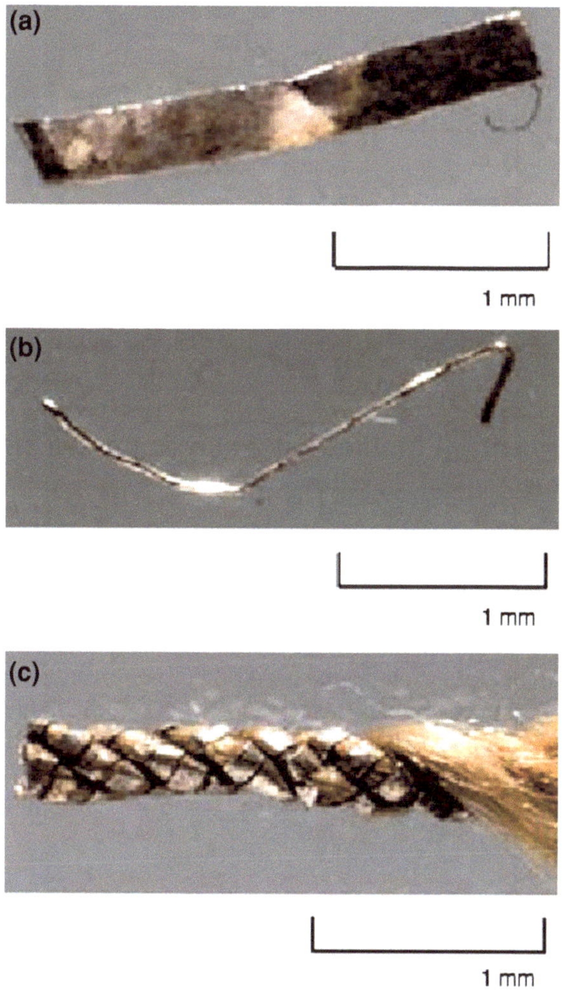

Fig. 5.28 Optical microscope images of metal threads of silver or gold: thin sheets (**a**), wire (**b**) and (**c**) thin sheets of gold spun around a silk or fine linen thread

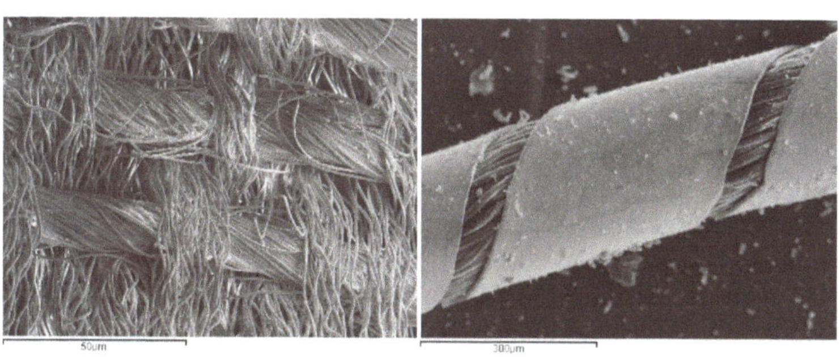

Fig. 5.29 SEM images of wool threads. *Left* a piece of wool textile. *Right* thin sheet of gold around a fine wool thread

Fig. 5.30 SEM images of cotton (*left*) and wool (*right*) threads

5.8 Physicochemical Methods of Analysis for the Study of Archaeological Textiles

Evangelia Varella and Maria Kyranoudi

Abstract Stereomicroscopy, Transmission Light Microscopy and sem–Energy-Dispersive Spectroscopy applied on 12 textile fragments found in ancient Macedonian graves in order to examine the condition of preservation, the nature of the fibres used, the spinning direction and the type of weave. Fibres of vegetable (cotton, flax) and animal (wool, silk) origin such as inorganic fibres (asbestos) appear in the samples. S and Z spinning was used for different fibres and the preferable type of weave is the plain weave.

5.8.1 Introduction

Archaeological textiles, as all organic materials, are hardly preserved in archaeological contexts due to biological attack and physicochemical alterations that occur before or during excavation [108].

Deterioration depends on many factors, the most important being the nature of the fibres [109] and the environment the material was kept in. They can be preserved in particular environmental conditions such as extreme dryness [110], extreme cold and absence of oxygen [111–113] or in association with metals such as copper or iron [111, 114–116].

When found, in spite of their fragmentary condition, they are an important testimony to everyday life, farming and trade, migration of nations, art and technical culture of different ethnological groups [117–119].

Application of analytical methods on archaeological textiles enriches the information received from indirect sources such as ancient writers and iconography. Moreover, it gives conservation scientists information about the conditions of

Table 5.13 Presentation of the samples

Sample number	Object in association	Chronology	Archaeological museum
1	Bronze little bell	Roman	Thessaloniki
2	Silver talisman	Early Byzantine	Thessaloniki
3	Bronze coin	Roman	Thessaloniki
4	Bronze/iron peroni	Fifth century BC	Thessaloniki
5a	Bronze krater	End of fourth century BC	Thessaloniki
5b	Bronze krater	End of fourth century BC	Thessaloniki
5c	Bronze krater	End of fourth century BC	Thessaloniki
5d	Bronze krater	End of fourth century BC	Thessaloniki
5e	Bronze krater	End of fourth century BC	Thessaloniki
6	Clay button	Hellenistic	Amphipolis
7	Bronze coin	Roman	Amphipolis
8	Bronze coin	Roman	Amphipolis

preservation and the physicochemical properties of the material, in order to choose the proper conservation method as well as the conditions of storage or display.

5.8.2 Experimental

Different methods of microscopy performed on 12 textile fragments found in Macedonian graves dated from Hellenistic to Byzantine period (Table 5.13). The information received regard the type of natural fibres (vegetable or animal origin), the spinning direction, the type of weave and natural colourants used for the fabrication and dyeing of the textiles.

5.8.2.1 Light Stereomicroscopy

All textile fragments studied, were found in association with a metal object (usually bronze), which was responsible for the preservation of the organic material (Fig. 5.31).

Before taking the samples, the fragments were photographed and observed under a Carl Zeiss Stemi 2000-C (for the objects kept in Archaeological Museum of Thessaloniki) or a Leica MZ6 (for the objects kept in Archaeological Museum of Amphipolis) light stereomicroscope (Figs. 5.32, 5.33). The aim was to find the macroscopical features and the degradation degree of the material and also to choose the area from where the samples would be taken.

Fig. 5.31 Roman textile fragment on a bronze coin

Fig. 5.32 Detail of a textile fragment on a roman bronze coin (Leica MZ 6, ×2)

Fig. 5.33 Organic fibres from a yarn on the back of a clay button (Leica MZ 6, ×2.5)

5.8.2.2 Transmission Light Microscopy

Single fibres of the sixth and eighth sample were mounted in 50 % glycerin and examined using a Carl Zeiss Axiostar Plus microscope for observing their anatomical characteristics (Fig. 5.34).

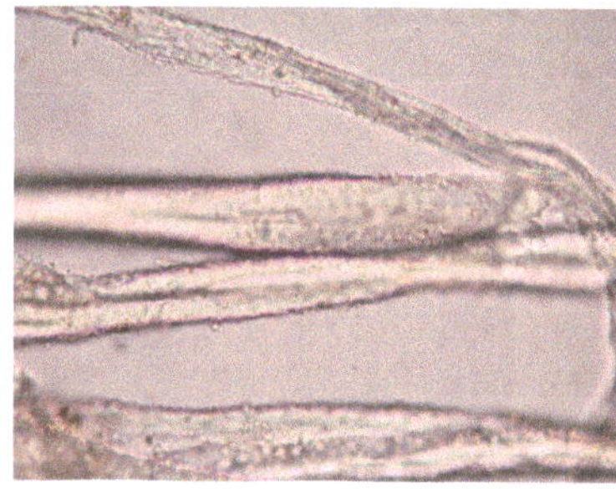

Fig. 5.34 Cotton fibres under a Carl Zeiss Axiostar plus light microscope

5.8.2.3 Scanning Electron Microscopy–Energy-Dispersive Spectroscopy

Scanning electron microscopy was carried out using a JEOL JMS-840A scanning microscope equipped with an Energy-Dispersive X-ray (EDS) INCA microanalytical system. Operating conditions were: accelerating voltage 20 kV, probe current 45 nA and counting time 60 s, with ZAF correction being provided online. The samples were coated with carbon, using a Jeol JEE-4X vacuum evaporator.

5.8.3 Results and Discussion

In the half of the samples, natural fibres are well preserved and easily identified in SEM photographs (Fig. 5.35a, b). The rest of the samples have undergone a different degree of mineralization as it appears in SEM photographs as well as in EDS spectra, making identification of the fibres more difficult. Flax, wool, silk, cotton and the inorganic fibre asbestos were identified (Fig. 5.36a–c). Both S and Z spinning appear in the samples whereas plain was the only type of weave used.

The analytical results of the study, concerning the type of the natural fibres used and the yarn characteristics of the textile fragments are displayed in Table 5.14.

5.8.4 Conclusions

Diffusion of metal ions is responsible for the preservation of the majority of the textile fragments studied in this project. The degree of the mineralization of the fibres depends on the burial microenvironment. In some cases, the organic material and its characteristics are completely lost, making fibre identification very difficult.

All types of natural fibres are used. Silk and flax are found in wealthier burials, while wool is connected with poorer roman burials. Wool and cotton yarns are

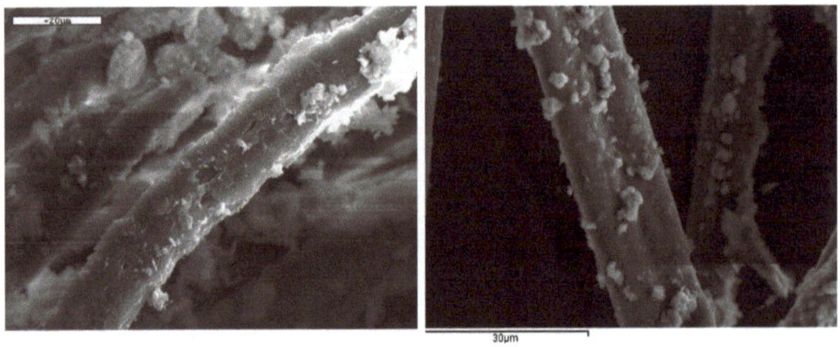

Fig. 5.35 **a** and **b** Wool and silk fibres under a JEOL JMS-840A scanning electron microscope

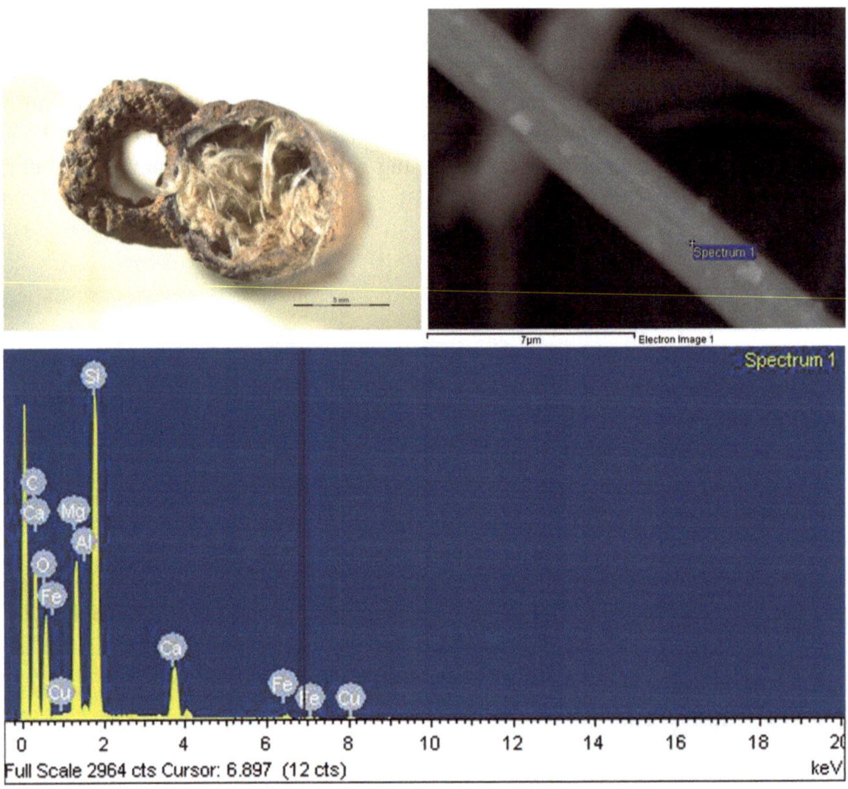

Fig. 5.36 **a–c** Early Byzantine talisman with asbestos fibres inside. SEM photograph and EDS spectrum of a single fibre (JEOL JMS-840A)

spun in S direction, silk yarns in Z direction and linen yarns in both directions. Plain is the type of weave used for the fabrication of the textiles in all cases.

Asbestos, an inorganic fibre known since antiquity for its strange properties [120, 121], is used in a talisman to protect its owner from the evil.

Table 5.14 A brief presentation of the results

Sample number	Fibre type	Fibre thickness (μm)	Yarn thickness (μm)	Spinning direction	Weaving type	Warp number/cm	Weft number/cm	Condition
1	Flax	15–20	250–300	S	Plain	24	24	Mineralized
2	Asbestos	1–5	–	–	–	–	–	Well preserved
3	Wool	10–13	150–600	S	Plain	11–12	26–27	Partly mineralized
4	Silk	5–6	250–330	Z	Plain	20–22	20–22	Partly mineralized
5a	Flax	13–18	240–270	Z	Plain	30	35	Mineralized
5b	–	–	200–250	–	Plain	16–17	16–17	Mineralized
5c	Cotton	20–25	250–300	S	Plain	12	16	Well preserved
5d	Flax	10–15	140–150	Z	Plain	20	60	Well preserved
5e	Silk	10–15	105–110	Z	Plain	25	65	Partly mineralized
6	Cotton	20–25	–	–	–	–	–	Well preserved
7	Wool	15–20	300–400	S	Plain	11	17	Well preserved
8	Wool	–	–	S	–	–	–	Well preserved

5.9 Identification of Purple by Means of Pyrolysis Gas Chromatography/Mass Spectrometry

Leopold Puchinger, Friedrich Sauter and Andreas Gössl

5.9.1 General Considerations

A multinational EU-sponsored research project [122] was dealing with ancient natural organic colouring materials applied to objects of art and archaeology, whereby the identification of these old dyes was a major aspect. The results on organic dyestuffs found on such antiquities are also of outstanding importance to help restorers with the application of adequate materials.

For the present studies, we chose Tyrian/Royal purple, as this was not only used for dyeing textiles, but also—upon conversion into a pigment—for painting purposes, thus being applied over centuries for a variety of works of art.

The precursors of purple substances can be found in several species of molluscs like *Bolinusbrandaris* and *Murex trunculus*. After isolation from the mollusc tissue the crude purpuric sap contains a variety of structurally related components (indigotin, indirubin, monobromoindigotin, monobromoindirubin and dibromoindirubin), the main colouring matter of course being 6,6′-dibromoindigotin [123].

Up to now a sample had to be taken irreversibly from the cultural object in question, being big enough to provide—upon extraction—enough material for (e.g.) HPLC analysis, a horror for everybody who wants to avoid even tiny destructions of a culturally valuable object [123].

To cope with such problems our studies were focused on pyrolysis capillary gas chromatography (Py-CGC), thus reducing right from the beginning the necessary quantity of the sample [124–137]. While in the beginning the pyrograms of the samples obtained by FID were straightforwardly identified by comparison with reference compounds [127–129], we only shortly later used a pyrolysis capillary gas chromatography/mass spectrometry (Py-CGC/MS) equipment (Py-CGC coupled to a mass spectrometer) [130–132].

Of course, this method was tested step-by-step with pure reference compounds (indigotin, indirubin, monobromoindigotin, monobromo- and dibromo-indirubin), with mixtures thereof, and with samples of the natural dyestuff in question, in particular with purple (*from Murex trunculus*).

Although all these tests yielded positive results allowing unambiguous identifications, one major problem still had to be overcome: up to this stage the sample demand of this method in fact was already very low, but it could not be considered as non-destructive.

So, we studied not only the pyrograms of the natural organic dyestuff, but also as a next step those of their possible carriers, e.g. of wool, cotton, silk, as well as of paper and parchment [133–136].

When applying the Py-CGC/MS method to purple simultaneously with its respective carrier we of course had to avoid interferences of the pyrograms. This was accomplished by using selected ion monitoring (SIM) instead of TIC mode (total ion current) of mass spectrometer. *Thus, we are able now to analyse the colouring materials by Py-CGC/MS while still fixed to their carriers,* i.e. without cutting out a large sample (since we now need less than 4 mg of some tiny fibres) and without applying any prior extraction.

Our studies showed that the Py-CGC/MS method proved to be useful and almost non-destructive for dealing with such analytical problems.

The present case study will explain the method a little more in detail and it will illustrate its efficiency by showing some examples, only for studies in which purple had to be identified.

In addition to purple we optimised the Py-CGC/MS method for altogether 20 natural organic dyes like indigo, weld, saffron, turmeric, brazilwood, logwood, madder, cochineal, dyer's bugloss, Persian berries and safflower.

5.9.2 Separation Technique

This case study was done by Py-CGC/MS. Using this technique the pure organic colouring material/the dyed textile and manuscript has to be heated in a pyrolyser unit, where degradation products from the dyeing compounds are formed or dyeing components volatise unmodified. As can be seen in Fig. 5.37 all products are transported by the carrier gas He, at first passing the hot evaporative injector and then running through a relative nonpolar fused silica column mounted in a temperature programmable oven, thus achieving their separation and after that the MS data are recorded using an appropriate chromatographic software. Some theoretical basics of chromatography and GC as well as the description of different injector and detector systems are not subject of this article and can be looked up elsewhere [134]. But what makes something the focus point of this study, the pyrolysis (Py) system and the MS, now will be illustrated a little more in detail:

To analyse a solid matter like an organic dyestuff or a dyed textile and manuscript—without its extraction —you only must combine a traditional GC instrument with a pyrolyser, which has to be mounted on a vaporising injector (Fig. 5.38a).

Analyses were carried out by a modern pyrolyser (Fig. 5.38b), if necessary suitable for "Double Shot Technique" with first thermal desorption (TD) and then followed by Py. In mode TD usually all volatile compounds are removed undecomposed running temperature programmes of the pyrolyser not higher than 300 °C. Py by constant temperature up to 800 °C happens alternatively either in "Single Shot Operation" without TD or in "Double Shot Technique (TD plus Py)".

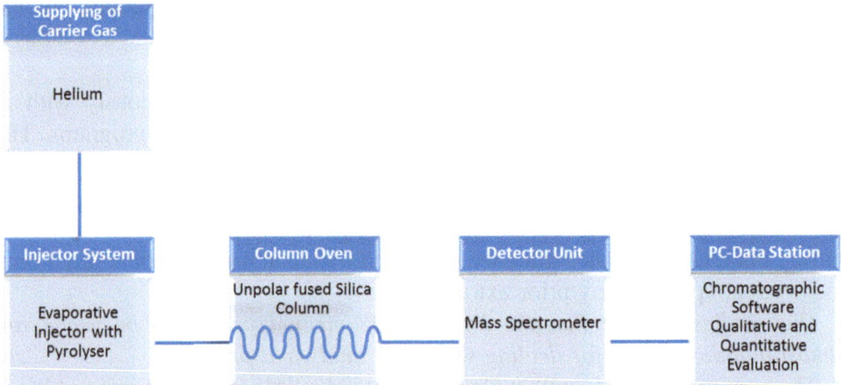

Fig. 5.37 Scheme of a pyrolysis capillary gas chromatographic working station

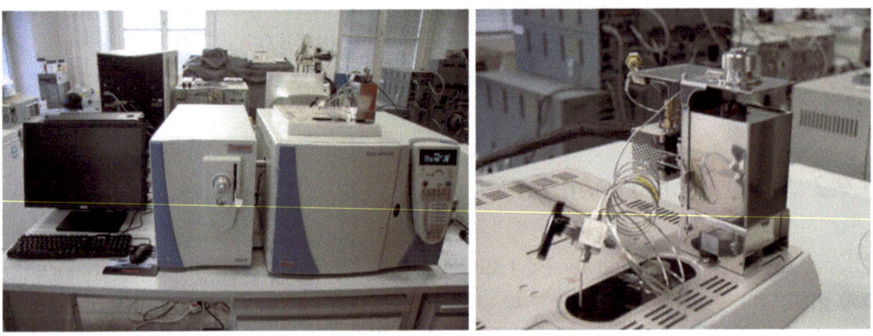

Fig. 5.38 FrontierLab-Thermo Ultra GC/MS, **a** Py-GC/MS equipment, **b** Pyrolyser unit

For TD and/or Py only a small cup has to be filled with the material of choice, so the pretreatment of a sample is less time-consuming (Fig. 5.39). In a further step the cup is connected with the sample holder by a stick and this unit is screwed onto the pyrolyser system. The combination cup-stick drops down vertically in a quartz Py tube, if a button on the top of the sampler has been pressed, and ends up in the micro-furnace (Py heater). There, immediately within a few milliseconds Py takes place. Compared to other techniques it is an advantage of micro-furnace method that evaporation, denaturation, curing and degradation do not occur during Py process.

One of the best practical options to detect the organic matter coming from the capillary column at different times due to its chemical–physical properties is the usage of a very specific detector, a quadrupole mass spectrometer (MS).

The MS process itself can be divided into for steps (Fig. 5.40):

- Ionisation of molecules by means of an electron beam produced with a filament
- Acceleration of the ions in an electric field, so that they all have the same kinetic energy

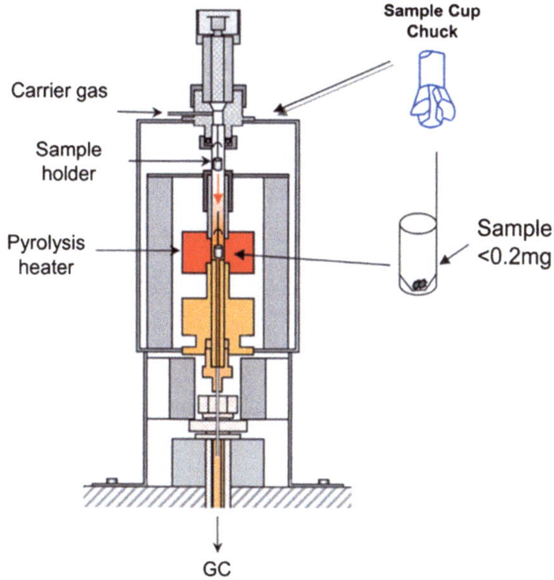

Fig. 5.39 Schematic diagram of the microfurnacepyrolyser (FrontierLab)

- Deflection of the ions by a magnetic field according to their masses (the lighter and more charged the ions are, the more they are deflected; Fig. 5.41)
- Detection of the mass/charge ratio and relative abundance of ions in a mass spectrum (stick diagram; Fig. 5.42), so that the molecule's structure and the elemental composition can be determined

For the identification of natural organic dyestuffs there should be adjusted the mode "total ion current (TIC)" of MS; however, it has to be used the mode "selected ion monitoring (SIM)" of MS, if a minor amount of colouring material must be analysed beside an excess of an organic supporting material, a situation, which can be found in dyed textiles and manuscripts.

5.9.3 Identification Technique

Samples

Purple from *Murex trunculus* delivered by a German company (Kremer Pigmente, Aichstetten, BRD: product number 36010)

Wool 1, wool 2, silk, cotton, paper and parchment, died in the vat of *Murex trunculus* and delivered by Inge Boesken Kanold (Lacoste, France)

In order to identify the components of Tyrian/Royal purple the following pure compounds were encompassed in the study (delivered by Mohammad S. Mubarak,

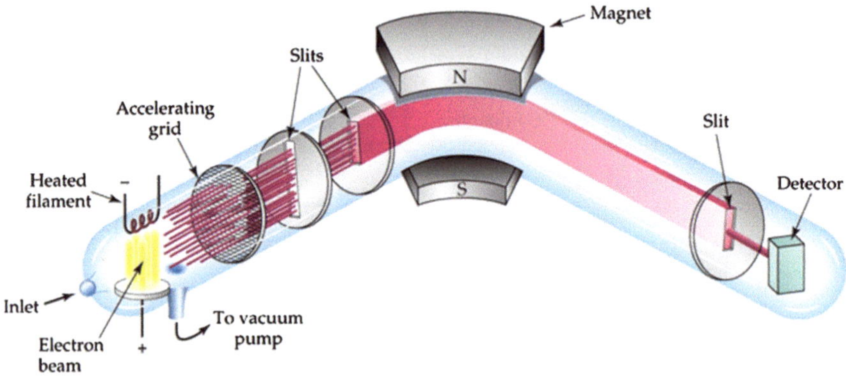

Fig. 5.40 Schematic quadrupol MS (source http://wps.prenhall.com/wps/media/objects/602/616516/Media_Assets/Chapter03/Text_Images/FG03_10.JPG)

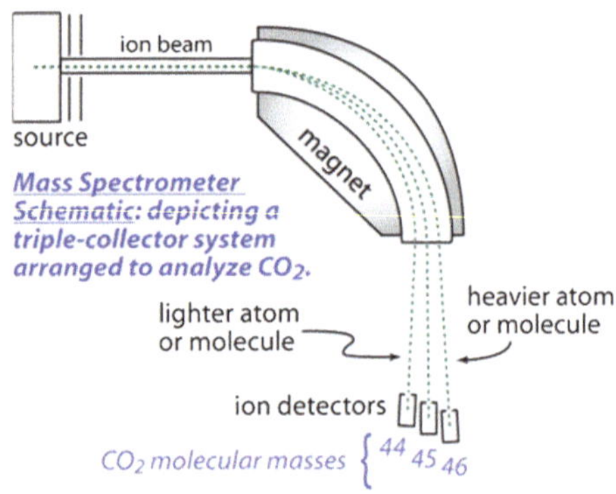

Fig. 5.41 Deflection of ions dependent on their masses (education/geochemsheets/techniques/gassourcemassspec.html)

University of Jordan, Amman): indigotin, indirubin, 6-bromoindigotin, 6,6′-dibromoindigotin, 6-bromoindirubin and 6,6′-dibromoindirubin (Fig. 5.43). The picture at the top right shows three vessels from the "Historical Wiesner Collection", which is in the ownership of the Vienna University of Technology.

Thermal Desorption (TD)/Pyrolysis (Py):

"Double-Shot Pyrolyser" (FrontierLab 2010D) in free fall technique

TD: programme: 50 °C (2 min), heating rate: up to 300 °C (25 °C/min), holding temperature: 300 °C (3 min); interface 330 °C

Py: temperature: 750 °C; time: 2 min; interface 330 °C

CGC:

"Thermo Trace GC Ultra"; column: Thermo TR-5 ms SQC, 15 m × 0.25 mm; film thickness: 0.25 μm; carrier gas: He; constant flow: 0.9 ml/min; injector temperature: 320 °C;

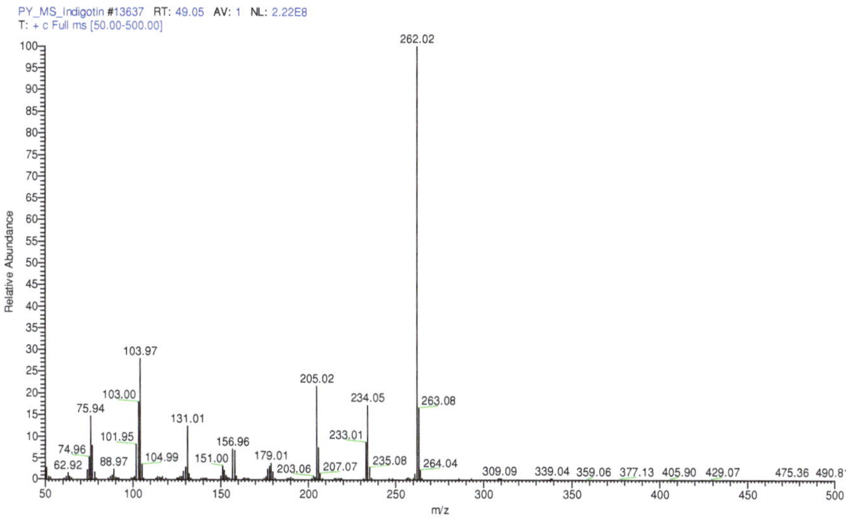

Fig. 5.42 Mass spectrum of indigotin

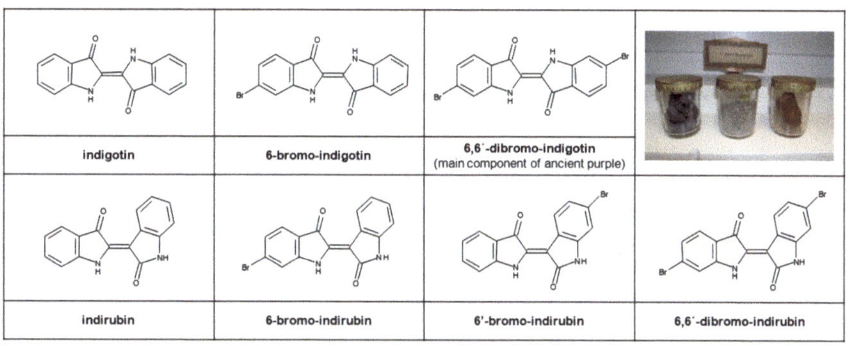

Fig. 5.43 Structure of studied indigoid components

oven programme *TD*: 35 °C (15 min), heating rate 1: up to 200 °C (25 °C/min), heating rate 2: up to 300 °C (10 °C/min), holding temperature: 300 °C (18.4 min)

oven programme *Py*: 50 °C (2 min), heating rate: up to 300 °C (5 °C/min), holding temperature: 300 °C (18 min)

MS: "Thermo DSQ II Single Quadrupole Mass Spectrometer"; EI: 70 eV; scan rate: 2.7 scan/s; TIC or SIM mode; ion source temperature: 250 °C; NIST library

Mass numbers adjusted in SIM mode:

TD: 5–20.5 min (89, 90, 198, 200), 20.5–22.5 min (115, 221, 223), 22.5–30 min (222, 237, 250, 265), 30–50 min (104, 205, 340, 342)

Py: 1.5–10 min (77, 156, 158), 10–20 min (102, 181, 183), 20–70 min (376, 390, 392, 406)

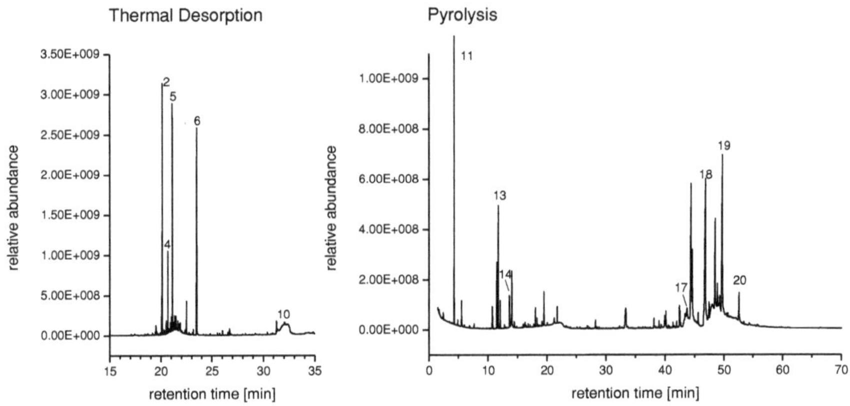

Fig. 5.44 TIC-chromatograms (TD and Py) of purple from *Murex trunculus*

Sample amount for Py-CGC/MS:
0.2–0.6 mg for pure dyestuffs
2–4 mg for dyed textiles and manuscripts

5.9.4 Applying the Techniques on Materials

5.9.4.1 TD/Py-CGC/MS Results of Purple from *Murex Trunculus*

By means of high pressure liquid chromatography (HPLC) indigotin, indirubin as well as the corresponding mono- and dibromo-compounds were found in purple [123].

Similar to HPLC TD/Py-CGC/MS allows an unambiguous identification of purple from other natural organic dyestuffs (Fig. 5.44).

Both modes of this method, TD and Py, are suitable to distinguish between all components occurring in purple.

But only by use of Py-bromobenzene (m/z 77/156/158; peak number 11), bromobenzonitrile (m/z 102/181/183; peak number 13), indole (m/z 90/117; peak number 14), indigotin (m/z 104/205/262; peak number 17) and dibromoindigotin (m/z 75/420; peak number 20) could clearly be identified in the sample of *Murex trunculus* all stemming from indigoid components (Table 5.15). 6, 6′-dibromoindigotin was the only one from all together five indigoid bromine containing reference compounds found in higher quantities in *Murex trunculus*.

The following substances were identified after TD of purple: bromonitrotoluene (m/z 89/90/198/200; peak number 2), bromoformanilide (m/z 171/173/199//201; peak number 4), bromomethylquinoline (m/z 115/221/223; peak number 5) and bromoindigotin (m/z 104/205/340/342; peak number 10; Table 5.16). The latter

Table 5.15 Components found in purple by pyrolysis (Py)

Studied materials	Substances and their *m/z* found by pyrolysis									
	Bromobenzene	Aniline	Bromobenzo-nitrile	Indole	Dihydro-indolone	Phenylimi-dazoisoindole	Indigotin	Unknown-bromine component	Unknownbromine-component	Dibromoin-digotin
	77/156/158	66/93	102/181/183	90/117	104/133	232	104/205/262	75/406	188/189/376	75/420
Peaknumber→	11	12	13	14	15	16	17	18	19	20
Dyestuffs										
Murex trunculus	*i*		*m*	*w*			*w*	*m*	*m*	*w*
Reference compounds										
Indigotin		*m*		*w*	*w*		*i*			
Indirubin		*i*		*i*	*w*	*w*	*w*			
6-Bromoindigotin	*w*	*w*	*w*	*w*	*w*		*w*	*m*		*w*
6,6′-Dibromoindigotin	*i*		*m*	*w*			*w*	*i*	*i*	*w*
6-Bromoindirubin	*w*		*w*	*m*	*w*				*w*	
6′-Bromoindirubin	*w*		*w*	*w*	*m*					
6,6′-Dibromoindirubin	*m*		*m*	*w*	*w*				*i*	

w weak, *m* medium, *i* intensive

Table 5.16 Components found in purple by thermal desorption (TD)

TStudied Materials	Substances and their m/z found by Thermal Desorption									
	Indole	Bromo-nitrotoluene	Dihydroindolo	Bromofor-manilide	Bromo-methylquinoline	Unknownbromine-component	Phenylimidaz-oisoindole	Indirubin	Indigotin	Bromo-indigotin
	90/117	89/90/198/200	104/133	171/173/199/201	115/221/223	222/237/250/265	232	205/234/262	104/205/262	104/205/340/342
Peaknumber	1	2	3	4	5	6	7	8	9	10
Dyestuffs										
Murextrunculus		*i*		*m*	*i*	*i*				*w*
Reference Compounds										
Indigotin	*w*		*w*						*i*	
Indirubin			*m*					*i*		
6-bromo-indigotin			*w*						*m*	*i*
6,6′-dibromo-indigotin									*i*	*m*
6-bromo-indirubin									*w*	
6′-bromo-indirubin			*w*						*w*	
6,6′-dibromo-indirubin										

w weak, *m* medium, *i* intensive

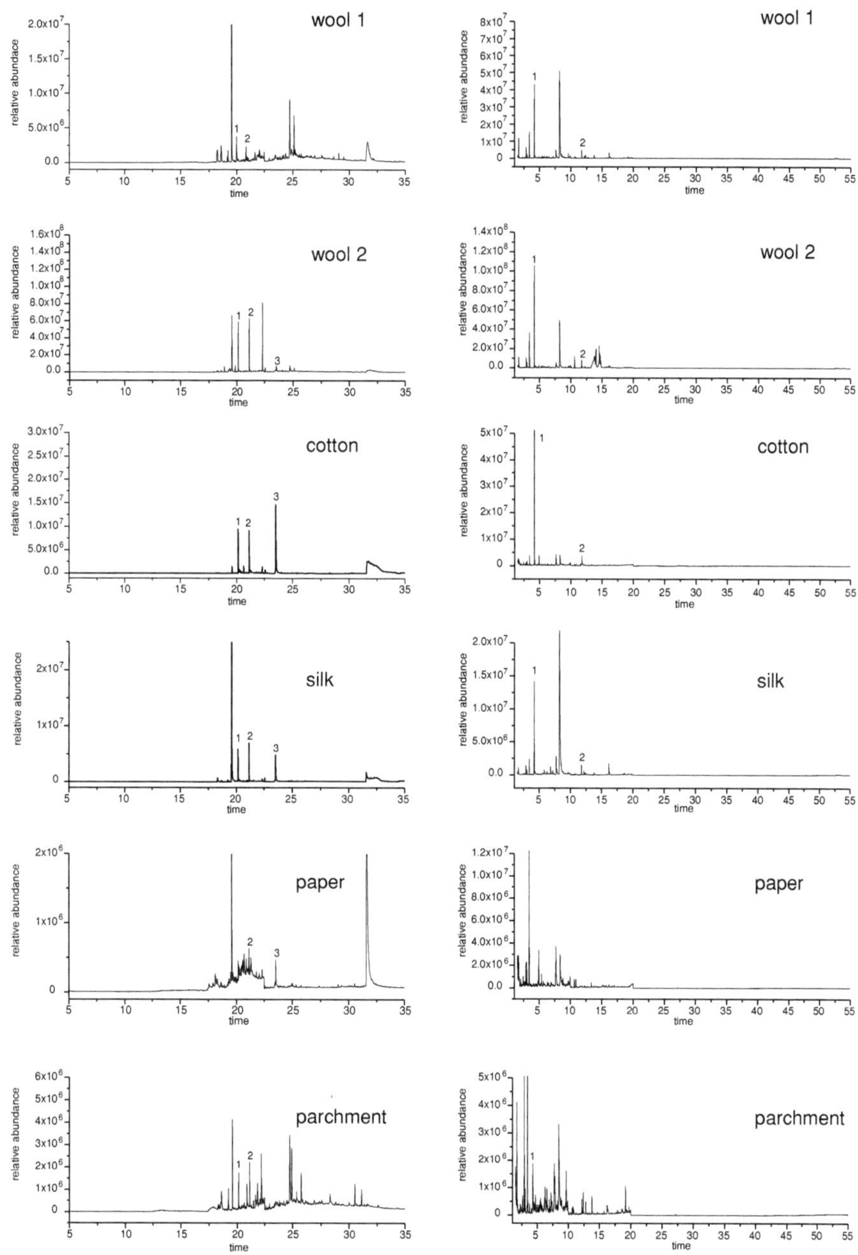

Fig. 5.45 Chromatograms from TD (on the *left*) and pyrograms from Py (on the *right*) of wools 1 and 2, cotton, silk, paper and parchment dyed with purple; SIM mode; TD: 1 bromonitrotoluene, 2 bromoethylquinoline, 3 bromoindigotin; Py: 1 bromobenzene, 2 bromobenzonitrile

may be a product of 6-bromoindigotin or 6, 6′-dibromoindigotin, while all other substances are non-indigoid components of purple.

5.9.4.2 TD/Py-CGC/MS Results from the Direct Identification of Purple on Textiles and Manuscripts

As can be shown in Fig. 5.45 even here two methods are practicable using the Frontier Lab pyrolyser for the detection of purple on textiles and manuscripts: TD and Py operating one after the other and making full use of the same sample.

Once more it has to be taken into account that purple will be found only in small quantities beside the dominant organic substances of the organic raw materials from wool, cotton, silk, paper and parchment. Therefore, the characteristic masses for purple components have to be adjusted in SIM mode of Py-CGC/MS (Chap. 4), otherwise the dyestuff pattern could not be distinguished from the complex organic matrix.

By means of TD the following pyrolysates were found: bromonitrotoluene (m/z 89/90/198,200), bromomethylquinoline (m/z 115/221/223) and 6-bromoindigotin (m/z 222/237/25/265).

Bromobenzene (m/z 77/156/158) and bromobenzonitrile (m/z 102/181/183) could clearly be identified by method of Py.

Considering Py-CGC/MS results from earlier studies in addition to our present analyses it has to be concluded that the synthetic purple used for dyeing of wool 1 was pure 6, 6′-dibromoindigotin, the main colouring component of Tyrian purple.

All other materials, cotton and silk as well as paper and parchment were dyed with natural purple.

Moreover, the chromatograms/pyrograms indicate differences in the intensity of purple components dependent on the materials probably referring to reactions of the dye with the supporting materials during TD and Py.

Thus, the efficiency of the described method could be ascertained.

References

1. Schweppe H (1979) Identification of dyes on old textiles. J Am Inst Conserv 19(1):14–23
2. Wouters J (1985) High-performance liquid chromatography anthraquinones: analysis of plant and insect extracts and dyed textiles. Stud Conserv 30:119–128
3. Wouters J, Verhecken A (1989) The coccid insect dyes: HPLC and computerized diode-array analysis of dyed yarn. Stud Conserv 34:189–200
4. Wouters J, Rosario-Chirinos N (1992) Dye analysis of pre-Columbian Peruvian textiles with high-performance liquid chromatography and diode-array detection. J Am Inst Conserv 31:237–255
5. Koren ZC (1994) HPLC analysis of the natural scale insect, madder and indigoid. J Soc Dyers Colour 110:273–277
6. Koren ZC (1995) An efficient HPLC analysis scheme for plant and animal red, blue and purple dyes. Dyes Hist Archaeol 13:27–37

7. Surowiec I, Quye A, Trojanowicz M (2006) Liquid chromatography determination of natural dyes in extracts from historical Scottish textiles excavated from peat bogs. J Chromatogr A 1112:209–217
8. Orska-Gawryś J, Surowiec I, Kehl J, Rejniak H, Urbaniak-Walczak K, Trojanowicz M (2003) Identification of natural dyes in archeological Coptic textiles by liquid chromatography with diode-array detection. J Chromatogr A 989:239–248
9. Karapanagiotis I, Lakka A, Valianou L, Chryssoulakis Y (2008) High-performance liquid chromatographic determination of colouring matters in historical garments from the Holy Mountain of Athos. Microchim Acta 160(4):477–483
10. Karapanagiotis I, Mantzouris D, Chryssoulakis Y, Saadeh HA, Alawi MA, Mubarek MS, Karadag R, Yurdun T, Alsaad Z, Abdel-Kareem O, Puchinger L, Sauter F (2009) Inter-laboratory chemical study of natural materials from the historical Wiesner collection. JJC 4(2):195–208
11. Deveoglu O, Karadag R, Yurdun T (2009) Preparation and HPLC analysis of the natural pigments obtained from Buckthorn (*Rhamnus petiolaris* Boiss) dye plants. JJC 4(4):377–385
12. Karadag R, Torgan E, Yurdun T (2010) Formation and HPLC analysis of the natural lake pigment obtained from Madder (Rubia tinctorum L.). Rev Anal Chem 29(1):1–12
13. Yurdun T, Karadag R, Dolen E, Mubarak MS (2011) Identification of natural yellow, blue, green and black dyes in 15th–17th centuries Ottoman silk and wool textiles by HPLC with diode array detection. Rev Anal Chem 30:153–164
14. Valianou L, Karapanagiotis I, Chryssoulakis Y (2009) Comparison of extraction methods for the analysis of natural dyes in historical textiles by high-performance liquid chromatography. Anal Bioanal Chem 395(7):2175–2189
15. Mantzouris D, Karapanagiotis I, Valianou L, Panayiotou C (2011) HPLC-DAD-MS analysis of dyes identified in textiles from Mount Athos. Anal Bioanal Chem 399(9):3065–3079
16. Puchalska M, Połeć-Pawlak K, Zadrozna I, Hryszko H, Jarosz M (2004) Identification of indigoid dyes in natural organic pigments used in historical art objects by high-performance liquid chromatography coupled to electrospray ionization mass spectrometry. J Mass Spectrom 39(12):1441–1449
17. Pawlak K, Puchalska M, Miszczak A, Rosłoniec E, Jarosz M (2006) Blue natural organic dyestuffs—from textile dyeing to mural painting. Separation and characterization of coloring matters present in elderberry, logwood and indigo. J Mass Spectrom 41(5):613–622
18. Zhang X, Good I, Laursen R (2008) Characterization of dyes in ancient textiles from Xinjiang. J Archaeol Sci 35:1095–1103
19. Szostek B, Orska-Gawryś J, Surowiec I, Trojanowicz M (2003) Investigation of natural dyes occurring in historical Coptic textiles by high-performance liquid chromatography with UV-Vis and mass spectrometric detection. J Chromatogr A 1012:179–192
20. Balakina GG, Vasiliev VG, Karpova EV, Mamatyuk VI (2006) HPLC and molecular spectroscopic investigations of the red dye obtained from an ancient Pazyryk textile. Dyes Pigment 71:54–60
21. Marques R, Sousa MM, Oliveira MC, Melo MJ (2009) Characterization of weld (Reseda luteola L.) and spurge flax (Daphne gnidium L.) by high-performance liquid chromatography-diode array detection-mass spectrometry in Arraiolos historical textiles. J Chromatogr A 1216(9):1395–1402
22. Blanc R, Espejo T, López-Montes A, Torres D, Crovetto G, Navalón A, Vílchez JL (2006) Sampling and identification of natural dyes en historical maps and drawings by liquid chromatography with diode-array detection. J Chromatogr A 1122:105–113
23. Sist. Daniilia, Minopoulou E, Andrikopoulos KS, Karapanagiotis I, Kourouklis GA (2008) Evaluating a Cumaean Sibyl:Domenichino or later? A multi analytical approach. Anal Chim Acta 611:239–249
24. Karapanagiotis I, sist. Daniilia, Tsakalof A, Chryssoulakis Y (2005) Identification of red natural dyes in Post-Byzantine icons by HPLC. J Liq Chromatogr R T 28:739–749

25. Karapanagiotis I, Valianou Y, sist. Daniilia, Chryssoulakis Y (2007) Organic dyes in Byzantine and post-Byzantine icons from Chalkidiki (Greece). J Cult Herit 8:294–298
26. Valianou L, Wei S, Mubarak MS, Farmakalidis H, Rosenberg E, Stassinopoulos S, Karapanagiotis I (2011) Identification of organic materials in icons of the Cretan School of iconography. J Archaeol Sci 38(2):246–254
27. Halpine SM (1996) An improved dye and lake pigment analysis method for high-performance liquid chromatography and diode-array detector. Stud Conserv 41:76–94
28. Ford BL (1992) Monitoring colour change in textiles on display. Stud Conserv 37(1):1–11
29. Cristea D, Vilarem G (2006) Improving light fastness of natural dyes on cotton yarn. Dyes Pigment 70:238–245
30. Cardon D (2007) Natural dyes-sources, tradition, technology and science. Archetype, London
31. Karapanagiotis I, Theologou J, Lakka A, Ozoline A, Panayiotou C (2011) Investigation of the colouring materials of Fustat carpet fragments. Archaeometry 53(3):587–599
32. Kumar JK, Sinha AK (2004) Resurgence of natural colourants: a holistic view. Nat Prod Lett 18(1):59–84
33. González M, Lobo MG, Méndez J, Carnero A (2005) Detection of colour adulteration in cochineals by spectrophotometric determination of yellow and red pigment groups. Food Control 16:105–112
34. Chengaiah B, Rao KM, Kumar KM, Alagusundaram M, Chetty CM (2010) Medicinal importance of natural dyes—a review. Int J PharmTech Res 2(1):144–154
35. Septhum C, Rattanaphani V, Rattanaphani S (2007) UV–Vis spectroscopic study of natural dyes with alum as a mordant. Suranaree J Sci Technol 14(1):91–97
36. Surowiec I, Orska-Gawryś J, Biesaga M, Trojanowicz M, Hutta M, Halko R, Urbaniak-Walczak K (2003) Identification of natural dyestuff in archeological coptic textiles by HPLC with fluorescence detection. Anal Lett 36(6):1211–1229
37. Cardon D (2007) Natural dyes. Archetype Publications, London
38. Ferreira ESB, Hulme AN, McNab H, Quye A (2004) The natural constituents of historical textile dyes. Chem Soc Rev 33:329–336
39. Karadag R (2007) Dogal Boyamacilik. T.C. Kultur ve Turizm Bakanligi, Ankara
40. De Santis D, Moresi M (2007) Production of alizarin extracts from *Rubia tinctorum* and assessment of their dyeing properties. Ind Crops Prod 26:151–162
41. Deveoglu O, Torgan E, Karadag R (2010) Identification by HPLC-DAD of dyestuffs in the natural pigments produced with Al^{+3}, Fe^{+2} and Sn^{+2} mordant metals from cochineal (*Dactylopius coccus* Costa) and walloon oak (*Quercus ithaburensis* Decaisne). Asian J Chem 22(9):7021–7030
42. Karadag R, Dolen E (2008) Identification of natural dyes in historical textiles by liquid chromatography. Chemistry and Conservation Science, Proceeding: second residential summer school Palermo (Italy)
43. Deveoglu O, Karadag R, Yurdun T (2011) Qualitative HPLC determination of main anthraquinone and lake pigment contents from *Dactylopius coccus* dye insect. Chem Nat Compd 47(1):103–104
44. Schweppe H (1993) Handbuch der Naturfarbstoff. Ecomed, Landsberg
45. Mikropoulou E, Tsatsaroni E, Varella EA (2009) Revival of traditional European dyeing techniques yellow and red colorants. J Cult Heritage 10:447–457
46. Zarkogianni M, Mikropoulou E, Varella EA (2010) Colour and fastness of natural dyes: revival of traditional dyeing techniques. Coloring Technol 127:18–27
47. Wouters J (1985) High-performance liquid chromatography anthraquinones: analysis of plant and insect extracts and dyed textiles. Stud Conserv 30:119–128
48. Wouters J, Verhecken A (1989) The coccid insect dyes: HPLC and computerized diode-array analysis of dyed yarn. Stud Conserv 34:189–200
49. Wouters J, Verhecken A (1989) The scale insect dyes (Homoptera:Coccoidea): species recognition by HPLC and diode-array analysis of the dyestuffs. Annales de la Sociét Entomologique de Française 25(4):393–410

50. Halpine SM (1996) An improved dye and lake pigment analysis method for high-performance liquid chromatography and diode-array detector. Stud Conserv 41:76–94
51. Karapanagiotis I, Daniilia S, Tsakalof A, Chryssoulakis Y (2005) Identification of red natural dyes in post-Byzantine icons by HPLC. J Liq Chromatogr Relat Technol 28:739–749
52. Karapanagiotis I, Mantzouris D, Chryssoulakis Y, Saadeh HA, Alawi MA, Mubarak MS, Karadag R, Yurdun T, AlSaad Z, Abdel-Kareem O, Puchinger L, Sauter F (2009) Interlaboratory chemical study of natural materials from the historical Wiesner collection. Jordan J Chem 4(2):195
53. Deveoglu O, Karadag R, Yurdun T (2009) Preparation and HPLC analysis of the natural pigments obtained from buckthorn (*Rhamnus petiolaris* Boiss) dye plants. Jordan J Chem 4(4):377–385
54. Deveoglu O, Torgan E, Karadag R (2010) Characterization of colouring matters by HPLC-DAD and colour measurements, preparation of lake pigments with *Ararat kermes* (*Porphyrophora hameli* Brand). Jordan J Chem 5(3):307–315
55. Deveoglu O, Sahinbaskan BY, Torgan E, Karadag R (2011) Dyeing properties and analysis by RP-HPLC-DAD of silk fibers dyed with weld (*Reseda luteola* L.) and walloon oak (*Quercus ithaburensis* Decaisne). Asian J Chem 23(12):5441–5446
56. Pollard A, Heron C (1996) Archaeological chemistry. The royal society of chemistry, RSC Paperbacks, Cambridge
57. Mills JS, White R (1994) The organic chemistry of museum objects, 2nd edn. Butterworth Heinemann, Oxford
58. Mathe C, Connan J, Archier P, Mouton M, Vieillescazes C (2007) Analysis of frankincense in archaeological samples by gas chromatography mass spectrometry. Anal Chim 97:433–445. doi:10.1002/adic.200790029
59. Mathe C, Culioli G, Archier P, Vieillescazes C (2004) High performance liquid chromatographic analysis of triterpenoids in commercial frankincense. Chromatographia 60:493–499
60. Mathe C, Archier P, Nehmé L, Vieillescazes C (2009) Study of Nabataean organic residues from Madâ'in Sâlih, ancient Hegra by gas chromatography mass spectrometry. Archaeometry 51:629–636. doi:10.1111/j.1475-4754.2008.00417
61. Lucas A (1989) Ancient Egyptian materials and industries. E. Arnold and Co., London
62. Forbes RJ (1993) Studies in ancient technology. Brill EJ, Leiden
63. de Morgan JJ (1903) Fouilles à Dashour, (I) mars juin 1894, Vienne (1895), (II) 1894–1895, Vienne
64. Vieillescazes C, Coen S (1993) Caractérisation de quelques résines utilisées en Egypte ancienne. Stud Conserv 38:255–264
65. Archier P, Vieillescazes C (2000) Characterisation of various geographical origin incense based on chemical criteria. Analusis 28:233–237
66. Martin P, Achier P, Vieillescazes C, Pistre MS (2001) HPLC coupled with fluorimetric detection for the identification of natural resins in archaeological materials. Chromatographia 53:380–384
67. Dev S, Misra R (1986) Handbook of terpenoids: diterpenoids, vol III. CRC, Boca Raton
68. Colombini MP, Modugno F (2009) Organic materials in art and archaeology. In: Colombini MP, Modugno F (eds) Organic mass spectrometry in art and archaeology. John Wiley and sons
69. Gough LG (1964) Conifer resin constituents. Soc Chem Ind 12:2059–2060
70. Scalarone D, Lazzari M, Chiantore O (2002) Ageing behaviour and pyrolytic characterisation of diterpenic resins used as art materials: colophony and Venice turpentine. J Anal Appl Pyrolysis 64:345–361
71. Van der Berg KJ (1996) State oxidation of diterpenoid Pinaceae resins in varnish, Max lining material, 18th century resin oil paint and a recent copper resinate glaze. ICOM Committee for Conservation, 11th triennial meeting, Edinbourg, vol 2, pp 930–937

72. Marchand-Geneste N, Carpy A (2003) Theoretical study of the thermal degradation pathways of abietane skeleton diterpenoids: aromatization to retene. J Mol Struct (Theochem) 635:55–82
73. Mahajan B, Taneja SC, Sethi VK, Dhar KL (1995) Two triterpenoids from *Boswellia serrata* gum resin. Phytochemistry 39:453–455
74. Proietti G, Strappaghetti G, Corsano S (1981) Triterpenes of Boswellia frereana. Plant Med 41:417–418
75. Mills JS (1966) The gas chromatographic examination of paint media. Part 1 fatty acid composition and identification of dried oil film. Stud Conserv 11:92–107
76. Colombini MP, Modugno F (2009) Organic mass spectrometry in art and archaeology. Wiley, Chichester
77. Colombini MP, Modugno F, Giacomelli M (1999) Characterisation of proteinaceous binders and drying oils in wall painting samples by gas chromatography-mass spectrometry. J Chromatogr A 846:113–124
78. Colombini MP, Modugno F, Menicagli E, Fuoco R, Giacomelli A (2000) GC-MS characterization of proteinaceous and lipid binders in UV aged polychrome artifacts. Microchem J 67:291–300
79. Colombini MP, Fuoco R, Giacomelli A, Muscatello B (1998) Characterization of proteinaceous binders in wall painting samples by microwave-assisted acid hydrolysis and GC-MS determination of amino acids. Stud Conserv 43:33–41
80. Tsakalof A, Bairachtari K, Sister Daniilia, Chrissoulakis I (2003) Identification of binding media in art objects: Protaton church and N.Gyzis' paintings, 4th Symposium on Archeometry, Athens, Greece, Book of Abstracts, p 67
81. Mills JS, White R (1994) The organic chemistry of museum objects. Butterworth-Heinemann, London
82. Grattan DW (1980) The oxidative degradation of organic materials and its importance in deterioration of artifacts. J IIC-CG 4(1):17–26
83. Van den Berg JDJ (2002) Analytical chemical studies on traditional linseed oil paints. PhD Thesis, University of Amsterdam
84. Casoli A, Musini PC, Palla G (1996) GC/MS approach to the problem of characterizing binding media in painting. J Chromatogr A 731:237–246
85. Schilling MR, Khanjian HP, Carson DM (1997) Fatty acids and glycerol content of lipids; effects of aging and solvent extraction on the composition of oil paints. Techne 5:71–78
86. Rasti F, Scott G (1980) The effects of some pigments on the Photooxidation of linseed oil-based paint media. Stud Conserv 25:145–146
87. Tsakalof A, Bairachtari K, Chryssoulakis I (2006) Pitfalls in drying oils identification in art objects by gas chromatography. J Sep Sci 29:1642–1646
88. Paci M, Federici C, Capitani D, Perenze N, Segre AL (1995) NMR Study Paper Carbohydr Polym 26(4):289–297
89. Odlyha M, Cohen NS, Foster GM, Aliev A, Verdonck E, Grandy D (2003) Dynamic mechanical analysis (DMA), 13C solid state nmr and micro-thermomechanical studies of historical parchment. J Therm Anal Calorim 71:939–950
90. Di Natale R (2003) Le Cinquecentine della Biblioteca dell'Archivio di Stato di Palermo, CRPR
91. Daniels V (2006) Conservation science: heritage materials. RSC Publishing, Cambridge
92. Paci M, Federici C, Capitani D, Perenze N, Segre AL (1995) NMR Study Paper. Carbohydr Polym 26:289–297
93. Capitani D, Segre AL, Attanasio D, Blicharska B, Focher B, Capretti G (1996) 1H NMR relaxation study of paper as a system of cellulose and water. Tappi J 79(6):113–122
94. Capitani D, Proietti N, Ziarelli F, Segre AL (2002) NMR Study of Water-Filled Pores in one of the most widely used polymeric material: the paper. Macromolecules 35(14):5536–5543
95. Vanderhart D, Atalla R (1984) Studies of microstructure in native celluloses using solid-state ^{13}C NMR. Macromolecules 17:1465–1472

96. Capitani D, Emanuele MC, Bella J, Segre AL, Attanasio D, Focher B, Capretti G (1999) 1H NMR relaxation study of cellulose and water interaction in paper. Tappi J 82(9):117–124
97. Capitani D, Segre AL, Attanasio D, Blicharska B, Focher B, Capretti G (1996) 1H NMR relaxation study of cellulose and water interaction in paper. Tappi J 79(6):113–122
98. Larsen R (2002) Microanalysis of Parchment. Archetype Publications Ltd, London
99. Miles CA, Ghelashvili M (1999) Polymer-in-a-box mechanism for the thermal stabilization of collagen molecules in fibres. Biophys J 76:3243–3252
100. Hartmann SR, Hahn EL (1962) Nuclear double resonance in the rotating frame. Phys Rev 128:2042–2053
101. Massiot D, Fayon F, Capron M, King I, LeCalve' S, Alonso B, Durand JO, Bujoli B, Gan Z, Hoatson G (2002) Modelling one- and two-dimensional solid-state NMR spectra. Magn Reson Chem 40:70–76
102. Maunu S, Liitia T, Kauliomaki S, Hortling B, Sundquist J (2000) 13C CPMAS NMR investigations of cellulose polymorphs in different pulps. Cellulose 7:147–159
103. Horii F, Hirai A, Kimaru R (1982) Solid-state high resolution 13C-NMR studies of regenerated cellulose samples with different crystallinities. Polym Bull 8:163–170
104. Aliev AE (2005) Solid-state NMR studies of collagen-based parchments and gelatin. Biopolymers 77(4):230–245
105. Ghioni C, Hilleri JC, Kennedy CJ, Aliev AE, Odlyha M, Boulton M, Wess TJ (2005) Evidence of a distinct lipid fraction in historical parchments: a potential role in degradation? J. Lipid Res 46:2726–2734
106. Müller M, Murphy B, Burghammer M, Snigireva I, Riekel C, Gunneweg J, Pantos E (2008) Identification of single archaeological textile fibres from the cave of letters using synchrotron radiation microbeam diffraction and microfluorescence. Appl Phys A Mater Sci Process 83(2):183–188
107. Karatzani A, Rehren T (2006) The use of metal threads and decorations in Byzantine-Greek Orthodox ecclesiastical textiles. J Miner Metals Mater Soc 58(5):34–37
108. Dimbleby GW (1967) Plants and archaeology, chapter 7: the preservation of plant materials. Royal Opera Arcade, London
109. Peacock EE (1996) Biodegradation and characterization of water-degraded archaeological textiles created for conservation research. Int Biodeterior Biodegradation 38(1):49–59
110. Wild JP (2003) Textiles in archaeology. Shire Publications, Buckinghamshire
111. Cronyn J, Pye E, Watson J (1985) The recognition and identification of traces of organic materials in association with metal artefacts, *Research Report* 58. Council for British Archaeology,
112. Caple C, Dungworth D (1997) Investigations into Waterlogged Burial Environments, Archaeological Sciences, Proceedings of a conference on the application of scientific techniques to the study of archaeology, Liverpool, July 1995. Oxbow Monograph 64:233–240
113. Tuross N, Dillehay TD (1995) The mechanism of organic preservation at Monte Verde, Chile, and one use of biomolecules in archaeological interpretation. J Field Archaeol 22:97–100
114. Chen HL, Jakes KA, Foreman DW (1998) Preservation of archaeological textiles through fibre mineralization. J Archaeol Sci 25:1015–1021
115. Anheuser K, Roumeliotou M (2003) Characterisation of mineralised archaeological textile fibres through chemical staining. Conservator 27:23–33
116. Janaway R (1983) Textile fibre characteristics preserved by metal corrosion: the potential of S.E.M. studies. Conservator 7:48–52
117. Batigne R, Bellinger L (1953) The significance and technical analysis of ancient textiles as historical documents. Proc Am Philos Soc 97(6):670–680
118. Rodman AO (1992) Textiles and ethnicity. Latin Am Antiq 3(4):316–340
119. Good I (2001) Archaeological textiles: a review of current research. Ann Rev Anthropol 30:209–226

120. Cook JG (1993) Handbook of textile fibres: natural fibres, vol I. Merrow Publishing Co., Wiltshire
121. Goffer Z (2007) Archaeological chemistry, 2nd edn. Wiley-Interscience, Hoboken
122. EU research project Investigation. Revival and optimisation of traditional mediterranean colouring technology for the conservation of the cultural heritage (Med-COLOUR-TECH; Inco CT 2005 015406), 2006–2009
123. Hofenk de Graaff JH (2004) The colourful past-origins, chemistry and identification of natural dyestuffs. Handbook, Abegg-Stiftung. Archetype Publications Ltd. 6 Fitzroy Square London W1T 5HJ, ISBN 3-905014 (Abegg-Stiftung), ISBN 1-873132-13-1 (Archetype)
124. Puchinger L, Leichtfried D, Stachelberger H (2002) Pyrolysis capillary gas chromatography (Py-CGC) of historical parchment samples. In: Rene L (ed) Microanalysis of parchment. Archetype Publications Ltd., London, ISBN 1-873132-68-9, pp 155–158
125. Puchinger L, Sauter F, Varmuza K, Leder S (2007) Studies in organic archaeometry VII—differentiation of wood and bark pitches by pyrolysis capillary gas chromatography (Py-CGC). Anal Chim 97(7):513–525
126. Moldoveanu SC (1998) Analytical pyrolysis of organic polymers-techniques and instrumentation in analytical chemistry: handbook, vol 20. Elsevier, Amsterdam
127. Puchinger L, Sauter F, Gössl A, Mubarak MS (2007) Pyrolysis capillary gas chromatography (Py-CGC) of Indigoid Dyestuffs. XII. Blue Danube symposium on heterocyclic chemistry, Club Tihany Hotel, Tihany (Hungary), Book of abstracts, ISBN 978-963-7067-15-0, PO-63, p 112
128. Puchinger L, Sauter F, Gössl A, Hofbauer N (2007) Chemische Identifizierung alter organischer Farbstoffe (untersucht an Farbstoffen des Indigo-Typs). Internationales ÖGUF-Symposium 2007—Zerstörungsfreie Archäologie, Aktuelle Methoden und Analysenverfahren im Einsatz der archäologischen Forschung, Naturhistorisches Museum Wien, Abstracts
129. Puchinger L, Sauter F, HofbauerN, Mubarak MS (2007) Studies in organic archaeometry VIII: microanalysis of indigoid dyestuffs by pyrolysis capillary gas chromatography. Petra international chemistry conference (PICC) and transmediterranean colloquium on heterocyclic chemistry (TRAMECH-5), Tafila (Jordan), Abstract book PO-52
130. Puchinger L, Sauter F,Gössl A (2009) PY-CGC/MS—a new method for analyzing organic dyestuffs. 28th meeting of dyes in history and archaeology (DHA 28). Institute of Natural Fibres & Medicinal Plants, Poznan (Polen), Book of abstracts, p 42
131. Sauter F, Puchinger L (2008) Analysis of indigo and related dyestuffs by PY-CGC/MS. 11th IbnSina international conference on pure and applied heterocyclic chemistry, Ain Shams University in Cairo (Egypt), Book of abstracts p 127, POII-21
132. Karapanagiotis I, Puchinger L, Sauter F (2009) Inter-laboratory chemical study of natural materials from the historical Wiesner collection. Jordan J Chem 4(2):195–208
133. Puchinger L, Sauter F (2008) Direct identification of indigo on textiles by means of PY-CGC/MS. 27th Meeting of dyes in history and archaeology (DHA 27), Marmara University in Istanbul (Turkey), book of abstracts, p 62; P-14
134. Puchinger L (2008) GC- and PY-CGC/MS for the determination of historical objects. In: Varella EA, Caponetti E (eds) Proceedings of the 2nd residential summer school: chemistry and conservation science 2008, Palermo. ISBN 978-88-86208-60-4, pp 199–214
135. Puchinger L, Sauter F, Gössl A, BoeskenKanold I (2009) PY-CGC/MS—a new method to identify purple on textiles and manuscripts. 13th Blue Danube symposium on heterocyclic chemistry. Program, abstract of papers, list of participants. Bled (Slovenia), p 116, PO-58
136. Puchinger L, Karapanagiotis I, Mantzouris D, Puchinger M (2010) Dyeing materials used for a Mediterranean Carpet and a traditional Greek child costume around 1900. Dyes in history & archaeology—DHA 29, Lisbon, Spain, Poster pp 78–79
137. Fabbri D, Chiavari G, Ling H (2000) Analysis of anthraquinoid and indigoid dyes used in ancient artistic works by thermally assisted hydrolysis and methylation in the presence of tetramethylammonium hydroxide. J Anal Appl Pyrolysis 56(2):167–178

Chapter 6
Applying the Techniques on Materials II

Nick Schiavon, Vasilios Melfos, Reiner Salzer, Renate Lunkwitz, K. Chrysafis, P. Spathis, D. Merachtsaki, K. Triantafyllidis, P. Giannakoudakis, P. Xidas, Giovanni Ettore Gigante, Stefano Ridolfi, R. Lahoz, L. A. Angurel, U. Brauch, L. C. Estepa and G. F. de la Fuente Leis

6.1 The Application of Back-Scattered Scanning Electron Microscopy to Unravel Building Stone Decay Mechanisms in Urban Environments

Nick Schiavon

Abstract Back-scattered Scanning Electron Microscopy (BSEM) has been used to identify weathering mechanisms occurring in two oolitic limestones from urban areas in London and Cambridge, United Kingdom. From a petrographical point of

N. Schiavon (✉)
Evora Geophysics Centre and Hercules Laboratory for the Study and Conservation of Cultural Heritage, University of Evora, Largo Marquês do Marialva 8, 7000-809 Evora, Portugal
e-mail: schiavon@uevora.pt

V. Melfos
Department of Mineralogy-Petrology-Economic Geology, School of Geology, Aristotle University of Thessaloniki, 54124 Thessaloniki, Greece
e-mail: melfosv@geo.auth.gr

R. Salzer · R. Lunkwitz
Department of Chemistry, Dresden University of Technology, 01062 Dresden, Germany
e-mail: reiner.salzer@tu-dresden.de

K. Chrysafis
Department of Physics, School of Science, Aristotle University of Thessaloniki, 54124 Thessaloniki, Greece
e-mail: hrisafis@physics.auth.gr

P. Spathis · D. Merachtsaki · K. Triantafyllidis · P. Giannakoudakis · P. Xidas
School of Chemistry, Aristotle University of Thessaloniki, 54124 Thessaloniki, Greece
e-mail: spathis@chem.auth.gr

E. A. Varella (ed.), *Conservation Science for the Cultural Heritage*,
Lecture Notes in Chemistry 79, DOI: 10.1007/978-3-642-30985-4_6,

view, the two stones can be described as oosparite and oomicrite, their main distinctive feature being the crystal size of the cement binding the limestone grains together. The sulphation mechanism, i.e. the replacement of calcium carbonate (calcite: $CaCO_3$) by calcium sulphate dehydrate (gypsum: $CaSO_4$ $2H_2O$), at the surface and within the stone fabric is confirmed as the general decay process. Differences in macroporosity/permeability distribution in the two limestones lead to different weathering patterns. BSEM provides evidence that gypsum patinas still commonly found on limestone facades in polluted urban locations are advancing inside the diseased stone and that their removal is urgently needed to arrest the growth of the in-growing weathering front.

6.1.1 Introduction

Rising levels of air pollution in modern urban agglomerates not only have deleterious effects on human health but may also be correlated with the onset of serious decay patterns on a variety of earth materials used in built heritage monuments. Stone is no exception. One of the most commonly used lithotypes is limestone, a high porosity sedimentary rock composed of a framework of grains (minerals, fossils and, as in the case illustrated in this chapter, ooids) bound together by a "cement" of calcium carbonate crystals (i.e. calcite: $CaCO_3$) either fine (<5 μm—micrite) or coarse grained (>5 μm—sparite). Ooids (also named oolites) are spherical or ellipsoidal grains which have regular concentric laminae of fine-grained carbonate developed around a nucleus, usually a quartz grain. Through laboratory and field based studies, studies focusing on the decay of limestones in polluted urban environments have

G. E. Gigante
Department SBAI Sapienza, University of Rome, Rome, Italy
e-mail: giovanni.gigante@uniroma1.it

S. Ridolfi
Ars Mensurae, Rome, Italy

R. Lahoz · L.A. Angurel · L. C. Estepa · G. F. de la Fuente Leis
ICMA (CSIC-University of Zaragoza), Maria de Luna, 3, E50018 Zaragoza, Spain

U. Brauch
Institut für Strahlwerkzeuge, Universität Stuttgart, Pfaffenwaldring 43,
D70569 Stuttgart, Germany

highlighted the importance of the sulphating mechanism, by which atmospheric SO_2, released mainly by the combustion of fossil fuels such as coal and oil, in reacting with the stone's $CaCO_3$ promotes the growth of decay-inducing sulphate patinas [1–7]. Despite the recent implementation of environmental laws setting strict limits to SO_2 levels in urban areas [8], which has now resulted in a significant reduction of SO_2 emissions in today's urban environments, gypsum patinas are still disfiguring building façades in several major European towns especially where high-sulphur coal had been extensively used in the past such it is the case in the UK [2, 7, 9] and Eastern Europe [10].

The growth of black gypsum-rich crusts on the stone surface in areas sheltered from episodes of rain wash-outs is therefore a well-known and still active phenomenon. However, the detailed growth mechanism and related chemical weathering pathways correlated with these patinas are still the subject of some debate. With regards to the progression of the weathering front inside the stone, for instance, some researchers suggest that gypsum crystals replacing calcite now seen at the outermost crust–atmosphere interface are the "oldest", i.e. the ones precipitated first. Others noted that that the distribution of aerosol particulate matter derived from fossil fuel burning imbedded within gypsum patinas seems to record an outward growth of the crusts during historic changes in urban pollution sources [2, 4, 11, 12]. Indeed, a "stratigraphical" pattern in the distribution of particulate pollutants in particularly thick gypsum crusts from central London (Westmister Palace) has been recognised [12]: in these studies, particles from historical coal burning activities were found concentrated close to the innermost crust–stone interface while particles typical of "recent" oil burning sources were most common near the outermost crust–atmosphere interface. This stratigraphical distribution, while supportive of the outward growth hypothesis, is, however, by no means present in all samples of black crusts from monument façades in urban areas. The scientific debate regarding the inward versus outward crust growth mechanism is of no trivial nature as it relates directly to a conservation open issue between conservators advocating a protective role of these crusts against further attack by environmental agents and those who strongly advise their prompt removal by different cleaning methods (wet chemical and/or laser assisted).

In this chapter, we examine the application of Scanning Electron Microscopy (SEM) interfaced with Energy Dispersive X-ray Spectroscopy (EDS) to examine the urban weathering behaviour of oolitic limestones in two British monuments (Fig. 6.1): King's College Chapel, Cambridge and St. Luke's Church, Chelsea, London. The SEM is used here in back-scattered mode (BSEM). BSEM is particularly useful in the examination of thin sections spanning the weathering patina–stone substrate interface as in this mode the brightness of individual mineral phases under SEM is directly correlated with the average atomic number of the elements forming the chemical composition of the same phases: in other words, the heavier the elemental composition of the mineral under view (for example Fe-containing minerals) the brighter the image in the SEM monitor.

Fig. 6.1 King' College Chapel, Cambridge (*left*) and St. Luke's Church, Chelsea, London (*right*)

Aim of this BSEM study was threefold: (a) assess the role played by petrographical differences in pollution-derived decay of apparently similar building limestones; (b) compare the relative soundness of two limestone lithologies subjected to different levels of past and present air pollution (Cambridge vs. London); (c) further investigate the detailed growth mechanisms of gypsum weathering patinas on limestone.

6.1.2 Petrographical Notes and Methodology

Both monuments are built in oolitic limestone. The main petrographical difference between the two limestones is the type of cement binding the constituent grains together: in the King's College stone, the cement is mainly composed of microcrystalline calcite (oomicrite) whereas in the St. Luke's stone the cement is largely composed of large interlocking calcite crystals (oosparite). Oolitic limestones, at least in the Cambridge case [13] were probably chosen because of their high degree of original cementation as both limestones show postdepositional features (such as interpenetrating grain boundaries) suggesting that they were subjected to considerable compaction during their geological history.

Polished surfaces of the samples were sputter-coated with gold and then examined with a JEOL 820L SEM interfaced with LINK system AN10000 EDS microanalysis system.

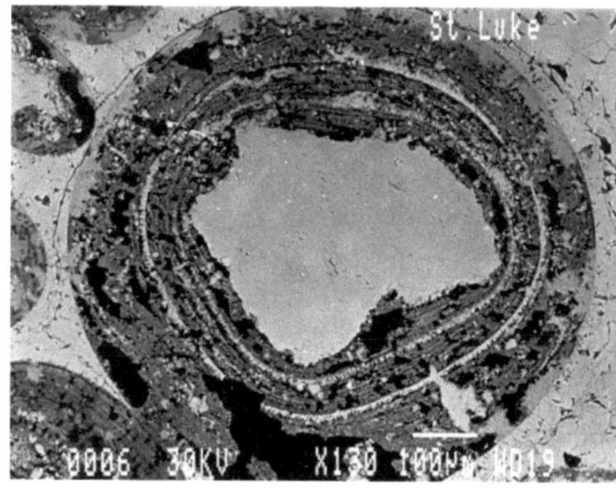

Fig. 6.2 St. Luke's Church. Former calcite oolite being replaced by gypsum. Some oolitic layers and nucleus still calcitic. Oolitic layered texture preserved

6.1.3 BSEM Observations

In both monuments, decay leads to the built-up of gypsum crusts of variable thickness (up to 1 cm). Thicker crusts in the London samples probably correlate with higher degrees of atmospheric pollution. Apart from gypsum crystals, other constituents include anthropogenic particles, calcite, silicate particles (quartz and feldspar) and fossil fragments. In particular, anhedral (i.e. with a not well-formed crystal habit) calcite grains are very common and evenly distributed throughout the crusts. Continuous calcite horizons are also present. Fossil grains can often be seen even near the crust–atmosphere interface. Aerosol particles from fossil fuel burning activities do sometimes show a "stratigraphic" distribution with a higher number of oil derived (i.e. recent pollution) carbonaceous particles at the outer edge and coal derived (i.e. historic pollution) aluminosilicate spherical particles (together with wood fragments) closer to the stone substrate interface. A typical feature within gypsum patinas is the presence of ooid "ghosts", i.e. areas now completely filled with acicular gypsum crystals still retaining a faint spherical oolitic outline (Fig. 6.2). These areas are free of imbedded particles and clearly represent areas formerly occupied by the stone substrate prior to sulphate attack.

Despite these apparent similarities, detailed BSEM investigation reveals how atmospheric chemical attack progressed differently in two building stones examined.

6.1.3.1 St. Luke's Church

The sulphation weathering front starts at the stone–crust interface and may reach a depth of several centimetres inside the stone. Calcitic oolites have been and are presently undergoing gypsum replacement. The preservation of calcitic remnants and the lower S content (as revealed by EDS analysis) at the centre of the oolites suggests that the replacement occurs from the outer edges of the oolites towards the centre (Fig. 6.2). The original ooid texture is commonly preserved suggesting that the gypsum replacement is a slow replacive process not always involving

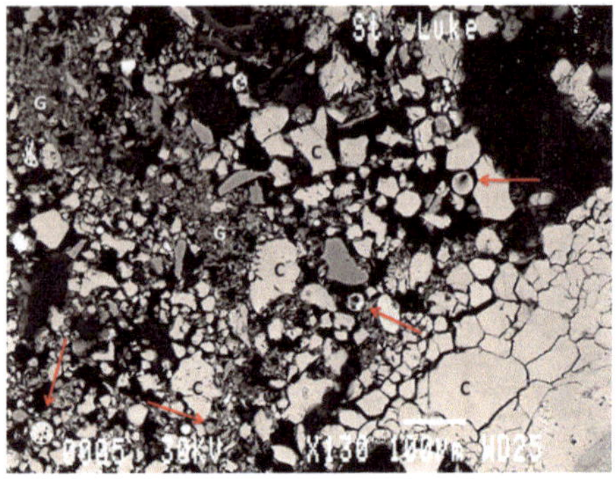

Fig. 6.3 St. Luke's Church: stone–crust interface. Gypsum crystals (*G*) crystallising within cleavage planes and intercrystalline boundaries split calcite cement into fragments (*C*) which are then incorporated into the growing crust. Anthropogenic particles from fossil fuel burning are also present (*red arrows*)

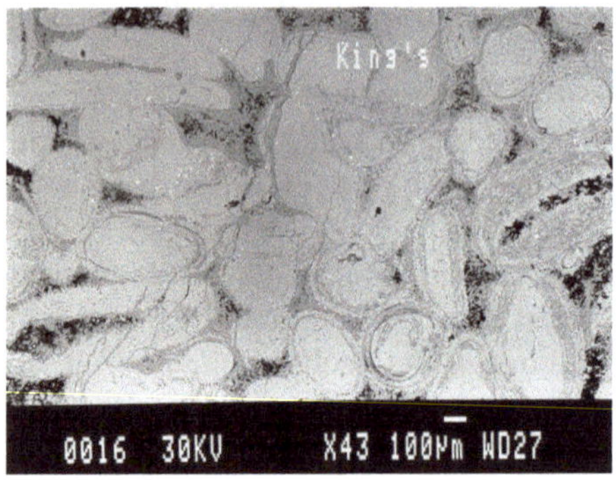

Fig. 6.4 King's College Chapel. Secondary acicular gypsum crystals precipitate within the pore spaces after dissolution of the original micrite cement. Stone porosity increases

previous dissolution of the calcite with an intervening void phase before gypsum precipitation. In some cases, however, the typical oolitic texture is destroyed although the spherical outline of the ooid is still visible. Occasionally, the space previously occupied by the ooid is now filled with anthropogenic particles and mineral fragments. The sulphation also affects the sparite cement but not through a replacement process: gypsum instead crystallises within fractures and cleavage planes splitting calcite crystals apart; detached calcite fragments are slowly incorporated into the growing crust (Fig 6.3).

6.1.3.2 King's College Chapel

Most of the oolites in Kings Chapel weathered stone samples, in contrast with St. Luke's ones, do not show extensive gypsum replacement, although crystallisation of gypsum within ooid laminae and microfractures has been observed (Figs. 6.4, 6.5, 6.6). The micritic cement is, however, almost completely replaced by acicular gypsum crystals up to a depth of several centimetres underneath the current stone–crust interface (Fig. 6.4). Authigenic dog-toothed calcite is also seen precipitating at the oolite edges as a fringe cement (Fig. 6.6). This type of decay may deceitfully lead to an overestimation of the stone soundness because of the

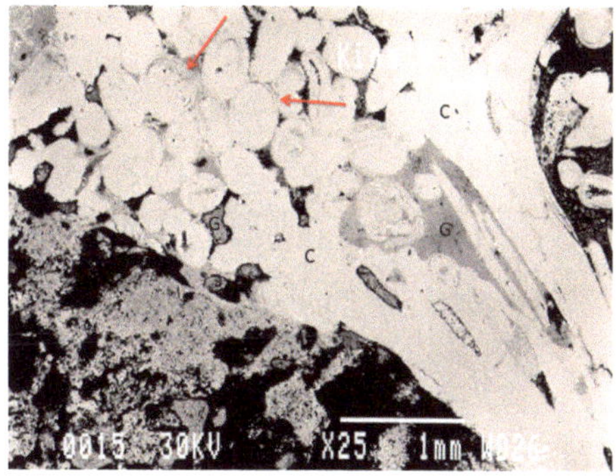

Fig. 6.5 King's College Chapel. Extensive areas of gypsum/replacement crystallisation (*G*) of micrtic cement (*C*) occur deep inside the stone fabric. Incipient oolite replacement is also present (*arrows*)

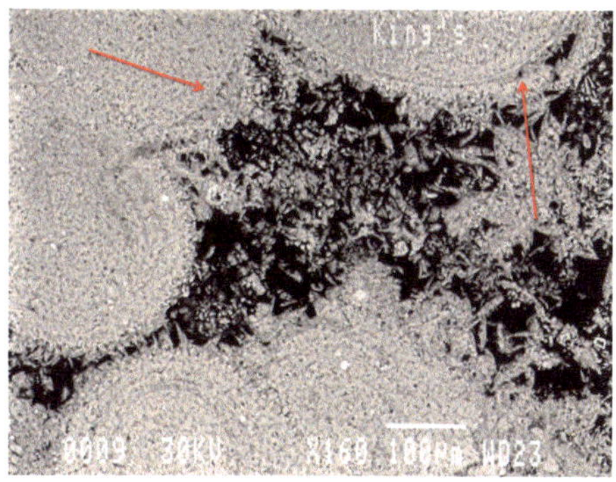

Fig. 6.6 King's College Chapel. Close-up of secondary acicular gypsum cement, showing also the reprecipitation of dog-toothed calcite as a fringe cement around oolitic grains as well as the detachment of layers of oolitic cortex due to gypsum crystallisation pressures (*arrows*)

apparent high degree of cementation. Widespread sulphation of the cement can also be seen in areas of the stone where exposure to episodes of rain wash-outs have prevented the development of fully grown gypsum patinas. The net effect of sulphation is a sharp increase in intergranular porosity: a fragile framework of oolite grains with intervening void spaces is thus created. Macroscopically, this leads to superficial crumbling of the stone into a fine "oolitic" powder and to a complete loss of cohesion. In many instances, a band of clear gypsum crystal (i.e. with no imbedded anthropogenic and/or soil dust particulate matter) is present near the stone–crust interface (Fig. 6.5).

6.1.4 Discussion and Decay Model

The ubiquitous presence of authigenic gypsum at the surface and within the pore spaces of the studied oolitic limestones confirms sulphation as the main decay process. The BSEM examination suggests that the following decay mechanisms have affected the building stones. Atmospheric pollutants, namely SO_2, are adsorbed on the stone surface either by wet or dry deposition. Wetting of the stone cause sulphation to begin and calcite starts to get replaced by gypsum. Weathering

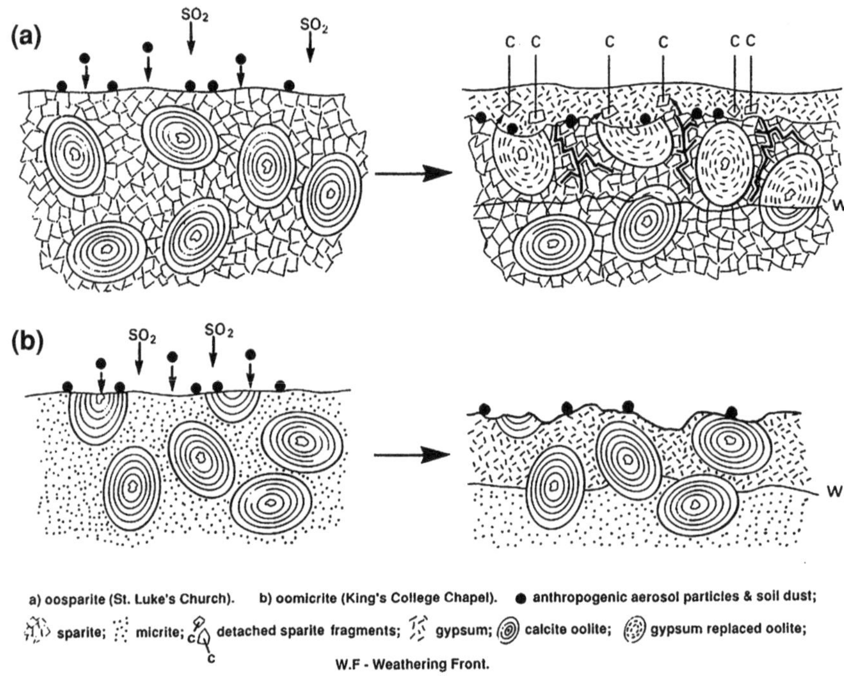

Fig. 6.7 Decay

solutions penetrate within the stone along cracks and intercrystalline boundary planes. At this stage, differences in stone microtexture lead to different decay patterns (Fig. 6.7).

Oosparite (*St. Luke's Church-London*). The oolitic grains are preferentially replaced with respect to the cement. The mechanism of replacement my follow one of two routes: (i) dissolution of the calcite followed by precipitation of gypsum in the void thus created; (ii) slower atom by atom replacement via a thin film mechanism similar to that reported for the transformation of the mineral aragonite ($CaCO_3$) into its polymorph (i.e. same chemistry, different crystalline structure) calcite [14]. The preservation of laminae within the cortex of many ooids (Fig. 6.2) would suggest the second mechanism to be predominant although episodes of dissolution and void creation are present. Gypsum crystals precipitating within cleavage planes and intercrystalline boundaries of calcite cement crystals results in their fragmentation (Fig. 6.3). Incorporation of these crystalline fragments into the growing crust is clearly visible and suggests that most of the anhedral calcite imbedded within the crust are not soil dust but remnants of the desegregated cement.

Oomicrite (*King's College-Cambridge*). Most of the oolite grains are still calcitic whereas the microcrystalline cement (micrite), despite its low permeability, is almost completely replaced by gypsum (Fig. 6.4). The replacement process is likely, in this case, to have involved early dissolution of the fine-grained micrite as suggested by the presence of intergranular microcavities in areas not yet subjected to sulphation. Microsplitting of ooid laminae due to gypsum growth is present but, in this case, does not greatly affect the overall stone soundness (Fig. 6.5).

Differences in micropore distributions between cement and ooid grains in oosparite and oomicrite are responsible for the observed selective replacement features. Limestone porosity studies indicate that in oosparites the microporosity is mainly located within ooid grains whereas in oomicrites, most of the interconnecting pores, affecting stone permeability are located within the micritic cement. The smaller crystal size of the micrite cement in King's College samples offers a greater specific area for chemical attack explaining why micrite is more easily replaced than large sparite calcite crystals. The fact that most of the ooid grains are still composed of calcite might also reflect the lower degree of SO_2 air pollution in Cambridge as opposed to London. The dissolution/replacement of micrite is probably occurring at a fairly early stage in the decay process: further decay would probably result in the in the replacement of ooids as seen in St. Luke's samples. In both examples, the presence of gypsum filled "ghosts" within the crusts near the stone–crust interface provides clear evidence that the chemically active weathering occurs inside the limestone and that crusts grow in thickness mainly at the expense of the stone substrate. Gypsum crystals get "younger" in an inward direction.

6.1.5 Conclusions

Compared to standard Optical Microscopy, BSEM is a useful tool in the investigation of building stone surface weathering in urban environments due to its higher magnification, higher lateral resolution and "instant" phase information based on changes in image brightness between minerals containing elements of various atomic numbers. In the case study illustrated in this chapter, for instance, the technique revealed slightly different decay pathways in the two similar but distinct lithological types examined. In oosparites, the main cause of decay is crystallisation pressures arising from gypsum crustallisation within cleavage planes and intercrystalline boundaries of the sparitic cement. In oomicrites, the main cause of decay is the dissolution of secondary gypsum cement replacing the original micrite.

In conservation terms, the oosparite seems to provide a better resistance to decay. The growth of gypsum crusts occurs in both directions, inwards and outwards, with respect to the original stone surface. The predominant direction is towards the interior of the stone, though, highlighting the need for an urgent removal of the patinas followed by the application of stone consolidants and water repellents.

6.2 Application of Microscopy, X-ray Diffractometry (XRD) and Stable-Isotope Geochemistry in Provenance Determination of the White Marbles Used in the Ancient Great Theatre of Larisa, Thessaly, Greece

Vasilios Melfos

Abstract The present work aims in identifying the exact source of the marbles used in the *koilon* from the ancient Great Theatre of Larisa, for maintenance purposes. Emphasis is given on the C–O isotopes of marble ($\delta^{13}C = +2.62$ to $+2.93$ ‰ and $\delta^{18}O = -4.75$ to -5.78 ‰). Mineralogical and petrographical investigations further refine the marble characteristics. Based on the comparison with seven different quarrying sites in Thessaly, it is concluded that the marble for the construction of the theatre was from the Kastri quarries.

6.2.1 Introduction

The Great Theatre of Larisa is one of the biggest well-preserved ancient theatres of Greece, totally built by stone. It was constructed in the beginning of the third century BC based on the Classical architecture [15, 16]. It consists of three major parts: the *Scene*, the *Orchestra* and the *Koilon*, a semicircular area of stepping seats, comprising the main part of the theatre. *Koilon* was built by fine-quality white-marble blocks. Fourteen series of seats are still preserved (Fig. 6.8), whereas the higher part (*epitheatron*) has been mostly removed or covered by modern buildings [15]. During the second century BC, a Doric style *proscenium* was added in front of the scene. Between the second and the first century BC, the theatre was subjected to extensive reconstructions due to a partial destruction attributed to a seismic event [17]. The Great Theatre was in use until the end of the third century A.D., when was probably flooded by the sediments of Pinios river which flows only 200 m to the West [18].

The first works for unearthing the theatre started in 1910, but the large revelation of the monument was due to the broader excavation programme between 1985 and 2000, by the fifteenth Ephorate of Prehistoric and Classical Antiquities, directed by Tziafalias. Today, a large project focusing on the preservation and restoration of the *koilon* is in process. The main problems are concentrated in damages, such as displacements and breaks of the large marble blocks, attributed to strong earthquakes [18]. For the maintenance purposes, it is of critical importance for the archaeologists to know the precise quarries in order to replace the broken and damaged slabs from the same quality marble with equivalent features.

The present work aims in identifying the exact ancient source of the marbles used in the *koilon* from the Great Theatre of Larisa, through a mineralogical, petrographical and isotopic study. In fact, the provenance studies contribute in the

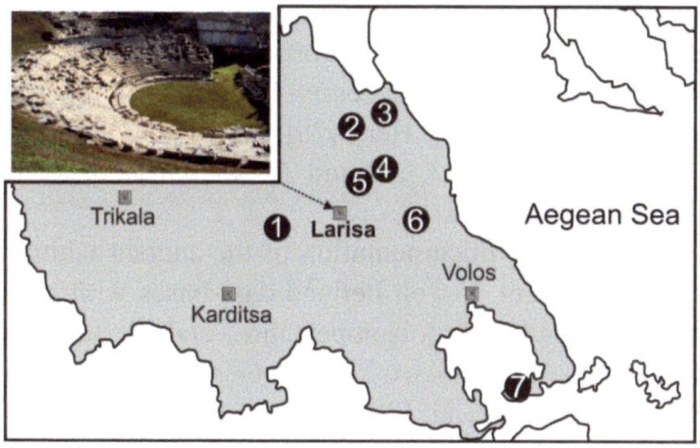

Fig. 6.8 Map of Thessaly in central Greece, showing the location of the ancient Great Theatre in Larisa (small photograph at the *upper left corner*) and the ancient white-marble quarrying sites: *1*. Tisaion mount (ancient Atrax), *2*. Olympos mount (ancient Gonnoi), *3*. Ossa mount (Tempi), *4*. Chasanbali, *5*. Kalochori, *6*. Mavrovounion mount (Kastri) and *7*. Magnesia–Tisaion mount

location of the best materials for restoration, but also offer valuable information in the ancient trading and communication routes during ancient times.

6.2.2 Sampling and Methodology

A total of six marble samples (AThL1-6) were collected from the *koilon* of the ancient theatre, especially from the main part, apart from one sample which is from the upper section, the *epitheatron*. It should be mentioned that all the samples were detached from hidden broken parts of damaged seats and they do not exceed $2 \times 2 \times 1$ cm in dimensions.

From each sample, a thin section was prepared for the purpose of mineralogical and petrographical study by polarising microscopy at the Department of Mineralogy-Petrology-Economic Geology, Aristotle University of Thessaloniki (AUTh). Microscopy was employed to determine the fabric of the mineral constituents, with particular reference to calcite, as well as to detect the frequency and distribution of the accessory grains. The maximum grain size of calcites (MGS) and the geometric relationships of the carbonate grains, such as the grain boundary shape, were also evaluated.

In addition, powders of the samples were processed by X-ray diffraction (XRD) in order to distinguish calcite from dolomite and to verify the related abundances in each sample. The XRD analyses were performed at the Department of Mineralogy-Petrology-Economic Geology, AUTh, with a PHILIPS PW1710 diffractometer (Ni-filtered CuKα/Ni: 40 kV, 0.01° 2θ, 3–63°, 0.02° 2θ/s).

Oxygen and carbon isotope analyses of the marble samples were carried out at the Department of Earth Sciences, Royal Holloway University of London. The oxygen and carbon isotope ratios are referred to the standard VDPB (*Belemnitella americana* from the Cretaceous Pee Dee Formation, South Carolina).

The applied methods have been successively used so far by many researchers, such as Coleman and Walker [19], Herz [20], Herrmann et al. [21], Capedri and Venturelli [22], Maniatis et al. [23, 24] and Al-Naddaf [25], for the source identification of ancient marble artefacts.

This study is based on the documentation of the ancient white-marble quarries of Thessaly, the development of well-defined data-bases with mineralogical, petrographical and oxygen and carbon isotope ratios, and the statistical treatment of the measured parameters. Previous works from Germann et al. [26], Capedri et al. [27], Melfos [28] and Melfos et al. [29] have defined the geological and isotopic characteristics of the marbles at seven ancient quarrying sites in Thessaly, located: at Titanos mountain (ancient Atrax), at Olympos mountain (ancient Gonnoi), at Ossa mountain (Tempi), at Mavrovouni mountain (Kastri), at Kalochori and Chasanbali, nearby the famous quarries of the ophicalcitic Green Thessalian stone [30], and at Tisaion mountain in Magnesia (Fig. 6.8). The mineralogical composition, the textural features, the MGS and a brief description of the marbles from each ancient quarry are depicted in Table 6.1, which also reports O and C isotope compositions.

6.2.3 Results

All the marble samples from the ancient theatre have a white colour and they are coarse grained, with visible calcite crystals. In some cases, grey-green coloured thin orientated stripes are observed parallel to the schistosity. The XRD analyses and the microscopic investigation showed that the mineralogical composition remains almost the same in the whole sample suite and consists mainly of calcite, with dolomite quartz, white mica, opaque minerals (sulphides) and Fe-oxides as accessory minerals. Traces of chlorite, when observed, are related to the grey-green orientated stripes.

The maximum grain size (MGS), a diagnostic feature for discriminating marbles, is 5–6 mm. This MGS is typical of the coarse-grained marbles from different localities in Thessaly, such as Gonnoi, Tempi and Kastri (Table 6.1). The marble samples exhibit a heteroblastic texture with small grains, up to 500 μm in diameter, coexisting with larger crystals with a size reaching 6 mm, and only occasionally they show a homeoblastic texture. This is typical of Atrax, Kalochori and Kastri marbles (Table 6.1). In some samples (AThL3, AThL4, AThL5), the marble is characterised by a high degree of preferred orientation of the calcite crystals which are considerably distorted by elongation, flattening and bending as a result of intensive strain. This fabric has been observed in some cases in marbles from Tempi and Kastri (Table 6.1). The shape of the grain boarders is sutured to dentate and frequently embayed (e.g. grains are interlocked) indicating non-equilibrated metamorphic conditions.

Table 6.1 Macroscopic characteristics, mineralogical composition, textures and isotopic compositions of the marbles in the ancient quarries of Thessaly

Locality	Macroscopic characteristics	Mineralogical composition	Texture	Shape of the crystal boarders	MGS[a]	C–O isotopes
Atrax	Coarse grained. White to greyish white colour	Ca, ap, slf, Fe-ox	Homeoblastic to slightly heteroblastic	Sutured dentate, embayed	4 mm	δ^{18}O: -1.77–3.36 ‰ δ^{13}C: 1.65–2.27 ‰
	Fine grained. White colour with pinkish tint	Ca, ap, slf, Fe-ox	Heteroblastic	Straight to curved, commonly embayed	2 mm	δ^{18}O: -4.60–7.00 ‰ δ^{13}C: 2.74–4.20 ‰
Gonnoi	Fine grained. Grey colour	Do, ca, wm, qu, slf, Fe-ox	Homeoblastic	Straight or curved	2 mm	δ^{18}O: -1.50–3.32 ‰ δ^{13}C: 2.00–2.70 ‰
	Coarse grained. Greyish white colour	Ca, do, chl, slf, Fe-ox	Homeoblastic	Curved	3 mm	δ^{18}O: -4.49–4.81 ‰ δ^{13}C: 1.77–2.02 ‰
	Very coarse grained. White colour	Ca, wm, Fe-ox	Homeoblastic	Dentate	5 mm	δ^{18}O: -7.63–8.31 ‰ δ13C: 2.00–2.70 ‰
Tempi	Middle to coarse grained. Greyish to white colour	Ca	Homeoblastic, rarely heteroblastic. Elongated crystals	Sutured dentate, embayed	5 mm	δ^{18}O: -1.10–6.00 ‰ δ^{13}C: 1.10–3.20 ‰
Kalochori	Fine to middle grained. White colour	Ca, do, wm	Heteroblastic	Straight, rarely curved	1 mm	δ^{18}O: -8.00–8.59 ‰ δ^{13}C: 0.84–0.91 ‰

(continued)

Table 6.1 (continued)

Locality	Macroscopic characteristics	Mineralogical composition	Texture	Shape of the crystal boarders	MGS[a]	C–O isotopes
Chasanbali	Fine grained. White colour	Ca, qu	Homeoblastic	Straight or curved	0.5 mm	δ^{18}O: -4.15–5.60 ‰ δ^{13}C: 2.74–2.94 ‰
Kastri	Coarse grained. White to greyish white colour with green stripes	Ca, do, qu, wm, chl, slf, Fe-ox	Heteroblastic, in some cases homeoblastic. Elongated crystals	Sutured to dentate, embayed	6 mm	δ^{18}O: -4.35–6.50 ‰ δ^{13}C: 2.10–2.85 ‰
Tisaion mount	Very fine grained. White colour with pinkish tint	ca, slf, Fe-ox	Homeoblastic	Straight to curved	100 μm	δ^{18}O: -1.79 δ^{13}C: 3.16 ‰
	Fine grained. White to greyish white colour. Dark coloured stripes	ca, slf, Fe-ox	Homeoblastic	Straight to curved	100 μm	δ^{18}O: -3.90 ‰ δ^{13}C: 4.13 ‰
	Fine grained. White to greyish white colour	ca	Homeoblastic	Straight to curved	100 μm	δ^{18}O: -0.45 ‰ δ^{13}C: 2.98 ‰×

Data combined from Germann et al. [26], Capedri et al. [27], Melfos [28] and Melfos et al. [29]
Ca: calcite, do: dolomite, qu: quartz, wm: white mica, chl.: Chlorite, ap: apatite, slf: sulphides, Fe-ox: Fe-oxides
[a] MGS: Maximum grain size of calcites

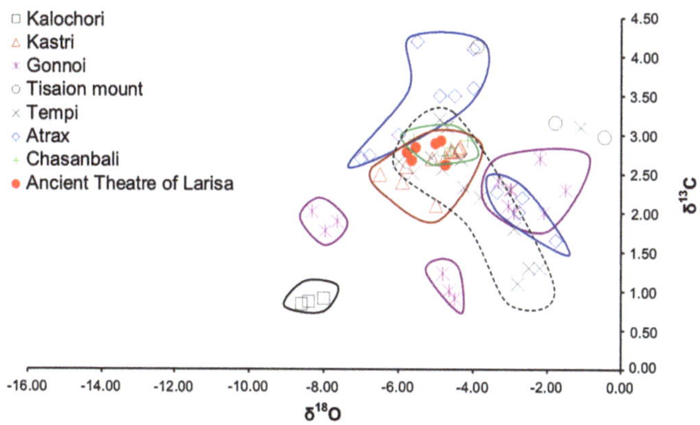

Fig. 6.9 A δ18O–δ13C diagram for the archaeological marble samples from the ancient Great Theatre of Larisa. Reference isotopic fields of the marbles from the ancient quarries in Thessaly are also shown: Kalochori, Kastri, Gonnoi, Atrax, Tisaion, Tempi, Atrax and Ghasanbali, published by Germann et al. [26], Capedri et al. [27], Melfos [28] and Melfos et al. [29]

All the samples contain small flakes of white mica and rounded quartz grains with a length up to 1 mm. Chlorite flakes with a maximum size 500 μm were observed in one only sample. Opaque minerals (probably sulphides) are seldom found among the calcite crystals and their size does not exceed 5 μm. This mineralogical composition resembles the Kastri and partly the Gonnoi marbles.

The C–O isotopic results of the marble samples from the ancient theatre are plotted in the diagram of Fig. 6.9. The δ^{13}C values range from +2.62 to +2.93 ‰ and the δ^{18}O values from −4.75 to −5.78 ‰. It is obvious that these values have a consistent signature and in the diagram of Fig. 6.9 they plot in a restricted field. Reference isotopic fields of the marbles from the ancient quarries in Thessaly are also shown in the same diagram.

6.2.4 Discussion

The microscopic investigation, the XRD and the stable-isotope analyses of six marble samples from the ancient Great Theatre of Larisa demonstrate consistent results with regard to the mineralogical composition, the petrographical features and mainly the C–O isotope ratios (Fig. 6.9).

Especially, the isotopic ratios of C and O provide usable signatures for determining the provenance of marbles. Isotopic analyses involve measuring of the ^{13}C/^{12}C and ^{18}O/^{16}O ratios in marble and the results are expressed in terms of the deviation from a conventional standard. This deviation, called δ, is expressed as δ^{13}C and δ^{18}O in parts per thousand (% or per mil) and forms the isotopic signature. The values exhibit a relatively restricted range in each quarry area or limited parts of a geological formation [31, 32]. On the $\delta^{13}C$–$\delta^{18}O$ correlation diagram, the geological samples of the marbles fall into well-defined groups and each quarrying site is therefore discriminated.

Comparing the isotopic results of the studied marbles from the theatre with the marbles from the Thessalian ancient quarries, it is evident that they plot in the groups of Tempi and Kastri and partly of Chasanbali marbles (Fig. 6.9). Consequently, all the other quarries are excluded as possible sources.

Based on mineralogical and petrographical features, discrimination among the possible sources (Tempi, Kastri or Chasanbali) is possible. The marble in Tempi consists only of calcite and its texture is mostly homeoblastic. The marble in Chasanbali contains calcite and quartz and is fine grained with homeoblastic texture. In contrast, the marbles from the theatre are made of calcite with traces of dolomite, white mica, quartz, opaque minerals, Fe-oxides and rarely chlorite and have a coarse-grained heteroblastic texture. Therefore both quarrying sites, Tempi and Chasanbali, are also excluded as possible marble sources for the theatre of Larisa. In addition the white-marble quarry in Chasanbali is small and unimportant and could not provide such a large amount of raw materials for the theatre construction.

Mineralogical composition obtained by XRD and microscopy showed that the Kastri marble is almost identical with the whole suite of the eleven studied marbles from the theatre. The textural features have also significant similarities being mainly coarse-grained heteroblastic, and in some cases elongated calcite crystals are observed. The large MGS, in the case of Kastri and the marbles from the theatre, are 5–6 mm.

It is evident that only the Kastri marble quarry demonstrates similar mineralogical and isotopic features with the studied marbles and shows that it is the most probable source for the marbles of the Great Theatre of Larisa (Fig. 6.8).

6.3 Case Studies: Investigation of Mortars by Infrared and Raman Spectroscopy

Reiner Salzer and Renate Lunkwitz

Abstract Infrared (IR) and Raman spectroscopy permit the identification of X-ray amorphous minerals. Mechanisms of material decay and principles of cause and effect can be elucidated. We discus three examples: (i) decay due to alkaline-rich mortar, (ii) effect of nitrate-rich mortar, (iii) identification of historic mortar. Damages to historic buildings may not only be caused by their actual utilisation or by contemporary pollutants. Natural influences acting for centuries even without any human interference may have put deteriorating strain on a historic structure as well. The impact of such strains rises if inappropriate materials were chosen or combined for construction, restoration or refurbishment. This kind of damages is known since many years. The underlying mechanisms can still not always satisfactorily be explained. Important sources of salt formation are (i) the historic

Table 6.2 Common ions in alternating composition and the corresponding solubility in water (g/100 g)

K_2SO_4	11.2	$MgSO_4$	35.6	$CaSO_4$	0.2
K_2CO_3	111.5	$MgCO_3$	0.01	$CaCO_3$	0.0015

building material itself, (ii) environmental pollution (air, soil), (iii) mistakes during restoration. Minerals may be mobilised in the presence of water, salt efflorescences subsequently become directly visible. Sometimes, these salts are merely recognisable by the damages they induced. Analytical data are needed to identify original sources of the material, to reveal the ancient technologies and to restore the works of art sustainably. Inorganic material is often characterised by its atomic composition, e.g. the content in Mg^{2+} and Ca^{2+} as well as and SO_4^{2-} and CO_3^{2-}. Under moist conditions, these ions may undergo ion exchange and form compounds of distinctly different properties. A crucial property is their solubility in water (Table 6.2): As solubility data indicate, single ion concentrations do not provide full information about the material. If local limestone contained appreciable amounts of dolomite ($CaCO_3$ * $MgCO_3$), the mortar may lose its strength gradually due to the enhanced mobility of Mg^{2+} components. Mobilisation is often indicated by salt coatings (efflorescences). Even non-crystalline samples can easily be identified by Raman or IR spectroscopy ([33], cf. also Sect. 3.3).

6.3.1 Choosing Mortar for Restoration

Salt coatings on stone were commonly called saltpetre in the past, nowadays they are usually considered sulphate depositions due to environmental pollution. Wall saltpetre is $Ca(NO_3)_2$, formed in staples everywhere in settlements. Coatings by wall saltpetre indicate high salt concentration inside the structure. Environmental sulphate depositions may be $CaSO_4$ * xH_2O (anhydrite, bassanite, gypsum), $Na_2Ca(SO_4)_2$ (glauberite) or various alkali sulphates. Salt coatings may appear rather localised on the surface due to varying permeability of different stone layers (Fig. 6.10). The chemical composition of the different salt bands may vary across the stone surface caused by a separation effect of the mobility mechanism.

The IR analysis of the efflorescences in Fig. 6.10 indicated natrite (Na_2CO_3 * $10H_2O$). Natrite would have been instantaneously transformed into sodium sulphate if environmental pollution played a role during formation of the salt bands. The formation of natrite indicates mortar of high alkali content used to position the sandstone slabs or the application of water glass to seal and strengthen the system. The recommendation based on the IR analysis was replacement by new sandstone and selection of an appropriate mortar type.

A high salt load may cause severe hazard to building material, particularly in case of high soluble salts like sodium sulphates. The solubility of thenardite

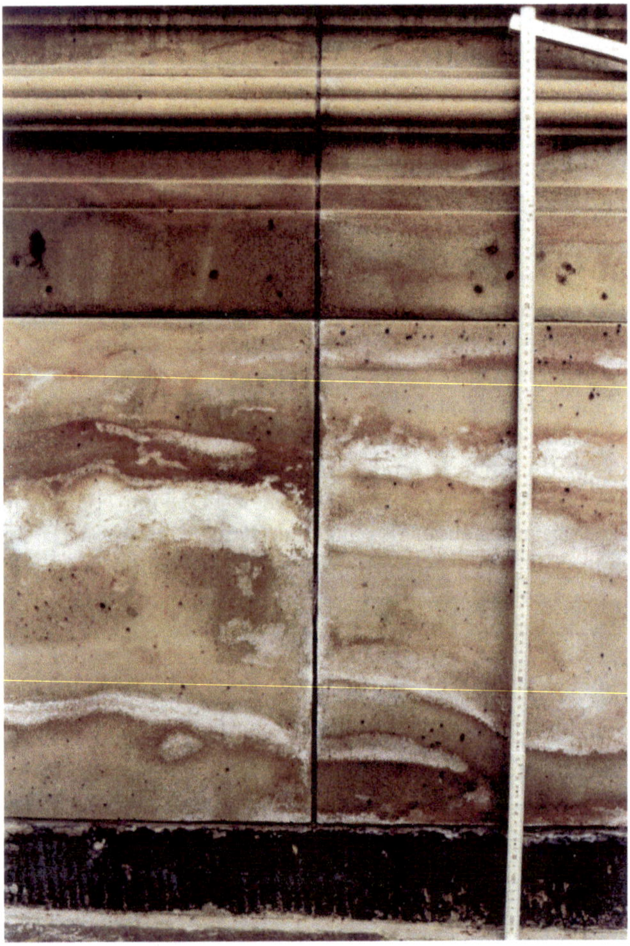

Fig. 6.10 Salt efflorescences on newly positioned sandstone slabs at the historic Schinkelwache Dresden/Germany

(Na_2SO_4) in water at 20 °C is 162 g/l, that of mirabilite ($Na_2SO_4 * 10H_2O$) is 900 g/l. The solutions migrate into the pores of the stone and equilibrate with the humidity in the surrounding air. The minerals crystallise/dissolve upon changes of the atmospheric humidity: mirabilite at 87 % relative humidity, thenardite at 81 % relative humidity. Thenardite requires a 4.2 times larger volume than that of mirabilite. The conversion from mirabilite to thenardite breaks up the pores inside the stone. The conversion temperature between the two minerals is at 32.5 °C. Both can be identified by details in their IR spectra.

Fig. 6.11 The remaining part of the wall of the destroyed Frauenkirche Dresden and the rebuilt body of the church fit well together. Incompatibilities of old and new material have to be fully excluded in order to ensure the stability of the building

Alkali content of mortar is crucial.
Efflorescences may be caused by raw material, diffusion or deposition.
Exact analysis is necessary to control the decay of stones

6.3.2 Analysis of Injuries in Historic Mortar

The Frauenkirche (Church of Our Lady) Dresden is an outstanding example of rebuilding a largely destroyed monument and preserving most of the old substance. New sandstones were obtained from historic quarries. Questions arose about the historic lime mortar (Fig. 6.11).

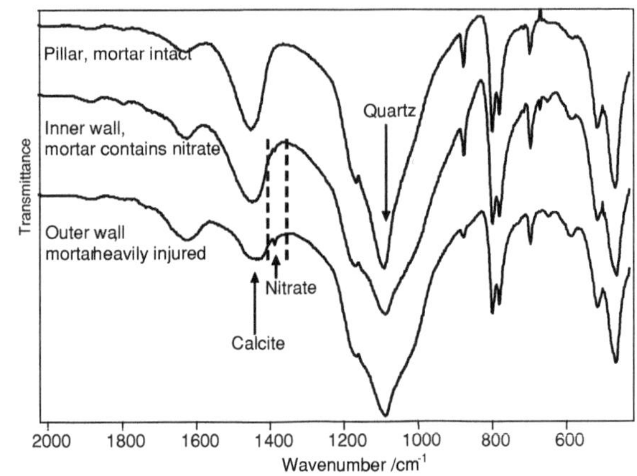

Fig. 6.12 IR spectra of mortar samples from the Frauenkirche Dresden. The formation of mobile nitrates caused the decay of the mortar in the outer wall

Local limestone is known to contain dolomite ($CaCO_3$ * $MgCO_3$). The solubility in water increases in the order $CaCO_3 < MgCO_3 * CaCO_3 < MgCO_3$. Acid (from rain or soil) dissolves magnesium ions, the corresponding salts migrate, the mortar inside the joints is weakened, and the stability of the building decreases.

After clearance of the debris, a preserved part of the outer wall just above ground showed intense signs of disintegration within its joints at the outside wall. Joints at the inside wall were weakened as well but to a much lesser degree. A nearby pillar did not exhibit any mortar decay within its joints. Samples were taken from the three sites and investigated by IR spectroscopy. All samples exhibited the characteristic features of lime mortar, lime and quartz (Fig. 6.12). Clay minerals would be indicated by a distinct band in the low-frequency wing of the quartz band. They cannot be identified in the actual samples.

The band sizes in the IR spectrum of mortar from the intact pillar indicate a normal ratio between lime (around 1,425 cm^{-1}) and quartz (below 1,100 cm^{-1}). The band assigned to lime becomes smaller for the inside wall and weak for the outside wall. This band indicates the amount of binding agent (lime) in the mortar. At the same time, a small feature below 1,400 cm^{-1} becomes visible. This little band indicates nitrate. It is generally weak but very characteristic. After this finding, a search in the archive revealed a former burial site near the outer wall.

6.3.3 Age Determination for Historic Mortar

The famous August the Strong, Elector of Saxony, ordered the construction of the Taschenberg palace. The palace next to Dresden Castle replaced earlier town houses, which dated back to around 1250. The building was destroyed in 1945 with only few parts remaining. We collected 120 mortar samples from the old ground walls. The sampling sites were recommended by archaeologists and cover the complete history of the historic building from the thirteenth to the eighteenth century. Minerals in the samples were analysed (i) by atomic absorption

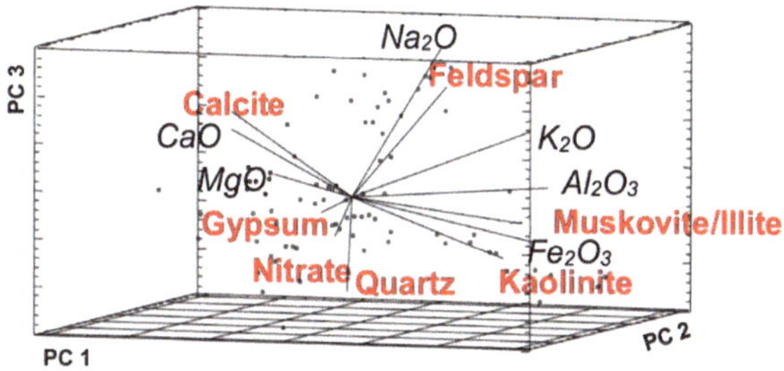

Fig. 6.13 PCA the AAS results (in *italics*) and the FT-IR results (*bold*) of 87 mortar samples (*dots*)

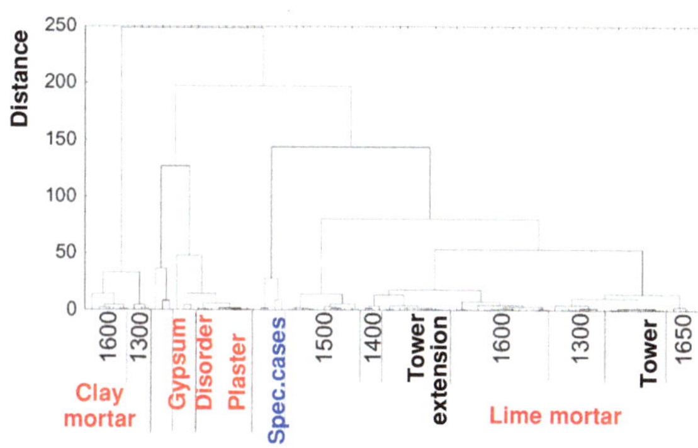

Fig. 6.14 Cluster analysis of the FT-IR data of the 87 mortar samples of Fig. 6.13. The assignment of the clusters to time periods of origin is based on reference data by archeaologists for approx. 50 % of the samples

spectrometry (AAS) after dissolution in HF/HCl/HNO_3 and (ii) by FT-IR spectroscopy after milling the samples and preparing KBr pellets [34].

Chemometric evaluations (i) may indicate and/or quantify relations between the samples (called objects) and the measured data (called variables) or (ii) may be used to model interdependencies between variables. Multivariate procedures evaluate several independent variables simultaneously. We employed Principal Component Analysis (PCA) and Cluster Analysis (CA). PCA reveals hidden, complex relations within the data set. CA merges the objects of a data set into homogeneous classes (clusters). Both kinds of evaluations are particularly helpful if no preliminary information is available, like in case of the historic mortar samples.

Figure 6.13 shows the PCA result for 87 mortar samples. AAS parameters are given in italics, FT-IR data are printed bold. The top three principal components, PC1–PC3, cover 71.30 % of the variance within the data set. The contribution of PC4 is distinctly lower than that of PC3. The samples (dots) are well distributed across the feature space. The orientation of the vectors within the cube is reasonable: the two aggregates feldspar and quartz point to opposite directions, as the mortar is either rich in feldspar or in quartz. The direction of lime (calcite) is orthogonal to (i.e. independent of) the feldspar/quartz ratio. Clay (illite/muscovite/kaolinite) is located opposite of calcite. Both clay and lime are binders, their content was well matched in historic mortar.

Subsequently, a hierarchical CA was performed for the FT-IR data of Fig. 6.13. The resulting dendrogram is shown in Fig. 6.14. The date of origin of roughly

50 % of the samples was assigned by archaeologists and could be taken as reference. The references were used to assign dates of origin to the obtained clusters. This information helped to gain new insights into the construction history of the palace. Inspections of the spectra indicate that the cluster "disorder" may be caused by a former storage of construction material, the "special cases" may indicated wall repairs in the early twentieth century.

6.4 Mortars-Thermal Analysis

K. Chrysafis

6.4.1 Introduction

Mortars and plasters in ancient structures are composite materials which have exhibited excellent durability through time and are constituted of a binder, such as lime (CaO) and/or gypsum ($CaSO_4 * 2H_2O$) and aggregates, such as sand or grit. The composition of mortars and plasters varies greatly and they are commonly divided into lime, gypsum and mixed, depending on the binding material and into hydraulic and non hydraulic depending on their ability to set under water. The main steps for lime processing are:

$$\text{the calcination process} \quad CaCO_3 \rightarrow CaO + CO_2 \tag{6.1}$$

$$\text{the hydration process} \quad CaO + H_2O \rightarrow Ca(OH)_2 \tag{6.2}$$

$$\text{the carbonation process} \quad Ca(OH)_2 + CO_2 \rightarrow CaCO_3 + H_2O \tag{6.3}$$

In the case of gypsum plaster, the plaster is produced by the addition of water and an aggregate to the hemihydrate ($CaSO_4 * 1/2H_2O$) and/or the soluble anhydrite ($CaSO_4$). In this way, interlocking crystal structures are developed corresponding to calcium sulphate dehydrate [35].

6.4.2 Experimental

Thermal analysis (TG–DTA) involves measuring the thermal variations associated with physical and chemical transformations (such as dehydration and decomposition) which occur during the heating of a sample. TG–DTA simultaneous analyses were performed using a SETARAM SETSYS 1750 TG–DTA system. Samples around 7.5 mg were placed in alumina crucibles. An empty alumina crucible was

Fig. 6.15 Some insights of the painted walls

used as reference. The samples were heated from ambient temperature to 900 °C, with heating rate 10 °C min^{-1}, in N_2 atmosphere.

The studied samples are from six monasteries and churches (Fig. 6.15). Most of them have been built after the tenth century but belong to different time periods. However, the majority of them are of the sixteenth century. Also, the studied samples in this work come from churches or monasteries from various Balkan countries. According to various sources, two of the studied wall paintings must have been done by the same iconographer. From one church, two types of samples were collected in order to examine the differences of the plaster between the two iconographers (two periods) (samples SA6, SB7). The plaster was taken from already damaged areas of the wall paintings, while from two other churches the plaster was removed, with the use of a microscalpel, beneath the pictorial layer (samples S1 and S2). Due to the destructive nature of sampling, the samples were carefully chosen from areas that had no aesthetic or iconographic value for future reconstruction.

6.4.3 Results and Discussion

For the examined samples, typical thermal curves obtained by DTG/TG analysis are shown in Fig. 6.16. For the thermal characterisation of the materials, the temperature range, can be divided into four regions corresponding to the mass loss in the thermal curve [36]: (i) <120 °C, (ii) 120–200 °C, (iii) 200–600 °C, and (iv) >600 °C. The first region is attributed to absorbed water evaporation (hygroscopic water), while the second to the evaporation of chemically bound water of the hydrated salts, such as gypsum. The third region refers to the evaporation of water chemically bound to hydraulic compounds—probably calcium silico-aluminate hydrates [36–39]. Finally, the fourth region corresponds to the carbon dioxide developed during the decomposition of carbonates. The endothermic peaks after 750 °C correspond to calcite decomposition (Eq. 6.4).

Fig. 6.16 Thermal curves of studied plaster samples: **a** S1, **b** SB7, **c** S5

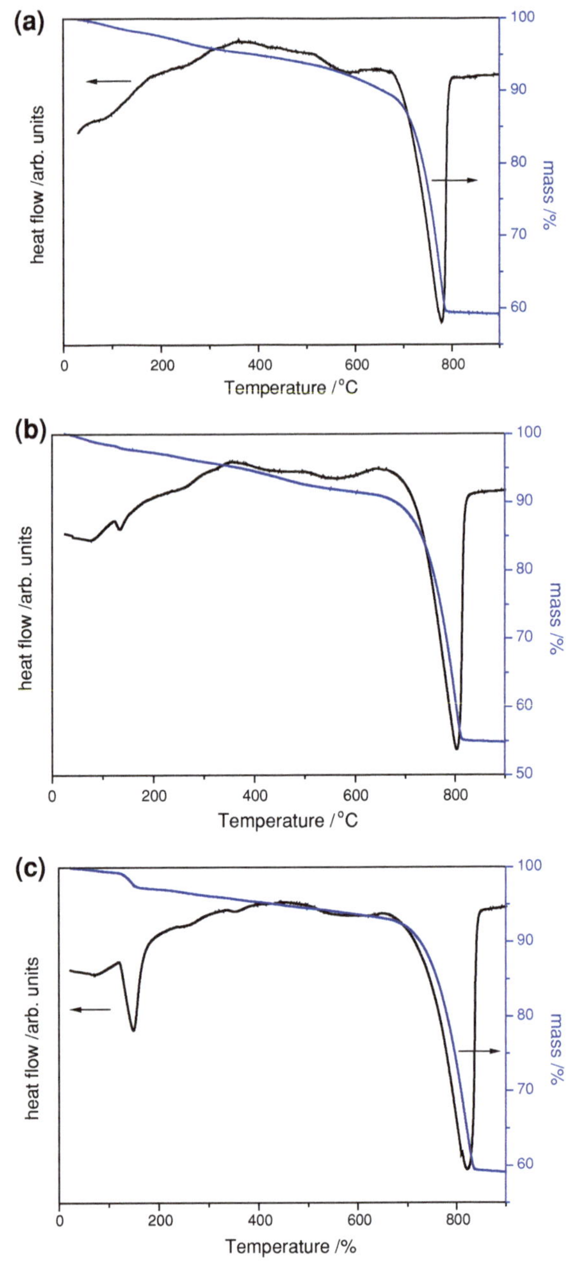

$$CaCO_3 \rightarrow CaO + CO_2 \quad (6.4)$$

While pure calcite decomposes near 840 °C, the lower degradation temperature for the plasters is characteristic of the CO_2 loss from $CaCO_3$ formed by recarbonation reaction of lime with atmospheric CO_2 (Eq. 6.3) [37, 38].

Figure 6.16a is representative of the samples taken from three churches, namely S1, S2 and S3. From that TG curve, a monotonous reduction of the sample mass was recorded up until 600 °C, which corresponds to the three of the four already discussed regions of mass loss. Above that temperature, the main mass loss was recorded as well as the corresponding endothermic peak.

The main difference between the other two examined samples which are presented in Fig. 6.16b, c respectively is located in the temperature range 120–200 °C. The samples S5 and SB7 from two different churches show endothermic peaks with minimum values at 132 and 149 °C, respectively. This peak is attributed, to a loss of moisture, and also to the dehydration of the gypsum ($CaSO_4$ * $2H_2O$), which takes place in two stages (Eqs. 6.5 and 6.6), between 130 and 160 °C [37, 38, 40].

$$CaSO_4 \cdot 2H_2O \rightarrow CaSO_4 \cdot 1/2H_2O + 3/2\ H_2O \quad (6.5)$$

$$CaSO_4 \cdot 1/2H_2O \rightarrow CaSO_4 + 1/2H_2O \quad (6.6)$$

It can be seen from Fig. 6.16b and c that the TG curves of these samples, mainly differ at the mass loss step of the area 130–160 °C. In Fig. 6.16b, it is difficult to see the step of mass loss, yet a small endothermic peak in the heat flow plot is visible, but the mass loss step is very clear in Fig. 6.16c. This difference means that the quantity of gypsum in the sample SB7 is larger than in the sample S5. Consequently, it is clear from the thermal curves that in samples S5 and SB7 there is gypsum participation. The presence of gypsum in these plaster samples was also verified by the FT-IR and EDS measurements which gave analogous results. Beyond the use of gypsum as a binding material of the plaster, it can also be formed during sulfation, as a result of Eq. 6.7, which involves the dry deposition reaction between limestone ($CaCO_3$) and sulphur dioxide (SO_2) gas, in the presence of high relative humidity, an oxidant and a catalyst (Fe_2O_3 or NO_2). In that case, gypsum is primarily detected close to the surface, especially in cracks and voids.

$$CaCO_3 + SO_2 + 2H_2O + 1/2O_2 \rightarrow CaSO_4 \cdot 2H_2O + CO_2 \quad (6.7)$$

Comparing the results of thermal analysis measurements for the samples S5 and SB7 with the results of other techniques, we can conclude that both frescoes painted by the same iconographer. SA6 which is a sample from the same church as of SB7 but from another wall had no traces of gypsum. This means that gypsum was not a deterioration result but the iconographer used it as a component in the initial mixture of the plaster. For this reason, we can conclude that the origin of the sample SA6 is from a wall painting which has been created by a different artist. Table 6.3 summarises the results from the thermal analysis studies. It is obvious that in all samples the main constituent is calcite as can be seen from the mass loss

Table 6.3 Percent mass loss in each temperature region and CO_2/H_2O ratio

Sample	<120 °C	120–200 °C	200–600 °C	>600 °C	CO_2/H_2O
S1	0.9	0.5	4.0	36.8	9.2
S2	1.3	1.0	5.8	32.8	5.6
S3	0.9	0.7	4.8	39.4	8.3
S4	1.9	1.1	5.3	37.3	7.1
S5	0.7	2.3	3.4	34.7	10.1
SA6	1.4	0.4	4.5	31.5	7.0
SB7	1.6	1.1	5.9	32.7	6.3

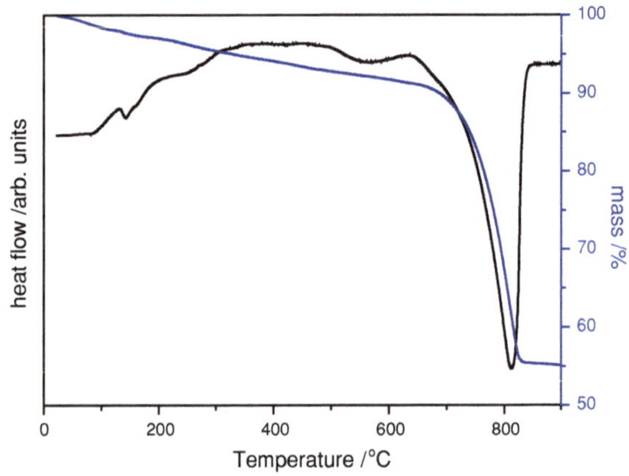

Fig. 6.17 Thermal curves of the sample S4, with the typical peak of gypsum

in the temperature range >600 °C. S3 has the greatest mass loss 39.4 % in this temperature area. SB7 and S5 perform in the second region mass losses of 1.1 and 2.3 % respectively due to gypsum dehydration, which means that gypsum was in greater proportion in the mixture of the plaster S5. S1 has a mass loss almost 1 % in the temperature range 120–200 °C, but the heat flow curve shows no peak corresponding to gypsum.

In Fig. 6.17, the thermal curves of the sample S4 are presented. The minimum value of the first endothermic peak was recorded at 142 °C, due to gypsum dehydration and the respective mass loss was 1.1 %. The sample was collected from an area where the painted layer was damaged, so the plaster was exposed to air. From the FT-IR and EDS measurements, there was no evidence of gypsum in the plaster studied below the pictorial layer in painted samples. On the contrary, gypsum was identified on the external surface of the painted layer. Therefore, the presence of gypsum can not be attributed to its inclusion in binder but is more likely to be related to the sulphation of the carbonate component, due to environmental pollution [39]. The FT-IR and EDS results are in good agreement with the TG–DTA analysis.

6.4.4 Conclusions

The historic plasters studied can be classified in two distinct groups depending on their thermal behaviour. From this set of measurements, the possibility that different wall paintings from different churches must have been made from one artist can be concluded. Also, the fact that different parts inside one church must have been done by different artists is proposed through the thermal analysis studies of each church. Thermal Analysis seems to be a reliable method for plasters' discrimination. The results give useful information on the understanding of the technology of historic plasters and on planning syntheses for restoration of wall paintings. Our work allows a consistent approach to the wall paintings conservation and repair to be achieved.

6.5 Electrochemical Impedance Spectroscopy Measurements for the Corrosion Behaviour Evaluation of Epoxy: (Organo) Clays Nanocomposite Coatings

P. Spathis, D. Merachtsaki, K. Triantafyllidis, P. Giannakoudakis and P. Xidas

Abstract Aim of the present work is the evaluation of corrosion behaviour of steel coated with epoxy–(organo) clay nanocomposite coatings by electrochemical impedance spectroscopy measurements. Electrochemical impedance basics and main characteristics of nanocomposite for protective coatings are presented. The investigation was carried out by electrochemical impedance measurements and the polarisation resistance for four types of specimens, non-coated, coated with pristine glassy epoxy polymer or with two types of epoxy–clay nanocomposites at various times of exposure in the corrosive environment was determined. Both epoxy–clay nanocomposite tested have improved anticorrosive properties in comparison with these of the pristine glassy epoxy polymer. The epoxy–montmorillonite clay modified with primary octadecylammonium ions, Nanomer I.30E, had a better behaviour than the modified with quaternary octadecylammonium ions, Nanomer I.28E.

6.5.1 Introduction

Organic coatings are widely used to prevent corrosion of metallic structures because they are easy to apply at a reasonable cost. It is generally accepted that the coating efficiency is dependent on the intrinsic properties of the organic film (barrier properties), on the substrate/coating interface in terms of adherence, on the inhibitive or sacrificial pigments used and on the degree of environment aggressiveness.

Over the years, polymeric coatings are developed due to their good barrier properties. However, these pristine polymeric coatings are still permeable to corroding agents such as water and oxygen. In order to enhance the barrier properties of polymeric coatings, various kinds of additives such as extenders and inorganic pigments which inhibit corrosion have been used.

During the last years, polymer clay nanocomposites have attracted a lot of attention. It has been reported that the incorporation of a small amount (1–5 %) of layered clay in organic polymers leads to significant improvements in mechanical performance, thermal stability, and barrier properties of organic coatings. These improvements are related to the morphology of the layered silicates and to the specific incorporation of the nanoparticules in the polymer [41–47].

EIS Basics

The development of electrochemical impedance methods in recent years has made possible the planning of electrochemical laboratory research into applications that traditional DC techniques could not be applied, such as corrosion measurements in cases where organic coatings and low conductivity materials are used. This technique allows analysis of the stages of a reaction when wide frequency ranges were used. Also, in the field of study of the materials is used to reveal relationships between electrical properties of materials and physical and chemical properties. The areas that have been demonstrated as appropriate for using EIS for corrosion measurements are:

Suspension or retardation of corrosion (uniform or local). Passivation of metals. Behaviour and properties of polymeric materials used as protective coatings against the corrosion of metals, which is very widespread and complex due to the high resistance of the coatings. Rapid estimation of corrosion rates. Estimation of extremely low corrosion rates and metal contamination rates ($<10^{-4}$ mm/year, <0.01 mpy) and in low conductivity media. Rapid assessment of corrosion inhibitor performance in aqueous and nonaqueous media.

Impedance

A resistance to direct current can be corresponded with an equivalent resistance to alternating current. This equivalent resistance is called "complex resistance" (Impedance). Measurements can be performed with dynamostatic control, i.e. when a voltage is applied and the respective current is measured. In this technique, typically, a small amplitude sinusoidal potential perturbation is applied to the working electrode at a number of discrete frequencies, ω. At each one of these frequencies, the resulting current waveform will exhibit a sinusoidal response that is out of phase with the applied potential signal by a certain amount (Φ) and has a current amplitude that is inversely proportional to the impedance of the interface. The electrochemical impedance, $Z(\omega)$, is the frequency-dependent proportionality factor that acts as a transfer function by establishing a relationship between the excitation voltage signal and the current response of the system:

$$Z(\omega) = E(\omega)/I(\omega),$$

where: $Z(\omega)$ = the impedance, ohm-cm^2; ω = frequency, radians-s^{-1}; E = the time varying voltage across the circuit, volts, $E = E_0 \sin(\omega t)$; I = the time varying current density through the circuit, amp-cm$^{-2;}$ for linear systems (i.e. where there is a continuous evolution of the phenomenon of corrosion: non-breaking of the coating of corrosion products), the response is defined as: $I = I_0 \sin(\omega t + \Phi)$; E_0 = the size of disturbance of the voltage; I_0: the size of disturbance of the current; t = time, s; Φ = phase angle, deg.

Only for linear systems the resulting current is pure sinusoidal. But in electrochemical systems, there is seldom a clean linear relationship between potential and current (Figs. 6.18 and 6.19).

The impedance of a single frequency can be represented by a vector of a length |Z| with an angle φ between the real axis of impedance Z' (real part of impedance) and a vector Z'' (imaginary part of impedance = the part of the capacity of the impedance) (Figs. 6.20 and 6.21).

These figures show that when a small amplitude sinusoidal potential perturbation is applied, there is a current response of small amplitude. When the amplitude remains small, it can be used the first term of Taylor series,

$$e^x = 1 + \frac{x}{1!} + \frac{x^2}{2!} + \frac{x^3}{3!} + \cdots, \quad -\infty < x < \infty$$

resulting in an approximately linear system and represents the response current. The voltage application and the resulting current can be written as a function: $E = E_0 \exp(j\omega t), I = I_0 \exp(j\omega\tau - \varphi)$, where j imaginary part $(-1)^{1/2}$.

The complex resistance (impedance) is:

$$Z(\omega) = E/I = (E_0/I_0)\ \exp(j\varphi) = |Z|\exp(j\varphi) = |Z|(\cos\varphi + j\sin\varphi) = Z' + jZ'';$$
$$Z' = |Z|\cos\varphi,\ Z'' = |Z|\sin\varphi.$$

Or: Real Current = I' = $|I|$ cos (ωt), (in-phase component); Imaginary Current = I'' = $|I|$ sin (ωt), (out-of-phase component); $E = E_{real} + E_{imag} = E' + jE''$; $I = I_{real} + I_{imag} = I' + I''$;

$$Z = E' + jE''/I' + jI'' = Z' + jZ'';\ |Z| = [(Z')^2 + (jZ'')^2]^{1/2}$$

-Grafic representation of impedance

Impedance of a system can be represented graphically by using Nyquist diagrams. The impedance is derived from the connection of the points between the edges of the impedance vectors as a function of frequency. An example of a Nyquist vector for an equivalent circuit of DC with a resistor in parallel with a capacitor is shown in Fig. 6.22.

Another method of expression of the impedance is Bode diagrams. In this expression, the logarithm of the absolute value of impedance enters in a diagram as a function of the logarithm of frequency. The phase angle enters in the same

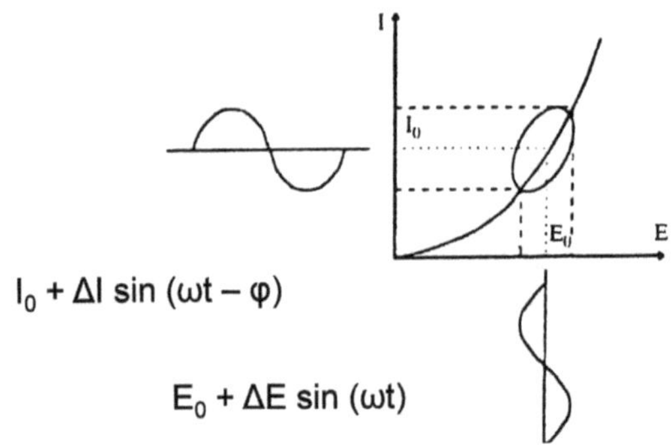

Fig. 6.18 Approximate linear relation I-E with the application of a small amplitude sinusoidal potential perturbation in a nonlinear system

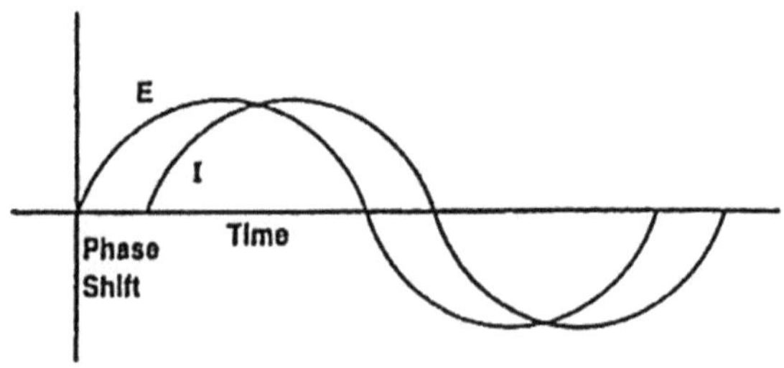

Fig. 6.19 Sinusoidal AC voltage and current signals

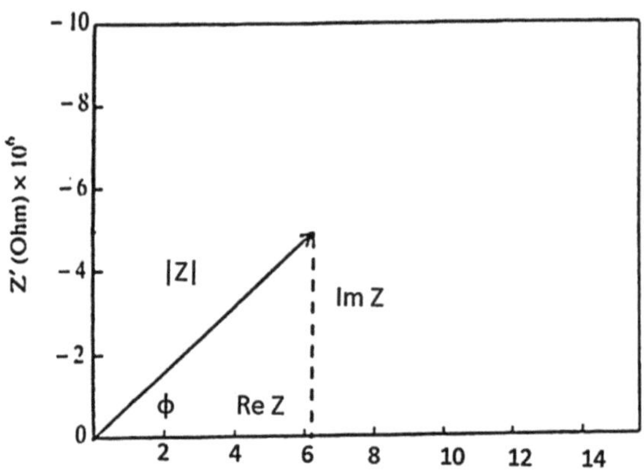

Fig. 6.20 Representation of impedance by a vector of a length |Z| with an angle φ between the real axis of impedance Z' (real part of impedance) and a vector Z'' (imaginary part of impedance)

diagram using an additional vertical axis on the right part of the Bode diagram. For the same circuit as the previous, the Bode diagram is shown in Fig. 6.23.

The electrochemical impedance is a fundamental characteristic of the electrochemical system it describes. Knowledge of the frequency dependence of impedance for a corroding system enables a determination of an appropriate equivalent electrical circuit describing that system.

The following expression describes the impedance for a system of Fig. 6.24: $Z(\omega) = \mathrm{Rs} + \mathrm{Rp}/\left(1 + \omega^2\mathrm{Rp}^2\mathrm{C}^2\right) - \mathrm{j}\omega\mathrm{CRp}^2/\left(1 + \omega^2\mathrm{Rp}^2\mathrm{C}^2\right)$, where ω = the frequency of the applied signal ($\omega = 2\pi\mathrm{f}$, rad-s^{-1}); f = the frequency of the applied

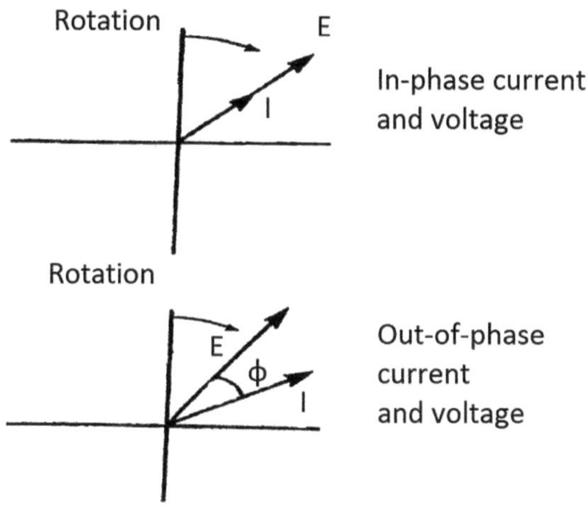

Fig. 6.21 In-Phase and out-of-phase rotation of current and voltage vectors

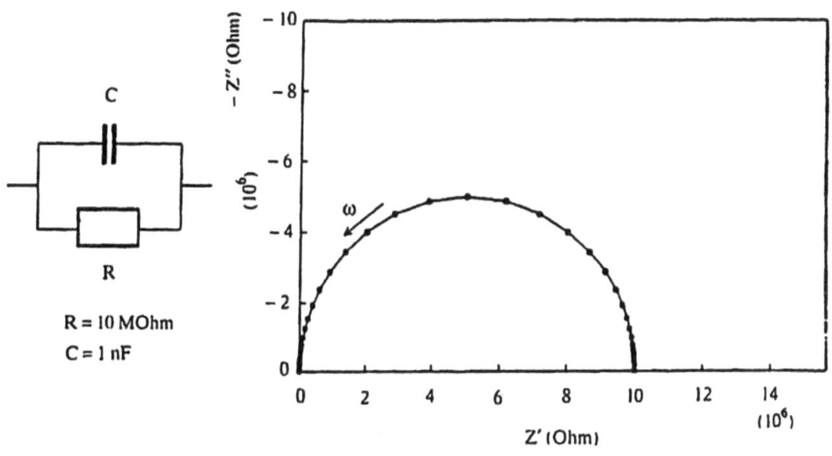

Fig. 6.22 **a** DC Circuit, consisting of a resistor and a capacitor, **b** Nyquist Diagram for the impedance to the AC equivalent circuit

signal, Hz (cycles-s^{-1}); Rs = the solution resistance, ohm-cm^2; Rp = the polarisation resistance, ohm-cm^2; C = the interfacial capacitance, Farad-cm^{-2} (Fig. 6.25).

The above equation and the Bode magnitude and phase information of Fig. 6.26 show that at very low frequencies:

$Z\omega_{\to 0}(\omega) = \mathrm{Rs} + \mathrm{Rp}$, while at very high frequencies $Z\omega_{\to\infty}(\omega) = \mathrm{Rs}$.

Determination of Rp is attainable in media of high resistivity because Rp can be mathematically separated from Rs by taking the difference between $Z(\omega)$ obtained at low and high ω

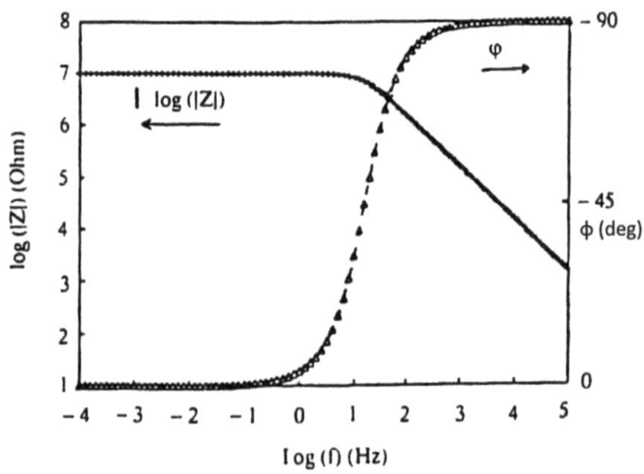

Fig. 6.23 Bode Diagram of impedance of the equivalent circuit of the previous shape

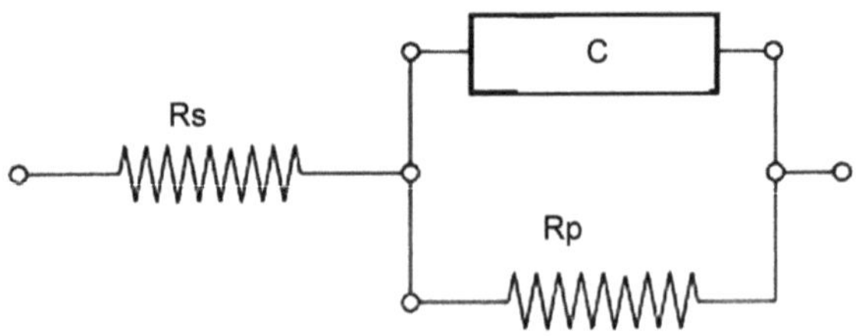

Fig. 6.24 Electrical equivalent circuit model simulating a simple corroding metal/electrolyte interface. Rs = solution resistance, Rp = polarisation resistance, C = interfacial capacitance, Rs = solution resistance, Rd = diffusion controlled resistance, Rct = charge transfer controlled resistance, Cd = coating capacitance, Cct = double layer coating-metal capacitance

$$\mathrm{Rp} = Z\omega_{\to 0} - Z\omega_{\to\infty}$$

Determination of the corrosion rate using the corresponding equation of the polarisation resistance method, also requires knowledge of Tafel slopes β_a, β_c and electrode area, which must be calculated according a suitable experiment as they are not obtained in the impedance experiment.

Either the anodic or cathodic half-cell reaction can become mass transport limited and restrict the rate of corrosion at Ecor. The presence of diffusion controlled corrosion processes does not invalid the EIS method, but does require extra precaution and a modification to the above presented circuit model. In this case, the finite diffusional impedance is added in series with the usual charge transfer parallel resistance (Fig. 6.25).

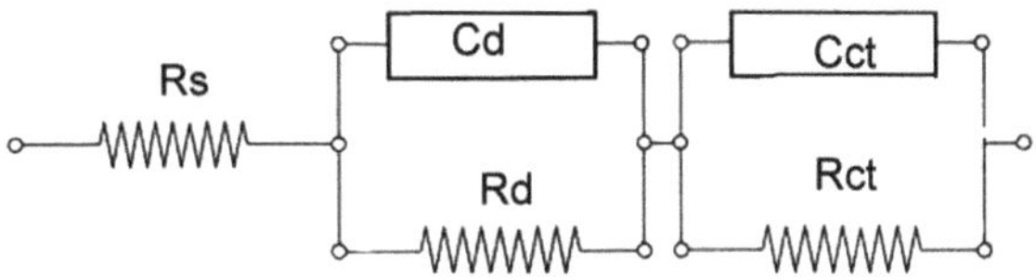

Fig. 6.25 Electrical equivalent circuit model simulating a corroding system metal/coating/electrolyte

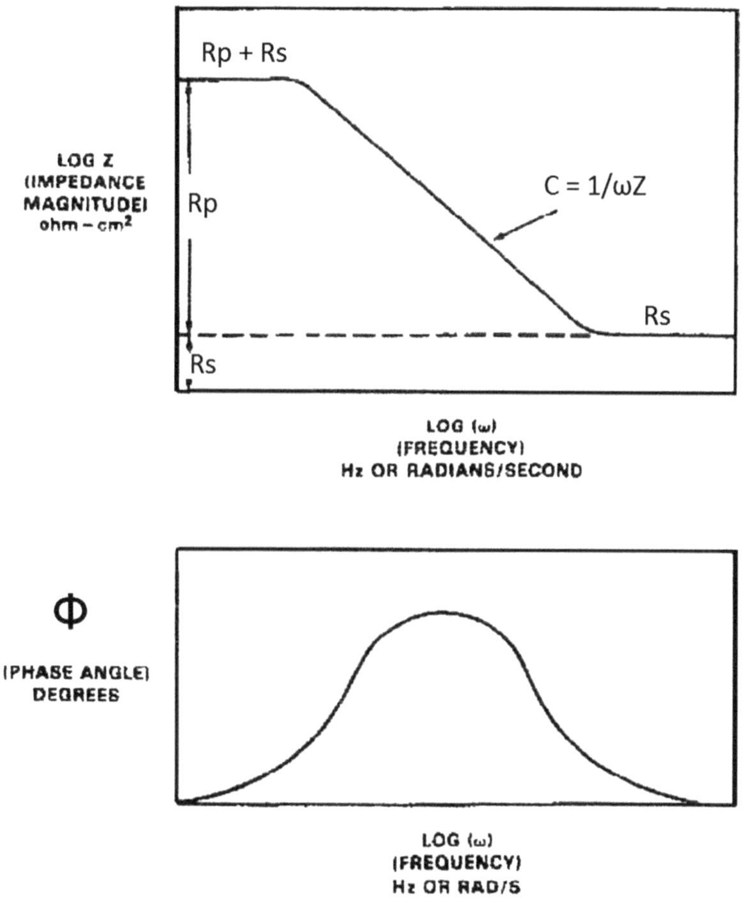

Fig. 6.26 Bode phase angle and magnitude plots demonstrating the frequency dependence of the impedance for the circuit model shown in Fig. 6.24

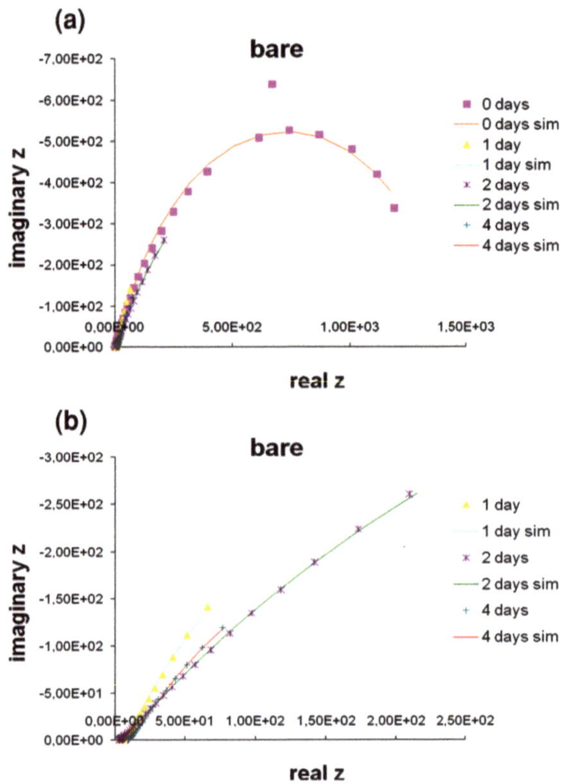

Fig. 6.27 Nyquist plots of impedance spectra for bare specimens

In a such case, Rp is the sum of the charge transfer controlled Rct, and diffusion controlled Rd contributions to the polarisation resistance assuming that Rd + Rct ≫ Rs → Rp = Rct + Rd [48–52].

6.5.2 Experimental Procedure

The steel tested was cold rolled steel (DC 01), the specimens were cut from a plate 0.3 cm thickness, the dimensions were 5 × 1.5 cm, the total exposed area was 6 cm^2.

The metallic specimens were coated with a ~20 μm thin film of pristine glassy epoxy polymer or epoxy–clay nanocomposites. The pristine liquid epoxy was diglycidyl ether of bisphenol A (DGEBA) (EPON 828RS, Hexion) with an average epoxide equivalent weight of ~187 (M.W. = 370) and was mixed at 50 °C with the appropriate amount of an aliphatic polyoxypropylene diamine (Jeffamine D-230, $M_w \approx 230$, Huntsman) which acted as curing agent. The metallic specimens were dipped in the liquid uncured mixture and where then kept vertically so that the excess of liquid was removed and left to cure at ambient conditions for 24 h. Finally, post-curing was performed at 75 °C for 3 h and

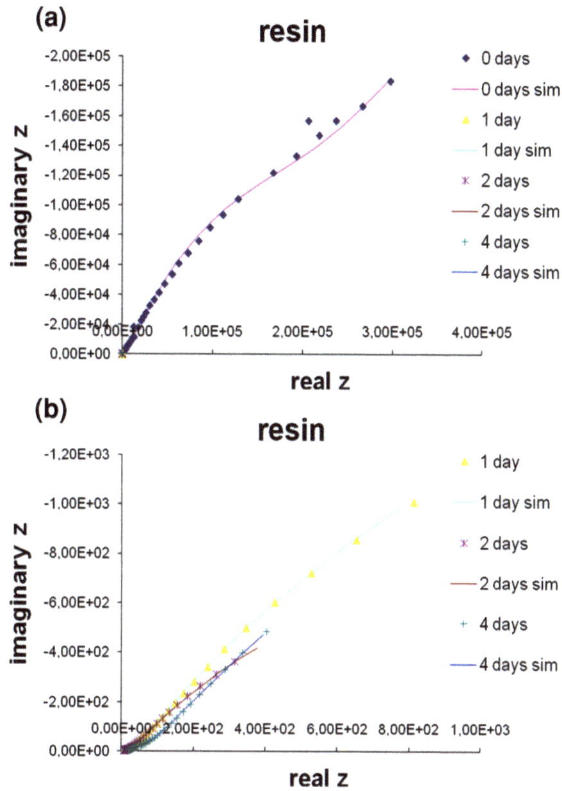

Fig. 6.28 Nyquist plots of impedance spectra for epoxy coated specimens

125 °C for another 3 h. The same procedure was applied for the coating of the specimens with the epoxy–clay nanocomposites, except that prior to adding the curing agent, the epoxy prepolymer was mixed with the (organo) clays for 1 h at 50 °C. The clays used were the Nanomer I.28E and the Nanomer I.30E (from Nanocor Inc.) which are montmorillonite clays that have been modified with quaternary and primary octadecylammonium ions, respectively.

Four types of specimens tested, blank (non-coated), coated with pristine glassy epoxy polymer, coated with the two types of epoxy–clay nanocomposites) and the times of exposure were 0 (measurement immediately after the immersion in the corrosive environment) 1, 2 or 4 days.

The electrochemical impedance spectroscopy measurements were carried out according standard methods [51, 52] in a corrosive environment of 3.5 % NaCl. They are realised in a typical electrochemical cell, in deaerated conditions and at the potential value of the corrosion potential (Ecor). The electrochemical impedance data is collected between 10,000 Hz (10 kHz) and 0.1 Hz (100 mHz) at 8–10 steps per frequency decade and the potential amplitude (E_0) was 10 mv.

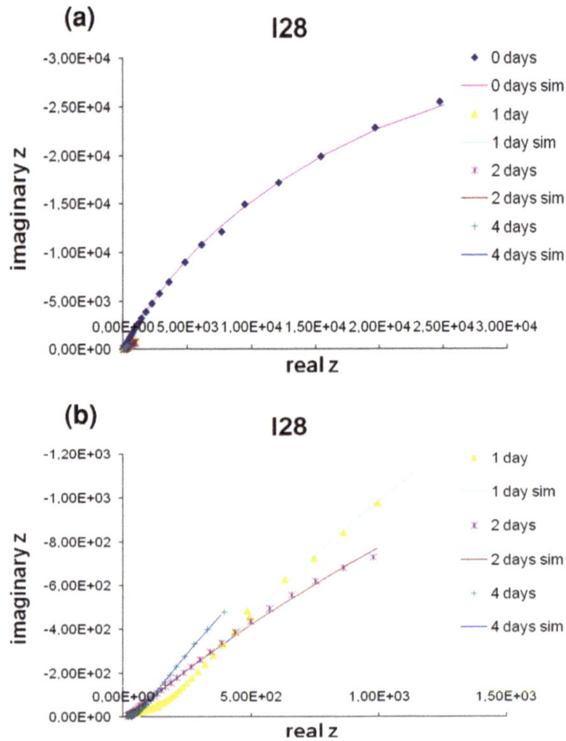

Fig. 6.29 Nyquist plots of impedance spectra for I28 nanocomposite coated specimens

6.5.3 Results and Discussion

The experimental data processing and fitting of the electrochemical impedance measurements was based on the equivalent circuits of Fig. 6.24 for bare steel and Fig. 6.25 for the coated steels and a suitable nonlinear regression analysis. The results are presented in Figs. 6.27, 6.28, 6.29, 6.30, 6.31 and 6.32 (Nyquist diagrams) and Table 6.4 (Polarisation Resistance values).

From the above results, it follows that all coatings used increase polarisation resistance and both nanocomposites have greater polarisation resistance values than pristine glassy epoxy polymer, indicating improved anticorrosive properties. The epoxy–montmorillonite clay modified with primary octadecylammonium ions, Nanomer I.30E, had a better behaviour than the modified with quaternary octadecylammonium ions, Nanomer I.28E.

A great polarisation resistance value in the case of bare steel in conditions of no prior exposure in the corrosive environment (0 days) is observed that decreases dramatically in any conditions of exposure, indicating great corrosion rate values and fast evolution of corrosion.

In the case of coated steel, the polarisation resistance values decrease for all coatings during the initial days of exposure, as it is expected due to corrosion

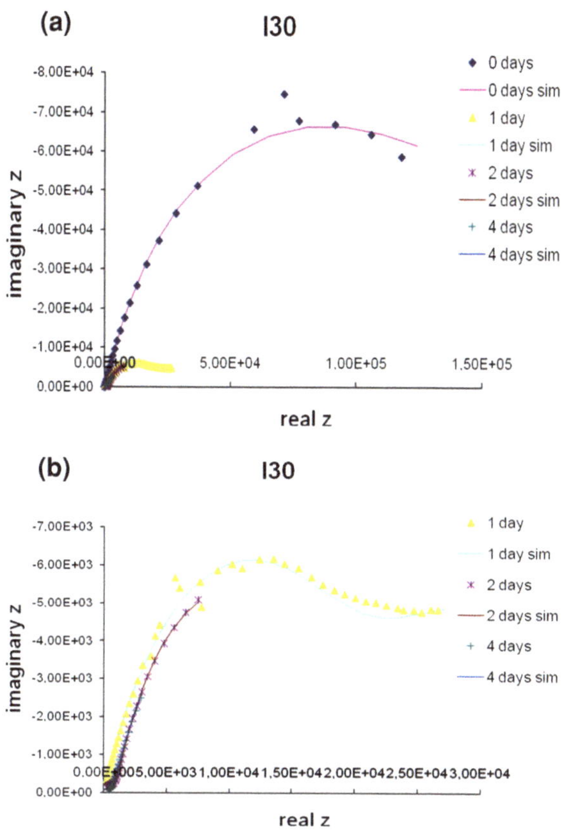

Fig. 6.30 Nyquist plots of impedance spectra for I30 nanocomposite coated specimens

initiation. But the decrease is much smaller and so corrosion evolution is too lower. Also, an increase of polarisation resistance values is observed in a longer exposure time of 4 days, more significant for nanocomposite coatings. This observation, indicating lower corrosion rates, may be is due to a passivation of the surface, but further examination is needed for a clear explanation.

A combination of Nyquist plots and also Bode plots of impedance spectra evaluation can give more precious information about corrosion reaction evolution and behaviour and properties of the polymeric and nanocomposite materials.

6.5.4 Conclusions

EIS method is a useful tool for the qualitative and quantitative study of corrosion, especially into applications that traditional DC techniques could not be successfully applied, such as corrosion measurements in cases where organic coatings are

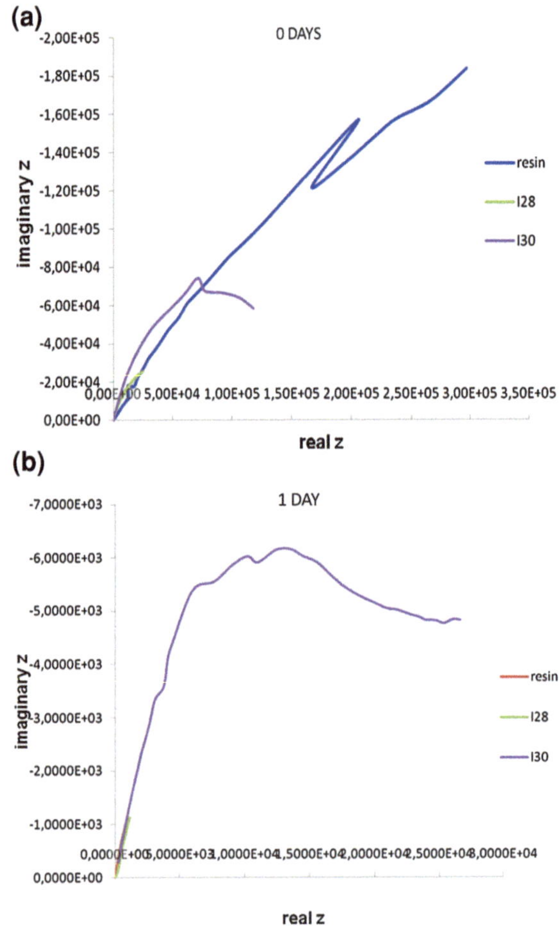

Fig. 6.31 Nyquist plots of impedance spectra for coated specimens (0 and 1 day of exposure)

used and low conductivity materials, suspension or retardation of corrosion (uniform or local), passivation of metals. The study of the behaviour and properties of polymeric materials used as protective coatings against the corrosion of metals, which is very widespread and complex due to the high resistance of the coatings and the estimation of extremely low corrosion rates is possible with EIS technique.

From the above electrochemical impedance measurements, it follows that both epoxy–clay nanocomposite tested have improved anticorrosive properties in comparison with these of the pristine glassy epoxy polymer, as it is shown from the electrochemical impedance spectroscopy measurements and the determined polarisation resistance values. The epoxy–montmorillonite clay modified with primary octadecylammonium ions, Nanomer I.30E, had a better behaviour than the modified with quaternary octadecylammonium ions, Nanomer I.28E.

Fig. 6.32 Nyquist plots of impedance spectra for coated specimens (2 and 4 day of exposure)

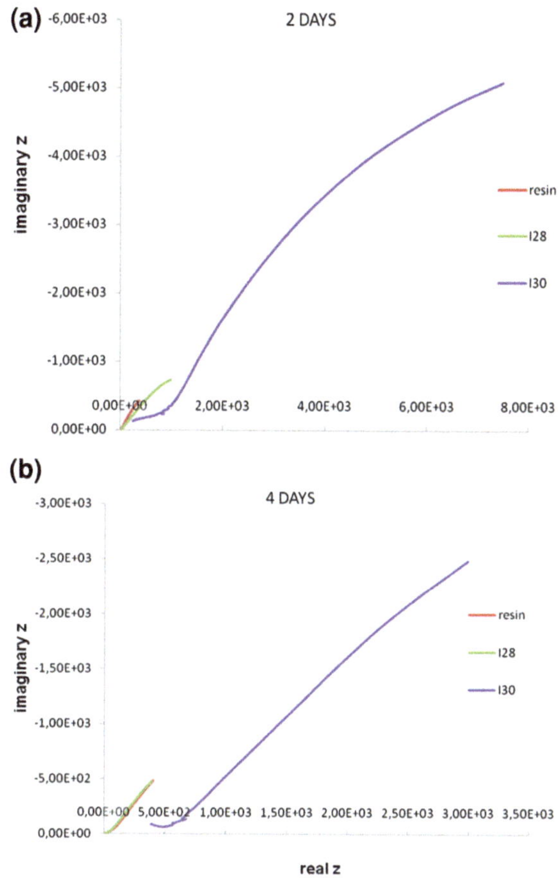

Table 6.4 Polarisation resistance values (ohm-cm^2) for bare and coated specimens

Bare	0 days	1 day	2 days	4 days
Rs	1,27E + 00	4,00E + 00	2,22E + 00	8,98E + 00
Rp	1,55E + 03	9,79E + 02	6,85E + 02	2,48E + 01
Resin	0 days	1 day	2 days	4 days
Rs	3,49E − 01	1,12E − 15	6,60E − 16	1,62E + 01
Rd	1,01E + 05	1,15E + 02	2,86E + 03	6,39E + 03
Rct	2,40E + 21	6,47E + 03	4,40E + 01	7,21E + 01
Rp	2,40E + 21	6,58E + 03	2,90E + 03	6,46E + 03
I 28	0 days	1 day	2 days	4 days
Rs	7,18E + 00	1,36E + 01	1,51E + 01	1,39E + 01
Rd	6,06E + 00	2,13E + 02	6,24E + 00	4,85E + 01
Rct	8,25E + 04	8,96E + 03	8,12E + 03	1,30E + 04
Rp	8,25E + 04	9,17E + 03	8,12E + 03	1,31E + 04
I 30	0 days	1 day	2 days	4 days
Rs	7,67E + 01	9,66E + 00	1,62E − 14	1,24E − 14
Rd	4,93E + 04	1,80E + 04	1,09E + 03	5,93E + 02
Rct	1,54E + 05	2,63E + 04	1,73E + 04	3,52E + 04
Rp	2,03E + 05	4,43E + 04	1,83E + 04	3,58E + 04

6.6 Study of Vatican Masterpieces

Giovanni Ettore Gigante and Stefano Ridolfi

6.6.1 Introduction

A number of masterpieces of the St. Peter's Basilica and its Museum have been studied, mainly during restoration works that were needed to improve the state of conservation and valourisation. The purpose of the investigation has been twofold, a better knowledge of the production technologies used and an assessment of the State of conservation of the objects. It has become a routine to fulfil diagnostic and archaeometric analyses and conservative investigations on metal works of great historical interest with the use of in situ tests and laboratory examinations. On previous occasions, as for the study of Perseus by Benvenuto Cellini and the statue of Bartolomeo Colleoni by Verrocchio [53] results have been of considerable utility for better planning the restoration operations and to better understanding of the work of art. In the last decade, the large availability of new portable instrumentations improved the diagnostic procedures that are becoming a rationale plan of measurements. A better understanding of the complementarity of the results obtained from the different techniques and the definition of the order to follow the experimental plans is an important result of these studies. The final goal is to develop protocols to obtain the maximum information with the minimum of scientific exams.

To some extent the purpose of our work is to provide, to avail non-destructive investigations and to get assisted by a limited number of laboratory tests, the basic experimental conservation plan, and to outline the subsequent conservation actions, the best way to control the degradation processes [54].

6.6.2 The Funeral Monument of Pope Sixtus IV: A Masterpiece of Italian Renaissance

The bronze funeral monument for Pope Sixtus IV (1471–1484) by Antonio del Pollaiolo, originally placed in the Choir Chapel in the Constantine Basilica of San Peter's in the Vatican and now in the Museum of the Vatican Basilica, has a complex composition with the Pope placed at the centre of an inner ring with the personification of the virtues and an outer one with ten allegories of Art and Science. The work was commissioned by the Cardinal Giuliano della Rovere, nephew of Pope Sixtus, himself a future pope later known as Giulio II. The work was created between 1484, the year in which the Pope died, and 1493, the date inscribed on the monument together with the artists signature by Antonio del Pollaiolo (1431–1498) one of the greatest painter, sculptor and goldsmith of the time.

The diagnostic campaign was [55] carried out by first making a series of non-invasive measurements using a transportable EDXRF and a portable Raman system, to map the deterioration processes that had altered the bronze surface and quantify the alloy used in the several pieces the monument was made of. As a consequence of the first non-invasive diagnostic campaign, the second campaign of micro-invasive tests was planned and carried out. The samples were analysed with SEM–EDS, NMR and XRD techniques. The procedure implemented was focused to maximise the diagnostic information gathered and minimise the microsampling on the work of art.

The EDXRF investigations were carried out in two different ways;

- without any removal of the patina, but after cleaning to remove the layer of wax, with a tube HV sufficient to reveal some light elements (S, Cl, Ca, K);
- by scratching an area of 2 mm of patina, in order to study the bulk alloy, and with the instrumental set-up so it would pick up also on tin and antimony.

On some areas of the bronze panels 2, 3, 4, examined in the first way, some control measurements were taken after cleaning with water and EDTA.

With the first method, 101 areas were examined (which were quite well distributed over all the panels) and nine control measurements were taken after a further cleaning.

Finally, 33 quantitative analysis were carried out on some of the parts put together to build the panels.

The portable EDXRF system used, is composed of an air cooled low power tube (35 kV and 0.2 mA) plus a Silicon Drift Detector (SDD) detector with 150 eV energy resolution at the iron *K* line. The tube worked with a 2 mm aluminium filter for the quantitative analysis on the bronze alloy while it worked without filter for the analysis on the patina.

In Fig. 6.33, we can see a moment of the non-invasive measurements. In Fig. 6.34, it is possible to see the analysed spots are shown on the "Music" panel.

The measures were divided in low energy and high energy. In the low-energy set-up, the elements from sulphur (2.3 keV) to iron (6.4 keV) were measured while in the high-energy set-up the elements from iron (6.4 keV) to antimony (26.4 keV) were analysed.

The systematic low-energy EDXRF measurements allowed us to analyse elements which are part of the surface treatment. These are: K, Ca, Mn and Fe. With the 111 surface areas studied on the ten panels, it has in fact been possible to demonstrate a good correlation between such elements. As a confirmation of the idea of a surface treatment, SEM/EDS and ^{1}H-NMR on samples were fulfilled. In the latter signals of an unsaturated waxy or oily substance, as well as signals relating to the use of cleaning products were found. While in the former the presence on all of the panels of the aforementioned elements agreed with a generalised surface treatment with coloured waxes and pigments such as bone black,

Fig. 6.33 Non-invasive measurements with a portable EDXRF on the funeral monument of Pope Sixtus IV

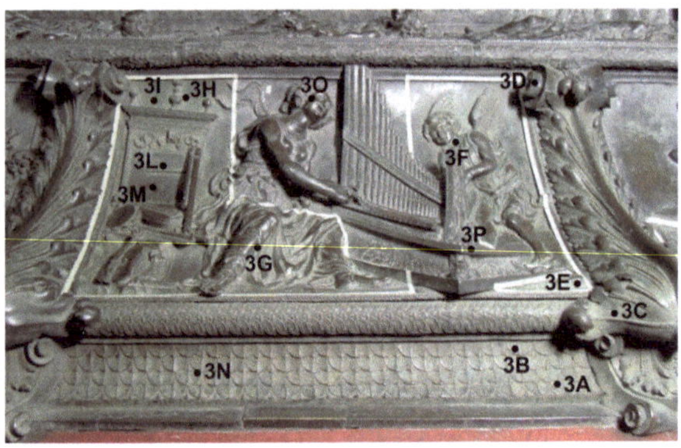

Fig. 6.34 Panel "the Music" and in black the spots where EDXRF non-destructive analyses were fulfilled

(given that in the EDS test on the swabs residual powder the phosphorous and calcium were found in stoichiometric ratios suggesting the presence of calcium phosphate) and Hearth pigment containing iron oxides and/or manganese which confer the characteristic dark colour of the statue.

In the high-energy set-up, 33 different bronze subparts of the burial monument were analysed to understand the bulk alloy composition. The quantitative analyses were fulfilled on the areas where the patina had been removed to avoid contamination of degradation products. In Table 6.5, the quantitative results are shown.

In the last two rows of the table, the mean composition and the standard deviation of the measurements are reported. The composition is typical of a bronze from that period.

The analysis of the individual elements gathered in the table allows us to verify whether Pollaiolo used alloys with different composition. The analysis of the variance for the four principle elements (copper, tin, lead, and antimony) has permitted us to demonstrate that the differences in composition of the alloys used to create the body of the masterpiece, in respect of those used for the panels and the floral decorations are significant.

Table 6.5 Composition of the different panels (EDXRF analyses)

		Cu	Sn	Pb	Ag	Sb	Fe	Ni	Zn	As
Panel 1	Foil	86.7	9.1	2.4	0.1	1.4	0.3	–	–	–
	Leaf	85.9	12	2.3	0.1	–	–	–	–	–
	Body	88	9.4	2.5	0.1	–	–	–	–	–
Panel 2	Body	87.7	8.7	2.7	0.1	0.2	0.5	–	–	–
	Foil	82.2	13	1.9	0.1	2.3	0.4	–	–	–
	Foil	83.5	3	3.2	0.1	–	0.4	3	6	–
	Leaf	86	12	2.5	–	–	–	–	–	–
Panel 3	Body	86.4	9.9	2.5	0.1	0.6	0.4	–	–	–
	Foil	80.9	13	2.8	0.1	3.2	–	–	–	–
	Leaf	86.8	9.9	2.2	0.1	0.9	–	–	–	–
	Leaf	87.7	9.2	2.4	–	0.7	–	–	–	–
Panel 4	Body	88.4	8.7	2.6	0.2	0.1	–	–	–	–
	Foil	79	14	2.6	0.1	3.9	0.4	–	–	–
	Leaf	84.1	12	2.8	0.1	1.5	–	–	–	–
Panel 5	Foil	84.6	10	1.6	0.1	3.3	–	–	–	–
	Leaf	81	14	2.4	0.1	1.7	0.4	–	–	–
Panel 6	Body	87.9	5.9	4.1	0.1	1.9	–	–	–	–
	Foil	87.5	8	1.5	0.1	2.5	–	–	–	1
	Leaf	90.2	7.9	1.8	–	0.1	–	–	–	–
	Body	89.5	7.5	2.7	0.2	0.1	–	–	–	–
Panel 7	Foil	86.4	8.5	2.3	0.1	2.6	–	–	–	–
	Leaf	87.8	9.9	2.3	–	–	–	–	–	–
	Body	90.9	6.6	2.3	0.2	–	–	–	–	–
Panel 8	Foil	86.4	9	2.5	0.1	2.1	–	–	–	–
	Leaf	88.1	9.8	2	0.1	–	–	–	–	–
Panel 9	Body	89.3	8.3	2.2	0.2	–	–	–	–	–
	Foil	87.9	8.5	2.3	0.1	1.2	–	–	–	–
	Body	87	7.8	3.2	0.1	1.4	0.5	–	–	–
	Leaf	86.6	9.3	2.5	0.2	1.1	0.3	–	–	–
	Leaf	89	8.2	1.7	0.1	1	–	–	–	–
Panel 10	Foil	86.4	8.5	4	0.1	0.6	0.4	–	–	–
	Leaf	88.5	9.4	2	0.1	–	–	–	–	–
	Body	90.6	7	2.2	0.2	–	–	–	–	–
	Mean	86.63	9.31	2.45	0.11	1.04	0.12			
	s.dev	2.80	2.30	0.57	0.06	1.14	0.19			

If we observe the table, we can see that the only area on a foil that has no presence of antimony is foil 10. It is also the only one with nickel and zinc, we can therefore hypothesise that we are dealing with later reparation.

Fig. 6.35 Inscription on the external gilded surface with the name of the author and date of production

Fig. 6.36 The Saint Peters Basilica and indicated the golden globe

6.6.3 The Golden Globe Placed on the Top of San Peter's Dome

The restoration and maintenance works of the famous golden globe placed on the top of San Peter's dome have been a unique opportunity to access the inside and outside of it and verify some aspects of the ancient production technologies and set-up of the big bronze artefacts. The analyses were fulfilled with a portable EDXRF system, maybe the only non-invasive technique that was possible to carry to such an inaccessible position. In Fig. 6.35, we can see the inscription present on the external gilded surface with the name of the architect Jacopo della Porta whom in 1593 moulded the globe.

The globe weighs 1,862 kg and his diameter measures 2.5 m. It is put right on top of the dome and just below the cross at 130 m from the floor of the Basilica. The globe is composed by 39 panels mutually soldered, numbered in longitudinal by letters and in latitudinal by numbers (starting from the 1 element top). The panels are composed of bronze and the external surface is gilded [56].

In Fig. 6.36, the Basilica of Saint Peter is shown and pinpointed is the golden globe.

The bronze alloy has a mean composition of 85.8 ± 0.5 % of copper, 10.1 ± 0.3 % of tin and 2.8 ± 0.1 % of lead. Minor elements, such as iron, silver and antimony, were as well quantified. The origin of these minor elements is related to impurities from the extraction processes for copper, tin and lead. Of major importance were the results related to the soldering processes that connected all the panels. The process was of autogenous welding. The alloy of the panels and the alloy of the welding material are the same.

In Table 6.6, the quantitative results plates and welding are shown.

The archaeometric analyses confirmed that the alloy has a very good mechanical resistance as it should be for a bronze object exposed to the weather in such a high position. The high homogeneity of the alloy of the different plates and welding material shows that the work was fulfilled with great attention and cure and proves the knowledge of the craftsmen in melting techniques. The welding is composed of approximately the same alloy of the plates indicating the use of autogenous welding that was not easy to fulfill in such an unreachable position.

6.6.4 The Bronze Statue of Saint Peters in the Saint Peters Basilica in Rome

The bronze statue of Saint Peters is positioned in the interiors of the Saint Peters Basilica in Rome. It was believed to belong the late roman period (fifth century A.D.) but some late studies, based on typological comparisons, changed its origin to Arnolfo di Cambio (thirteenth century A.D.); but discussion it is still open.

The scientific analyses conducted on the statue were mainly non-invasive. Too important is the votive aspect of the statue to intervene in more destructive ways. The analyses conducted were:

- Determination of the alloy composition availing of a portable non-invasive XRF system
- International Annealed Copper Standard (IACS) on the surface of the statue for non-invasive qualitative valuation of the homogeneity of the alloy.
- Spectrocolorimetic non-invasive analyses of the surface of the statue.
- Metallographic destructive examination on samples
- SEM/EDS destructive analyses of samples
- XRD on surface powder
- Lead isotopic ratio on a sample

About the EDXRF measures, nine different points were cleaned. In Table 6.7, the XRF results are shown.

In Fig. 6.37, a moment of the XRF measures is shown.

The XRF results show a very high homogeneity of the statue except at the right hand and the fingers of the right hand that are supposed to be of restoration and so different from the original alloy of the statue.

Table 6.6 Composition of the different panels (EDXRF analyses)

	Zn	Pb	Sn	Fe	Ag	Sb	Cu
4a	0.9	2.9	9.3	0.2	0.1	0.5	86.2
5a	0.5	3.1	10.5	0.1	0.1	0.6	85.1
Welding between 5a and 6a	0.3	2.3	8.6	0.0	0.1	0.4	88.2
6a	0.4	2.9	9.5	0.1	0.1	0.6	86.5
3b	0.5	2.2	8.5	0.1	0.1	0.5	88.1
Welding between 3b and 4b	0.9	2.8	8.5	0.2	0.1	0.5	87.1
4b	0.4	4.3	13.0	0.1	0.1	0.9	81.3
Welding between 4b and 5b	0.2	2.5	9.4	0.1	0.1	0.5	87.3
5b	0.4	2.0	9.1	0.1	0.1	0.6	87.8
Welding between 5b and 6b	0.4	2.0	7.0	0.1	0.1	0.4	90.0
6b	0.3	3.4	10.7	0.1	0.1	0.6	84.8
Restoration spot in 3b	0.5	2.4	9.8	0.1	0.1	0.7	86.5
3d	0.7	3.0	10.9	0.2	0.1	0.6	84.5
4d	0.5	3.4	11.8	0.2	0.1	0.7	83.4
5d	0.3	3.6	13.5	0.1	0.1	0.6	81.7
Welding between 5d and 6d	0.3	2.3	8.8	0.1	0.1	0.5	87.9
6d	0.4	3.7	11.7	0.1	0.1	1.1	83.0
Restoration spot in 5d	0.5	2.5	11.1	0.1	0.1	0.6	85.1
5e	0.3	2.9	10.7	0.1	0.1	0.8	85.0
Welding between 5e and 6e	0.3	2.1	8.8	0.1	0.1	0.6	88.0
6e	0.3	2.5	9.0	0.1	0.1	0.5	87.5
4 g	0.5	2.6	9.7	0.2	0.1	0.6	86.3
5 g	0.4	2.3	9.0	0.1	0.1	0.5	87.6
4 h	0.4	2.1	9.6	0.1	0.1	0.5	87.2
5 h	0.4	2.6	9.0	0.1	0.1	0.6	87.4
Restoration spot in 5 h	0.5	0.7	0.5	0.1	0.1	0.7	97.4
4 l	0.4	2.1	8.6	0.1	0.1	0.5	88.2
5 l	0.4	3.0	9.9	0.1	0.1	0.6	85.9
4 m	0.5	2.3	8.4	0.1	0.1	0.5	88.2
Welding between 4 and 5 m	0.5	1.8	8.4	0.1	0.1	0.5	88.7
5 m	1.4	4.2	22.9	0.3	0.2	1.2	69.9

Table 6.7 Composition of the different parts (EDXRF analyses)

	Pb	Sn	Fe	Ag	Sb	Cu
Left thigh	5.6	24.8	0.6	0.1	1.5	67.5
Key	4.8	26.3	0.9	0.1	1.9	66.1
Right thigh	4.3	25.8	0.4	–	1.5	68.1
Hair	6.2	21.6	0.3	–	1.1	70.8
Neck	5.7	26.5	0.4	0.1	1.3	66.0
Back	4.0	22.8	0.3	0.1	1.4	71.4
Right hand forefinger	6.4	30.0	0.5	0.1	1.9	61.2
Right hand middle finger	7.1	29.2	0.4	0.1	1.8	61.4
Right hand palm	4.1	34.2	0.3	0.1	1.9	59.3

Fig. 6.37 The statue of Saint Peter in Vatican, EDXRF analyses

The results of the non-invasive techniques and the metallographic analysis indicate that:

- The alloy is very rich in Tin, that is typical of a bell bronze, than can be casted by artisans doing this work. They were present in the area from the fifth to the thirteenth century.
- The colorimetric measurement raised the suspect of a different alloy for the legs.
- The results of metallographic and SEM studies show the presence of large Tin and Lead segregation zone and the presence of dendritic structure, typical of an alloy melted with already casted metals.

6.7 Laser Applications in the Preservation of Cultural Heritage: An Overview of Fundamentals and Applications of Lasers in the Preservation of Cultural Heritage

R. Lahoz, L. A. Angurel, U. Brauch, L. C. Estepa and G. F. de la Fuente Leis

6.7.1 Laser Fundamentals for Non-Specialists

Lasers do emit a special type of light which can be useful in an extremely wide range of applications. To understand what makes laser light special and how it is produced, we will begin our introduction to lasers with a short discussion of what is the nature of light and what is different with laser light. From here, it will become quite obvious that what is basically needed to build a laser producing this kind of light. We will then discuss a very simple and intuitive set of equations, the so-called rate equations, which allow describing many laser phenomena quantitatively. Finally, some specific laser systems will be looked at, the disc and fibre laser as two versions of solid-state lasers, and the diode laser.

6.7.1.1 The Nature of Light

People have always been wondering what light is made of. The dispute became famous between Huygens, who considered light as a kind of wave, and Newton, who thought of light as particles, propagating in straight lines (light beams). This dispute seemed to be finally settled when *Maxwell* came up with his famous set of equations unifying the description of both electric and magnetic phenomena:

$$\begin{aligned} \text{rot } H &= \dot{D} + j \quad & \text{div } D &= \rho \\ \text{rot } E &= -\dot{B} \quad & \text{div } B &= 0 \end{aligned}$$

Here, E and D are the electric field and electric displacement vectors, H and B are the magnetising and magnetic field vectors, and ρ and j the macroscopic charge and current densities, respectively. Maxwell could show that his equations include the propagation of electromagnetic fields through space as waves described by the *wave equations*

$$\Delta E = \frac{1}{c^2}\ddot{E} \quad \Delta B = \frac{1}{c^2}\ddot{B}$$

They can be derived using some mathematical transformations from the Maxwell equations assuming vacuum ($\rho = j = 0$; $D = \varepsilon_0 E$; $B = \mu_0 H$)

The propagation speed of these electromagnetic waves

$$c = \frac{1}{\sqrt{\varepsilon_0 \mu_0}} = 2.997 \cdot 10^8 \frac{m}{s}$$

happens to be identical to the *speed of light*. From then on, it was clear that light is just a (small) part of the electromagnetic spectrum which ranges from radiowaves and microwaves via Infrared, Visible and UV, all the way to X and γ rays (Fig. 6.38).

There are many different classes of solutions to the wave equation. Perhaps, best known are plane harmonic waves in space and time (and their linear combinations):

$$E = E_0 \sin 2\pi \left(\frac{r}{\lambda} - vt\right)$$

$$B = B_0 \sin 2\pi \left(\frac{r}{\lambda} - vt\right)$$

The relation between wavelength λ and frequency v is given by the speed of light: $c = \lambda v$ (Fig. 6.39).

In principle, plane waves extend from $-\infty$ to $+\infty$. Laser beams with a well-defined beam diameter are better described by "Gaussian beams" which have a Gaussian field and intensity distribution in the cross-section perpendicular to their propagation direction.

In the beginning of the twentieth century, people became aware that there might be more to it than what is described by Maxwell's equations: An important observation was the *photo electric effect*. Lenard noticed that electrons escaping from a surface which is hit by light have a kinetic energy that does not depend on the light intensity or power density—which is proportional to the product of E_0 and B_0—but rather depends on the light frequency v:

$$W_{\text{kin}}^{\text{el}} = \frac{m_0}{2} v^2 = hv - \chi$$

χ is a material constant describing the energy necessary to get the electron out of material (work function). Einstein noticed in 1905 that this equation is simply the energy conservation law if one assumes that light comes in energy packages of hv or $\hbar\omega$, so-called light quanta or photons. $h = 6.626 \times 10^{-34}$ Js is a constant which had to be introduced by Planck a couple of years earlier to correctly describe the blackbody radiation spectra (Fig. 6.40).

Perhaps, even more puzzling was the *double-slit experiment* that could be done with light or with matter particles, electrons, for example. In both cases, one did not find a simple superposition of the images of the two slits but rather interference effects—of course not so much surprising in case of light (electromagnetic waves!) but rather strange in case of what were thought to be classical particles. These interference effects also showed up if the intensity was so much reduced that only

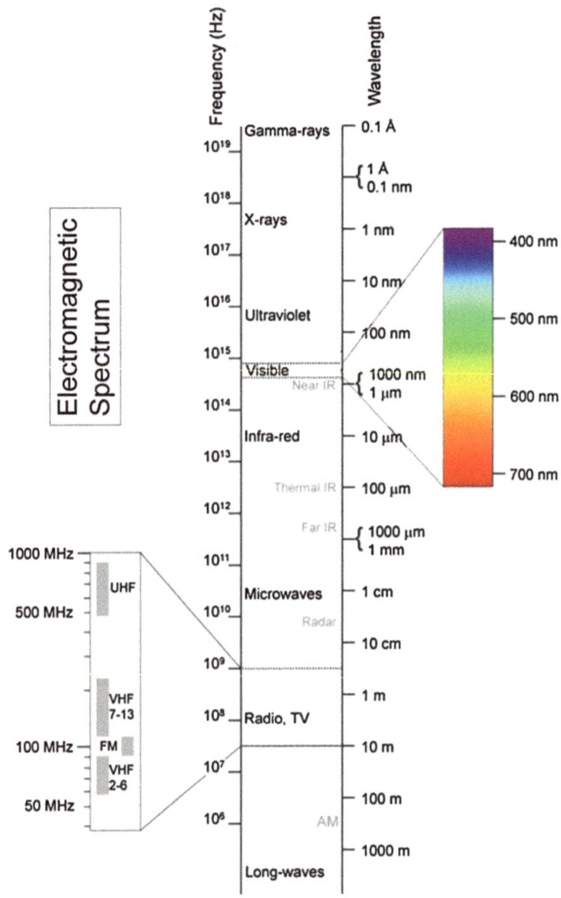

Fig. 6.38 The electromagnetic spectrum. Only a very small section with wavelengths between 400 and 700 nm is visible to the human eye [57]

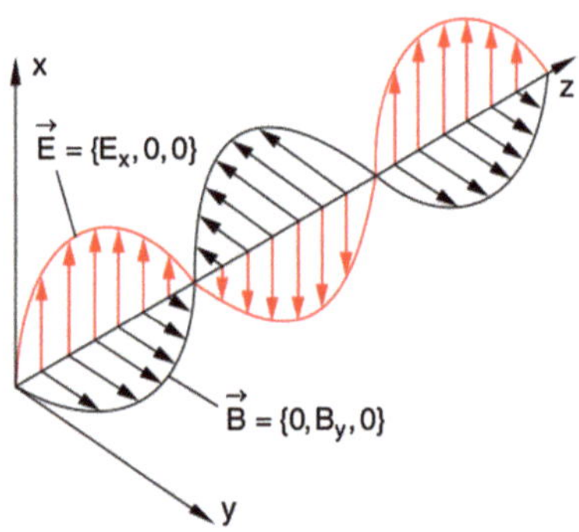

Fig. 6.39 Electromagnetic plane wave propagating along z [58]

one electron, photon, etc. passed the slits at a time. Adding up all these single-photon/single-electron experiments again resulted in interference effects. So, each single particle (electron or photon) must have "known" of the existence of both slits or might have passed through both slits (Fig. 6.41).

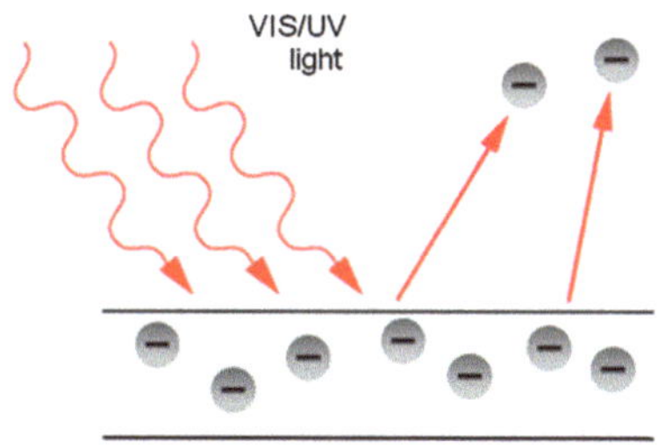

Fig. 6.40 Visualisation of the photo-electric effect [59]

In the 1920s, a formalism was developed by Schrödinger, Heisenberg and others that could handle the wave–particle dualism of classical (mechanical) particles (*quantum mechanics*); somewhat later a similar formalism was developed for classical fields (quantum field theory; in case of the electromagnetic field: *quantum electrodynamics, QED*). In QED, the proven classical solutions of the Maxwell equations (consistent with the boundary conditions, e.g. given by a laser resonator)—the so-called "modes"—form the basis. The state of the system is then characterised by the number of photons that occupy each mode. This means the energy that can be found in each of the electromagnetic modes cannot be chosen arbitrarily but can have discrete numbers—multiples of $h\nu$—only. In most practical cases, there is no need to use the full QED formalism. Instead, one can consider light either as classical electromagnetic field (characterised by frequency ν, wavelength λ, electric and magnetic field amplitudes E and H) or alternatively think of it as classical particles moving with the speed of light c, having an energy $E = h\nu$ and a momentum $p = h/\lambda$—whichever is more convenient.

Now, back to the question at the beginning: What makes laser light special? In "high-quality" laser light ideally all photons occupy one *single mode* while in light from thermal sources like the sun or tungsten filament lamps the photons will be distributed over all available modes with a distribution over the frequencies or wavelengths according to Planck's formula for the blackbody radiation (Fig. 6.42).

6.7.1.2 Stimulated Emission

There is another famous paper by Einstein—on the absorption and emission process—that gives a hint what we can do to favour the emission of identical photons. He considered the simplest atomic system, consisting of two energy levels only, the ground state E_1 and the excited state E_2.

First, assume the atom is in the ground state. If a photon arrives with $h\nu = E_2 - E_1$ (energy conservation!), the photon can be absorbed by lifting the atom in its excited state [*case c, (stimulated) absorption*]. With some delay which varies statistically—the spontaneous lifetime τ—the atom returns to its ground state by emitting a photon of energy $h\nu$ (*case a, spontaneous emission*).

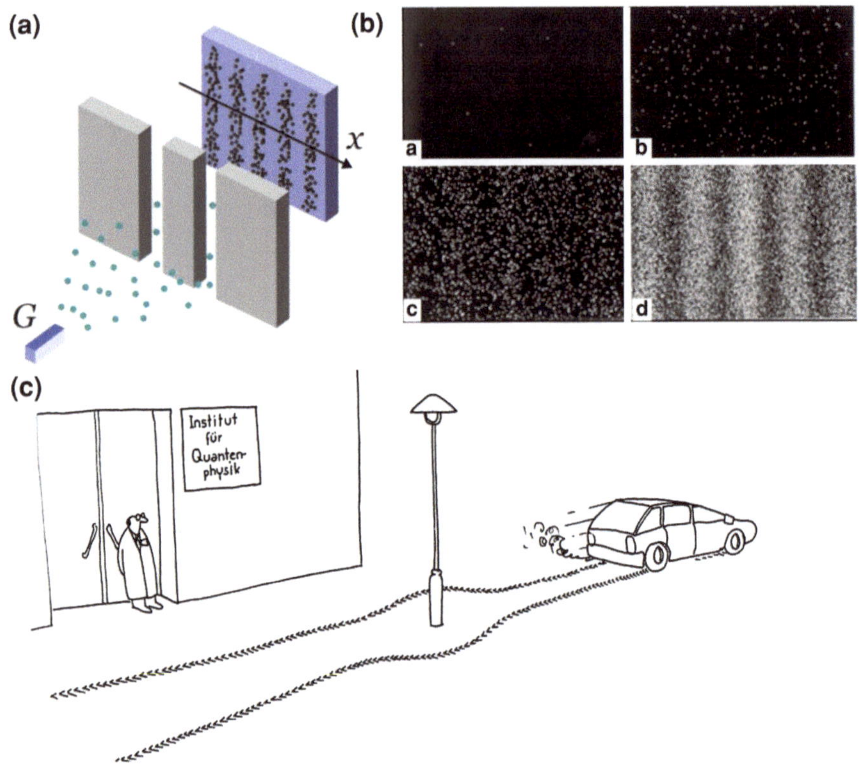

Fig. 6.41 **a** The double-slit experiments with particles resulting in interference effects known from similar experiments with light [60]. **b** These effects show up even if only one particle (electron or photon) passes the double slit at a time. Shown here are the interference patterns of 8/270/2000/60000 (single!) electrons [61]. **c** The macroscopic analogue to the double-slit experiment [62]

However, there is another type of emission process that is the absorption process (case **c**) in reverse: Again, a photon of energy $h\nu$ is arriving, but now the atom is in its excited state. As in the absorption case, a transition will be stimulated from the occupied level (in this case E_2) to the empty level (in this case E_1) and for energy conservation another photon of energy $h\nu$ will be emitted (*case b*; *stimulated emission*). Either there is net (stimulated) absorption or net stimulated emission depends solely on the occupation difference of the two energy levels. If the higher level E_2 more strongly populated than the lower level E_1, i.e., $N^* = N_2 - N_1 > 0$, *inversion*, there is stimulated emission or gain, if $N^* < 0$ there is (stimulated) absorption, if $N^* = 0$ the material is transparent.

The stimulating and the stimulated photon have the same properties, i.e., they belong to the same mode. This stimulated-emission process is exactly the mechanism needed to increase the number of photons in a certain mode and hence gave the *LASER* its name: *Light Amplification by Stimulated Emission of Radiation* (Fig. 6.43).

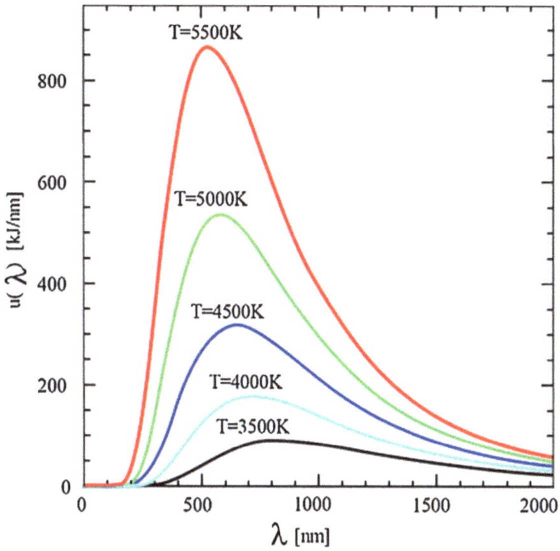

Fig. 6.42 Thermal or blackbody radiation with all available electromagnetic modes occupied by photons according to Planck's law [63]

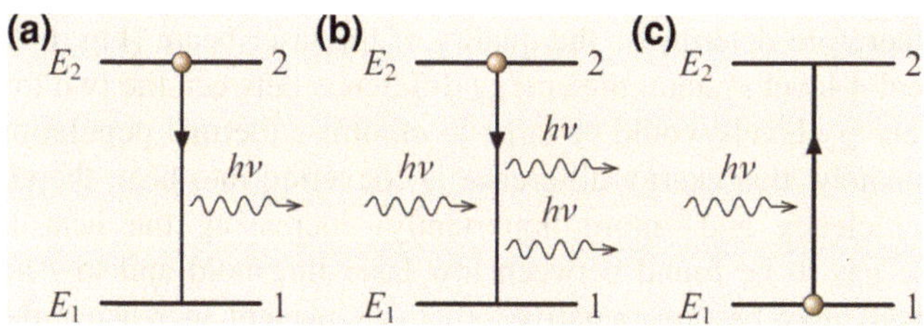

Fig. 6.43 The three types of optical transitions **a** spontaneous emission, **b** stimulated emission, and **c** (stimulated) absorption in a 2-level system with $E_2 - E_1 = h\nu$ [64]

6.7.1.3 Laser Set-up

In order to favour the stimulated-emission process, an optical material is needed which has a partly occupied higher energy level E_2 and a mostly empty lower energy level E_1 (inversion $N^* > 0$). This can be accomplished best by using a (more or less) *ideal 4-level system*. Here, the assumed fast (non-radiative) relaxation process between E_1 and E_0 and between E_3 and E_2 ensures (almost) empty levels E_3 and E_1. In this way, (almost) perfect absorption occurs when photons of energy $h\nu_{\text{pump}} = E_3 - E_0$ hit the material. Remember, if E_3 was partially occupied there would be also the reverse process, reducing the absorption efficiency. Similarly, if laser photons of $h\nu_{\text{laser}} = E_2 - E_1$ hit the sample with the final, i.e. lower laser level E_1 being mostly empty, and E_2 partially filled, there will be mainly stimulated emission (Fig. 6.44).

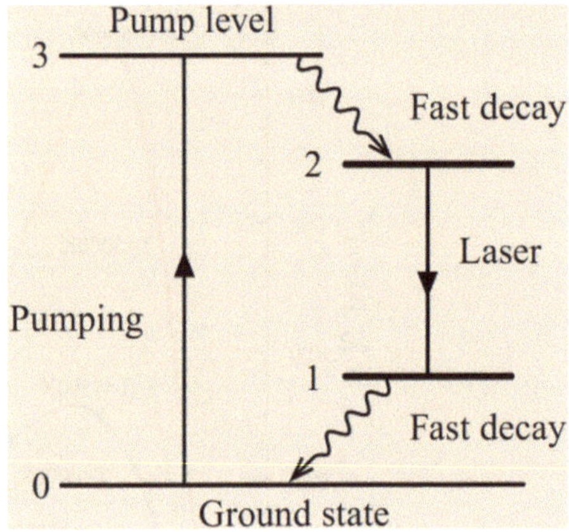

Fig. 6.44 Energy-level diagram of a 4-level laser [64]

To operate this 4-level system as laser, one only has to make sure that a sufficient number of atoms is in the excited state, for example by *pumping* the active material with $h\nu_{\text{pump}}$ of an intense light source, and that enough stimulated photons $h\nu_{\text{laser}}$ are around by recycling the already emitted laser photons with the help of mirrors, more precisely by putting the active material in an *optical cavity* or a *resonator*. The resonator defines the mode(s) where stimulated emission takes place and therefore determines the quality of the laser beam (Fig. 6.45).

In an ideal 4-level system, the energy difference between the two lower and the two upper energy levels would be large to minimise thermal population of E_1 and E_3. Unfortunately, this energy difference is converted into heat, thereby reducing the laser efficiency and—more importantly—increasing the heat load. So, a compromise has to be found between low laser threshold and low heat load. In addition, there may be non-radiative processes present increasing the heat load further. In any case, a *cooling* system is needed which efficiently removes the heat from the active material.

Putting everything together, one ends up with a set-up like this one (Fig. 6.46):

6.7.1.4 The Rate Equations

Let us consider what processes do influence the *population of the upper laser level* N_2 of an ideal 4-level system. These are:

1. *Pumping*

$$W_{\text{gen}}^{\text{opt}} = \eta_{\text{gen}}^{\text{opt}} \frac{\phi_{\text{pump}}}{s} \qquad W_{\text{gen}}^{\text{inj}} = \eta_{\text{gen}}^{\text{inj}} \frac{I}{q}$$

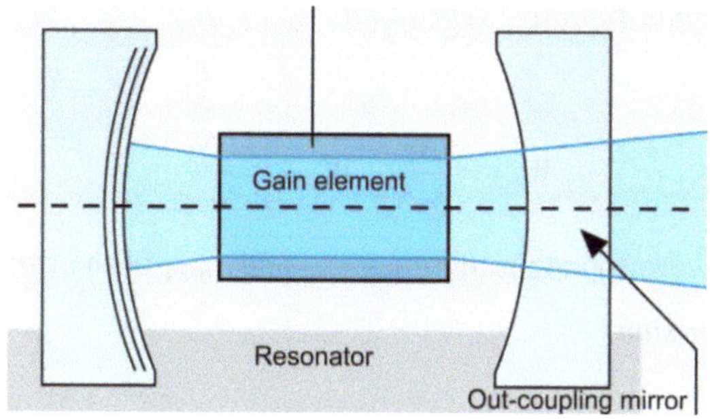

Fig. 6.45 Optical cavity or resonator consisting of the gain medium amplifying the (laser) photons and two mirrors (highly reflective on the *left* and semitransparent on the *right*) defining the laser mode, favouring stimulated emission into this mode by optical feedback, and finally coupling out the stimulated photons as laser beam [65]

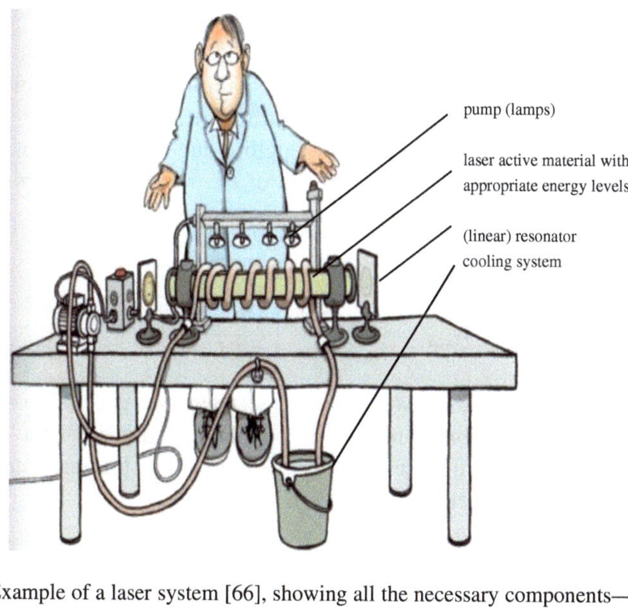

Fig. 6.46 Example of a laser system [66], showing all the necessary components—including the operator

In case of optical pumping, the pump rate is the number of pump photons Φ_{pump} absorbed per second; in case of electrical pumping (laser diodes), it is the number of electrons injected per second I/q, in either case modified by the efficiency of transferring this into an excited state

2. *Radiative and non-radiative relaxation*

$$W_{\text{rad}} = \frac{N_2}{\tau_{\text{rad}}} \quad W_{\text{nrad}} = \frac{N_2}{\tau_{\text{nrad}}}$$

according to the corresponding lifetimes τ_{rad} and τ_{nrad} of the excited state

3. *Stimulated emission*

$$W_{\text{stim}} = BN_2\,\phi = c_{\text{gr}}\,g(N_2)\phi$$

proportional to the Einstein coefficient of stimulated emission B, population of the upper laser level N_2, or—in more experimental terms—the gain $g(N_2)$, and the number of laser photons (Φ)

Now, let us do the same for the *number of photons in the laser mode* Φ. Here, we have the following three processes:

1. *Stimulated emission*. This is exactly the term above
2. *Radiative relaxation*. In this case, only those photons count that accidentally get into the laser mode, usually only a small fraction β.
3. *Finite cavity lifetime* τ_{cav} in a cavity of length L due to resonator losses *Abs* and outcoupling of laser photons through one of the resonator mirrors with transmittance T

$$W_{\text{cav}} = \frac{\phi}{\tau_{\text{cav}}} = \frac{\phi}{c_{\text{gr}}}\left[\frac{\text{Abs} + T}{2L}\right]$$

Putting everything together (with the correct sign) we get two coupled equations, the laser *rate equations*, one for the population number of the upper laser level, one for the number of laser photons in the cavity, that are very useful to model many aspects of lasers:

$$\frac{\mathrm{d}}{\mathrm{d}t}N_2 = W_{\text{gen}} - W_{\text{stim}} - W_{\text{rad}} - W_{\text{nrad}} \tag{6.8}$$

$$\frac{\mathrm{d}}{\mathrm{d}t}\phi = W_{\text{stim}} + \beta\,W_{\text{rad}} - W_{\text{cav}} \tag{6.9}$$

6.7.1.5 Continuous-Wave Operation

In general, the rate equations have to be solved numerically. In case of continuous-wave (cw) operation (time derivatives equal zero), however, the rate equations can be solved analytically:

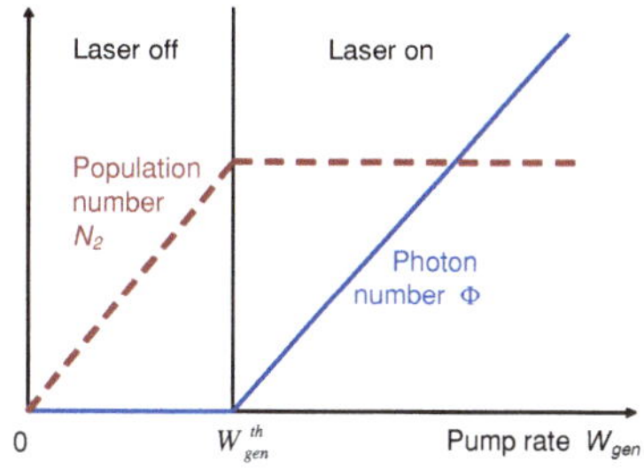

Fig. 6.47 Photon number Φ and population number N_2 or gain G as function of the pump rate W_{gen} with onset of lasing at the threshold pump rate W^{th}_{gen}

From Eq. 6.8, the population of the upper laser level follows which is needed for lasing, i.e., the population at and above (!) the laser threshold:

$$c_{gr}\ g(N_2) = \frac{1}{\tau_{cav}} \quad \Rightarrow \quad N_2^{th}$$

From Eq. 6.9, we get the two solutions

$$\phi = 0 \text{ the non-lasing case with } N_2 \leq N_2^{th}$$

and

$$\phi = \tau_{cav}(W_{gen} - W^{th}_{gen}) \text{ the case of lasing with } N_2 = N_2^{th}$$

So, up to the laser threshold the population number increases with the pump rate but there are no stimulated photons in the laser cavity because the resonator losses are still higher than the gain at the given population number. Increasing the pump rate further opens another relaxation channel: the stimulated emission which exactly balances the increasing pump rate, thereby clamping the population number at its threshold value. This means, theoretically all absorbed pump photons (minus those needed to reach threshold) will be transformed into stimulated cavity photons (Fig. 6.47).

To be useful, part of the circulating photons will be coupled out (transmittance T) with an efficiency

$$\eta_{oc} = \frac{T}{\text{Abs} + T}$$

Hence, the available laser power is

$$P_{out} = \eta_{gen}\, \eta_{oc} \frac{h\nu_{out}}{h\nu_{pump}} \left(P_{pump} - P^{th}_{pump}\right)$$

An optimum between minimising the laser threshold by minimising T and maximising the outcoupling efficiency by maximising T has to be found.

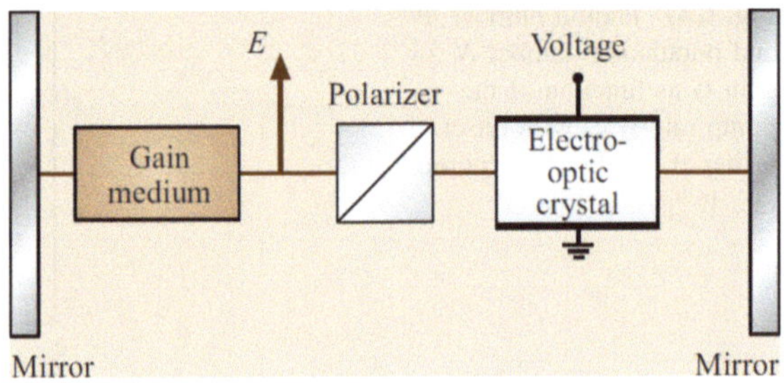

Fig. 6.48 Laser resonator with electro-optic modulator for Q-switching [64]

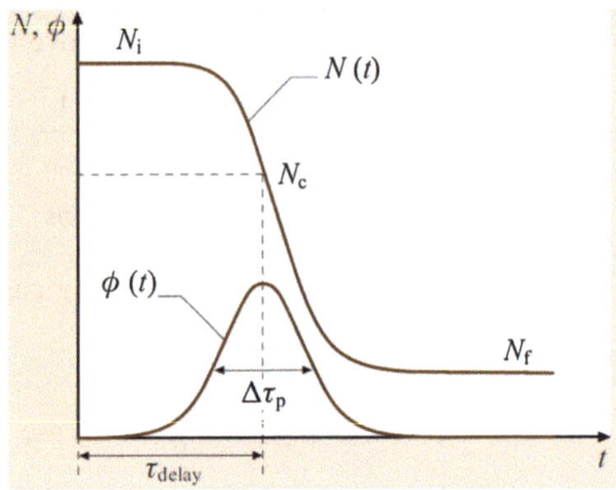

Fig. 6.49 Inversion N and photon number Φ after switch-off of the extra cavity loss (at $t = 0$). At the beginning, Φ grows exponentially to a number much higher than the equilibrium number, thereby reducing the inversion within a very short timespan $\Delta\tau_p$ to its final value N_f. Ideally, $N_f - N_i$ is converted into (laser) photons Φ [64]

6.7.1.6 Pulsed Operation

Q-Switch

In many solid-state laser materials, the lifetime τ_{rad} of the excited state is of the order of milliseconds and can be used to store the pump energy which, at the end of the pump period, can be released in a short pulse with correspondingly high peak power. For this, a Q-switch—for example an electro- or acousto-optic modulator—is placed inside the resonator (Fig. 6.48).

During the pump phase, the transmission losses of this modulator are set to high, stimulated emission cannot take place, therefore the population of the upper laser level can grow much higher than it would do in cw laser operation where it is clamped at its threshold value. At the end of the pump phase, the modulator losses are switched off, the gain is now much higher than the (remaining) losses, the photon number grows very fast to a number which is orders of magnitude higher than during cw operation, converting in a very short time, $\tau_{pulse} \approx 10$ ns, most of the energy stored in the upper laser level into laser energy. Ideally, one can get a peak power which is by a factor of $\tau_{rad}/\tau_{pulse} \approx 10^4 \ldots 10^5$ higher than the corresponding cw output power (Fig. 6.49).

Mode Locking

For even shorter pulses, one has to take into account that according to the Fourier theorem a pulse of a certain length $\Delta\tau$ is composed of a set of frequencies of width $\Delta f = 1/\Delta\tau$. So, production of extremely short pulses does not work with single-frequency (mono-mode) lasers but several modes have to be superimposed with the correct phase (phase-coupled, mode locked). A 100-fs pulse, for example, requires a frequency spectrum of 10 THz, corresponding to a wavelength spectrum of 30 nm at a centre wavelength of 1 μm. This means the gain of the active materials has to cover at least this frequency range. In addition, a modulator is needed to phase-couple these modes. This could be an active modulator, running with the modulation period equal to the photon round-trip time. Another, often preferred option is a *saturable absorber*, a passive element that bleaches at high intensites and hence favours pulses with the highest peak power that have the lowest transmission losses. Because of the high peak power (density) one has to make sure that non-linear effects do not destroy the laser (by self-focusing) or broaden the laser pulses.

6.7.1.7 Solid-State Lasers

Already within a couple of years after the first realisation of the laser by Maiman in 1960, many more *laser materials* have been discovered. Just to name a few types: rare-earth and transition-metal ion doped insolating crystals and glasses (Nd^{3+}:YAG, Cr^{3+}:Al_2O_3, summarised as solid-state lasers), noble-gas atom (HeNe) and ion (Ar^+) vapour, molecular vapour based on electronic energy levels (N_2) or on vibrational or rotational energy levels (CO_2), excimers (short form of excited dimers, ArF), dye molecules, mostly in solution (Rhodamin 6G), electrons in semiconductors (GaAs, often as diode), free electrons (FEL), etc. *Excitation* was mostly either optical using lamps (later also other lasers), via electric discharge (often using a mixture of gases, one for excitation, one for energy transfer, one for lasing, one for deexcitation etc.) or using a p–n junction in case of semiconductor lasers (diode lasers).

Ruby, a Cr^{3+} doped sapphire crystal, was the very first laser. A cylindrical rod was placed in the centre of a helical flashlamp used for pumping. The resonator mirrors have been directly coated onto the endfaces of the rod. Since ruby is a 3-level system ($E_0 \equiv E_1$), its laser threshold is very high and it is mostly used in pulsed mode if laser emission in the deep red (693 nm) is needed. In many applications, ruby was soon replaced by *Nd^{3+} doped Yttrium–Aluminium–Garnet (YAG)*, which is an almost ideal 4-level laser system. It became the standard laser material for continuous-wave (cw) and pulsed operation for many years. Its emission is in the infrared at 1,064 nm.

With discharge lamps being more and more replaced by diode lasers, another crystal came into focus: *Yb^{3+} doped YAG*. This is still a 4-level system but the splitting of the lower (and upper) levels is so small that people often call it a quasi-3-level laser system (Fig. 6.50).

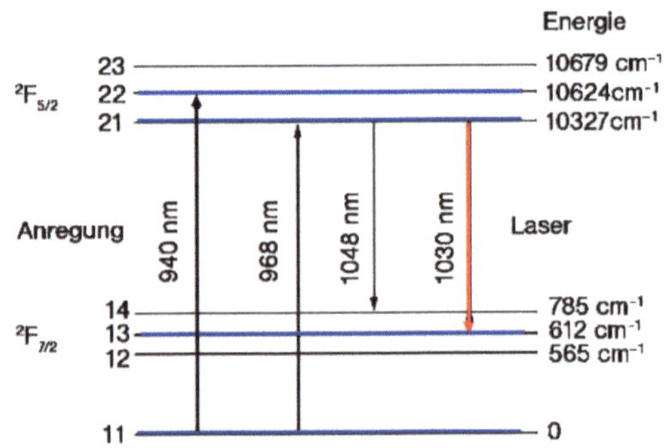

Fig. 6.50 Energy level diagram of Yb:YAG and main absorption and emission wavelengths

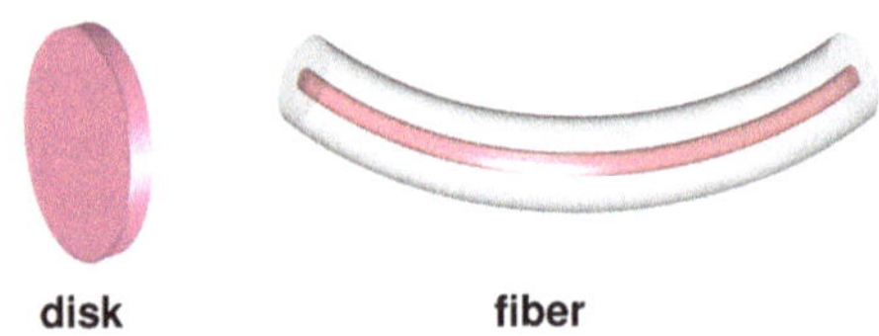

Fig. 6.51 Geometries of the laser-active medium with high surface-to-volume ratio [67]

This small splitting increases potentially the efficiency and reduces the heat by roughly a factor of 3 compared to Nd:YAG, where approximately 30 % of the pump energy goes into heat. The disadvantage is the relatively high threshold that allows efficient cw operation at high pump-power-density levels only, which in turn require efficient cooling, i.e. a gain element with large surface-to-volume ratio: the thin-disc as extremely short rod and the fibre as extremely thin and long rod (Fig. 6.51).

Thin-disc laser

The laser crystal is typically a disc 200 μm thick and 10 mm in diameter, highly reflective coated on its back side and mounted with this side on a heat sink. The resonator axis is basically collinear to the heat flow, resulting in low thermally induced wave front distortions. Reducing the thickness of the disc not only improves the cooling efficiency but also reduces that part of the laser threshold that is needed to achieve transparency, simply because there is less material to be inverted (Fig. 6.52).

The small absorption length in the disc is usually compensated by an external pump optics which allows reimaging several times the pump photons that are not absorbed during the first pass. A very nice feature of the disc laser is that the output power scales with the area of the pump spot at constant pump (and cooling) power densities. Output powers of several Kilowatts per disc are commercially available; presently up to four discs can be operated in series in one resonator, quadrupling the maximum output power. The laser efficiency (laser output related to the electrical input power) is between 20 and 30 % (Fig. 6.53).

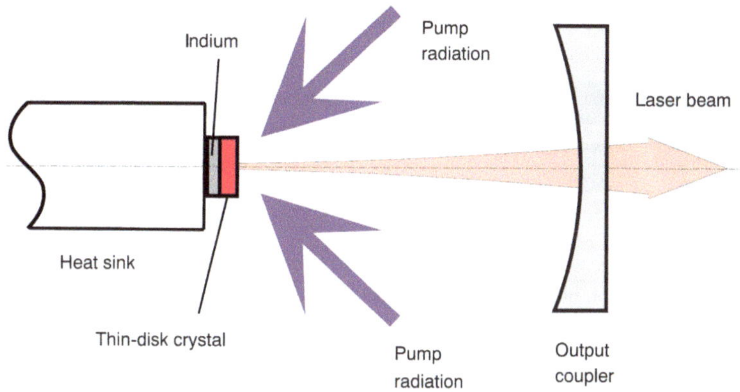

Fig. 6.52 Cross-section of thin-disc laser. The thin disc serves as gain element and—HR coated on its back side—as rear laser mirror. The pump optics is not shown [68]

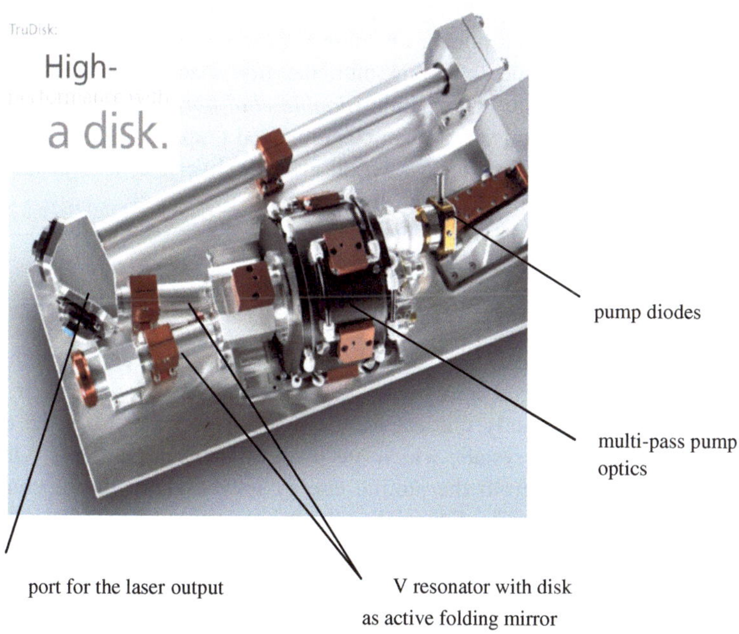

Fig. 6.53 Example of a commercial thin-disc laser [69]

Fibre Laser

In a fibre (laser), the light is guided by the difference in index of refraction of the core region and the surrounding cladding region. Core diameters of typically 10 times the wavelength allow to guide the (laser) light in a single transverse mode.

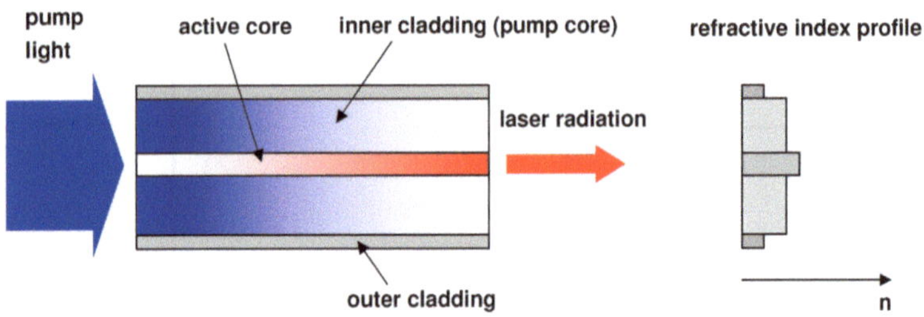

Fig. 6.54 Cross-section of a short piece of a double-clad fibre laser [67]

Power scaling is by increasing the fibre length (typically 10 m). Especially, in single-mode fibres this leads to extremely high-power densities and nonlinearities in the core which can easily damage the fibre. Increasing the core diameter without losing the single-mode behaviour of the fibre and broadening the laser spectrum is one of the measures to shift the power limit further up. Presently, the maximum output power from a single-mode cw fibre laser is 10 kW (IPG). The small core diameter makes it difficult to get the pump light directly into the active region. The trick here is to use the much thicker cladding region as core region for the pump radiation and to put another cladding onto the first cladding which serves as cladding (light guide) for the pump light (*double-clad fibre*). A big advantage of the fibre laser is that the active fibre can be connected ("spliced") easily and with minimal losses to a transport fibre which allows to deliver the laser beam to the place where it is actually needed (Fig. 6.54).

6.7.1.8 Diode Laser

In what is traditionally called "solid-state laser", the active elements are individual rare-earth or transition-metal ions being placed in a solid-state matrix. In semiconductor lasers, it is exactly this crystalline matrix, more precisely the energy bands of semiconductor crystals, which are used for the laser process. One can derive the energy bands from the atomic energy levels taking into account the coupling between these levels. Especially, the higher energy levels belonging to less localised atomic states couple strongly, thereby forming broad energy bands that extend over the whole crystal (see 6.72.1). The optical transitions take place between the valence band (the energetically highest filled band) and the conduction band (the lowest empty band). The energy of the emitted photons corresponds approximately to the width of the energy gap E_g between valence and conduction band. Inversion in a semiconductor is achieved by putting together an n-doped and a p-doped semiconductor and applying a voltage. N-doping (with electron donors) results in free electrons in the conduction band, p-doping (with electron acceptors) results in free holes (missing electrons) in the valence band. Applying a voltage across this p–n junction (diode) of approximately E_g/q pushes the electrons and

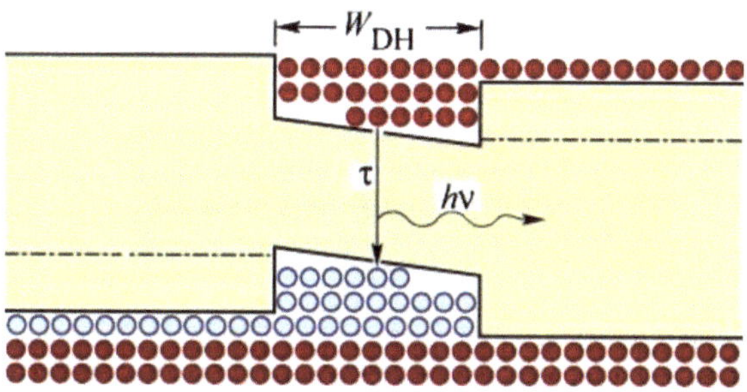

Fig. 6.55 *p–n* junction with a voltage applied that pushes the conduction-band electrons from the n-doped side and the valence-band holes from the p-doped side into the junction area. In all modern diode lasers, there are extremely thin layers of semiconductor crystals with a smaller band gap near the junction plane where the electrons and holes get trapped and can recombine more efficiently [70]

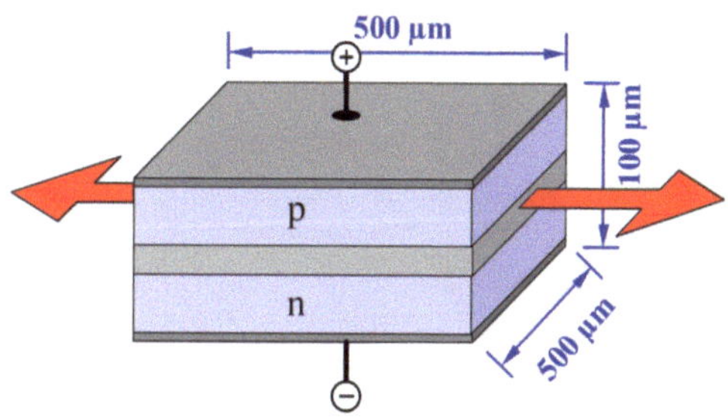

Fig. 6.56 A simple edge-emitting diode laser [71]

holes into the junction area, thereby creating an inversion layer where stimulated emission can take place (Figs. 6.55 and 6.56).

So, the simplest (edge emitting) diode laser looks like this:

The laser photons propagate mostly along the p–n junction (in the direction of the red arrows) where they get the maximum gain. Often, there is an additional optical waveguide along the p–n junction, optimising the interaction between the free carriers and the (stimulated) photons. The endfaces serve as resonator mirrors, one side HR, and the other side AR coated. Power scaling is mostly via the lateral width of the junction, which results in poor beam quality (focus ability) of high-power diodes in the lateral direction (highly multi-mode), whereas the beam perpendicular to the p–n junction single mode but highly divergent. The maximum power such a diode can deliver is of the order of 100 mW/μm emitter width. The maximum output power of a bar of 1 cm width is above 100 W; several bars can be stacked. Here, essential is a very efficient cooling of the bars (Fig. 6.57).

The output of single emitters, bars and stacks on the one hand and the efficiency of incoherent beam combining on the other hand have made considerable progress. Meanwhile, these lasers can replace lamp-pumped Nd:YAG rod lasers, but with by far better efficiencies. Further progress is likely, e.g. by spectrally combining

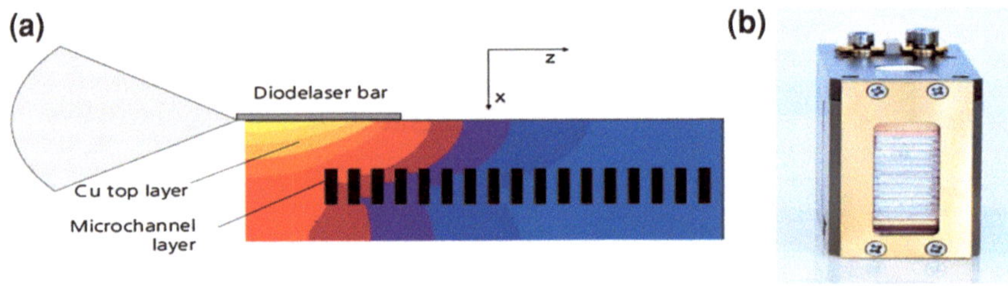

Fig. 6.57 **a** Diode laser bar on microchannel cooler. The colours indicate the temperature distribution [72]. **b** Commercial stack of diode laser bars [73]

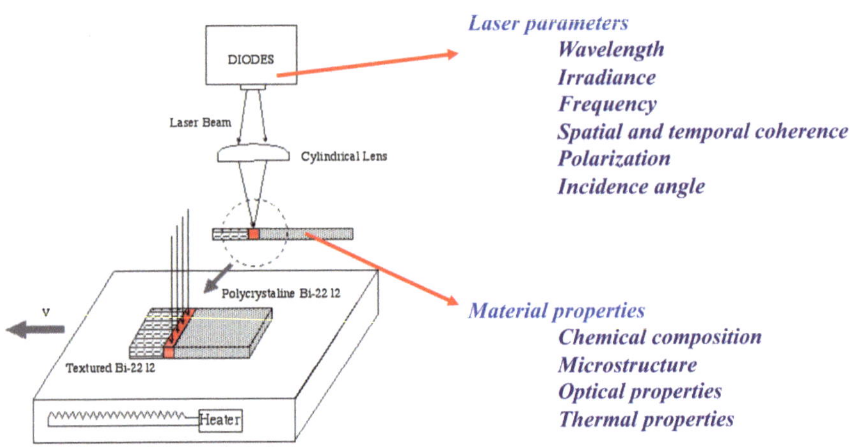

Fig. 6.58 Laser parameters and material properties that are relevant in the interaction between lasers and materials

grating-wavelength-stabilised bars of slightly different wavelength. So, in the future we may see more and more diode lasers replacing cw diode-pump solid-state lasers.

6.7.2 *Laser–Matter Interaction Basics for Nonspecialists*

This section of the chapter reviews the basic principles of the interaction between lasers and materials and how they affect what phenomena may be observed or expected while attempting modification of materials surfaces with the types of lasers presented in the previous section. The previous literature explains deeply both laser interaction mechanisms and surface modification under different approaches [74–79]. This section makes a quick review regarding exclusively the main interaction mechanisms related to laser ablation applied to cleaning and conservation of cultural heritage.

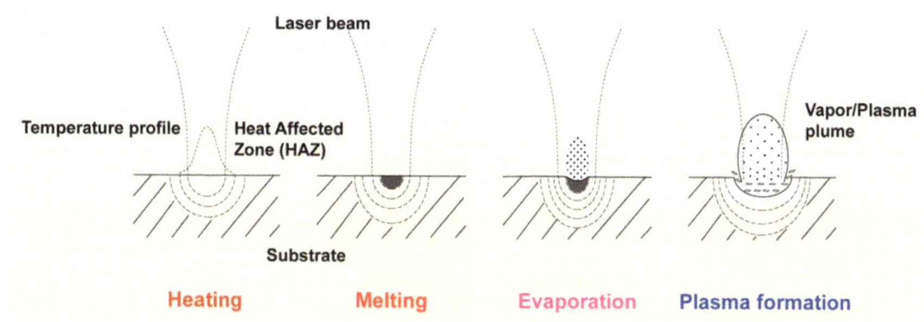

Fig. 6.59 Physicochemical phenomena that can be induced in an irradiated surface

When the surface of a material is irradiated with a laser beam, several mechanisms may be induced within the interaction volume, depending essentially, although not exclusively, on the laser emission parameters (wavelength, cw or pulsed operation mode, etc.) and on the properties of the material. Figure 6.58 illustrates what parameters and properties must be generally taken into consideration when a given material is irradiated with a laser. Laser parameters include the wavelength, irradiance, frequency, spatial and temporal coherence, polarisation or the incidence angle. The material is characterised by its chemical composition and microstructure, which determine the type of elementary excitations present, as well as their interactions. They are also responsible for the optical and thermal properties of the material, which are also important for the understanding of the different interactions that take place between lasers and materials.

The combination of laser parameters and material properties will induce different physicochemical phenomena on the irradiated surface. These are governed by the type of photon–electron interaction induced mechanisms, and, from the physical point of view, are largely determined by the intensity and/or the emission mode of the laser. The different physical processes that usually appear on the irradiated surface include heating, melting, evaporation and plasma formation, as illustrated in Fig. 6.59.

6.7.2.1 Influence of the Electronic Energy Band Structure

Let us first of all consider how the properties of a material influence its behaviour when irradiated with a laser beam. As an initial idea, photons of laser irradiation interact with electrons in solid and the understanding of this phenomenon requires knowing the energy band in the solid. For each individual atom, there exist discrete energy levels that may be occupied by electrons, arranged into shells and subshells. The electrons in most atoms fill just the states having the lowest energies. Following the Pauli exclusion principle, in each state only two electrons with opposite spin can be placed.

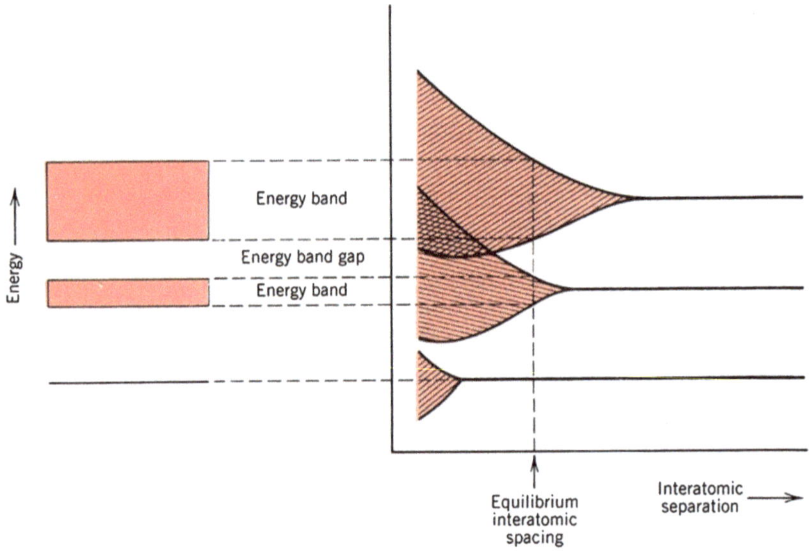

Fig. 6.60 Splitting of the atomic electronics energy levels in energy band when a solid is formed [80]

A solid contains a very large number of atoms. When atoms are separated long distances, they are independent from each other and do not interact. When they come within close proximity of one another, however, the electrons from adjacent nuclei or atoms perturb electrons from one atom. Due to this electronic perturbation or interaction, the energy levels split into many closely spaced electron states to form what is called an electron energy band. This is schematically illustrated in Fig. 6.60.

At the equilibrium interatomic spacing, energy gaps may exist between adjacent bands. Energies lying within these band gaps are not available for electron occupancy. In addition, some degree of overlap between the energy bands can take place in some cases.

With this in mind, another important aspect must be considered within this interaction scheme. This is related to the fact that the shape of the energy bands depends on the sample microstructure, thus causing many of the material's electronic and other properties to be anisotropic. That is, to depend strongly on grain or crystal orientation.

The band that contains the highest-energy electrons is called the valence band. The conduction band is the next higher energy band, which is virtually unoccupied by electrons in most circumstances.

Four different types of band structures are possible at OK, as exemplified in Fig. 6.61. In the first case, on the left side of the figure, the valence band is only partially filled with electrons. This is the case that corresponds to elements with an odd atomic number, elements that show typical metallic or conductor behaviour, representative of the metals.

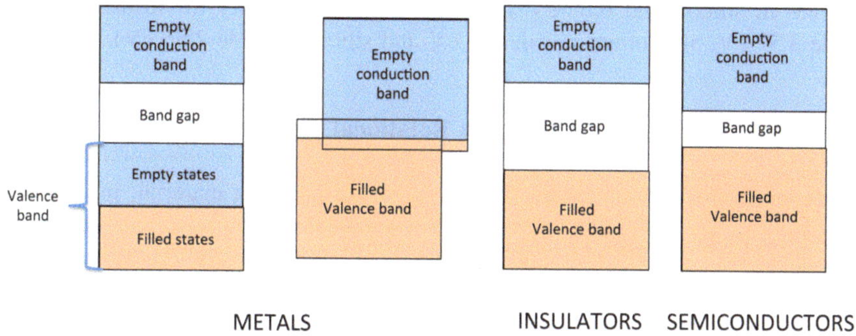

Fig. 6.61 Diagram of the electronic energy band structure in metals with an odd atomic number, metals with an even atomic number, insulators and semiconductors

Fig. 6.62 Scheme showing the different the energy transitions involved in electronic, vibrational and rotational transitions [81]

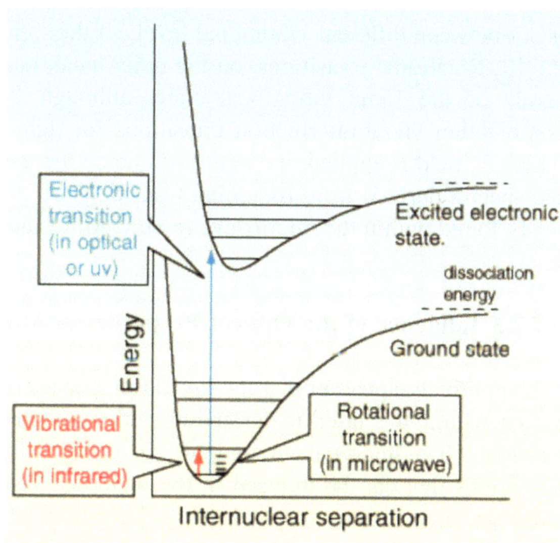

Another situation observed in metals may be described with that illustrated for the second illustration from left to right. The valence band is full in this case, but it overlaps with the conduction band. This is the case of some metals with an even atomic number.

The final two cases, on the third and fourth illustrations to the right side of Fig. 6.61, correspond to the situation in which the valence band is completely filled with electrons, but there is an energy band gap between this valence band and the empty conduction band above it. This is the electronic condition used to define insulators and semiconductors. The difference between these two types of materials, semiconductors and insulators, is the magnitude of the energy gap. Typical values may be considered around 1 eV in the case of semiconductors

(1.1 eV in silicon and 0.67 eV in Germanium) and higher for insulators (zinc oxide 3.37 eV, aluminium nitride 6.3 eV and silicon carbide 2.86 eV).

6.7.2.2 Influence of Vibrational and Rotational Transitions

When incident electromagnetic waves hit the surface of a material, in addition, they may excite individual atoms and/or molecules present at the outermost surface layers. The most commonly observed atomic or molecular excitation phenomena involve electronic, vibrational or rotational transitions.

Let us consider a diatomic molecule. The electronic states can be represented by plots of potential energy as a function of internuclear distance, as illustrated in Fig. 6.62. Electronic transitions are vertical or almost vertical lines on such a plot, since the electronic transition occurs so rapidly that the internuclear distance cannot change much in the process. The involved energies correspond to UV, visible and near-IR part of the electromagnetic spectrum. Vibrational transitions occur between different vibrational levels of the same electronic state, typically in the IR. Rotational transitions, on the other hand, occur mostly between rotational levels of the same vibrational state, although there are many examples of combination vibration–rotation transitions for light molecules. When an electromagnetic field is applied, it exerts a torque on the atoms or molecules. This causes the spectra derived from rotational transitions of atoms and molecules to be typically found within the microwave region. All of these are shown in the Fig. 6.62.

6.7.2.3 Influence of the Optical Properties of Materials

It is intuitively apparent that the optical properties of a material are a determinant factor within the laser–material interaction scheme. When light interacts with materials, it is important to compare the energy of the photons with the possible transitions that can be induced in the material. Visible light can be absorbed by two basic mechanisms: (i) electronic polarisation, where the electric field shifts the electronic cloud relative to the nucleus—only important near the relaxation frequency; (ii) electron excitations.

Materials may be opaque or transparent to radiation. This produces the phenomenology that it is usually observed in the interaction of the material with light, the absorption (*A*), reflectance (*R*) and transmittance (*T*) of an incident (*I*) beam on the surface of a bulk material (Fig. 6.63). In opaque materials, radiation is absorbed or reflected while in transparent ones the absorbed part is very small allowing transmitting part of the energy.

In a metal, the radiation can be absorbed because electrons have free states to be excited in a broad range of energies. Most of the absorbed radiation is re-emitted from the surface. A small amount of the energy associated with the electron decay is dissipated as heat. This explains why metals are opaque in the visible range of the

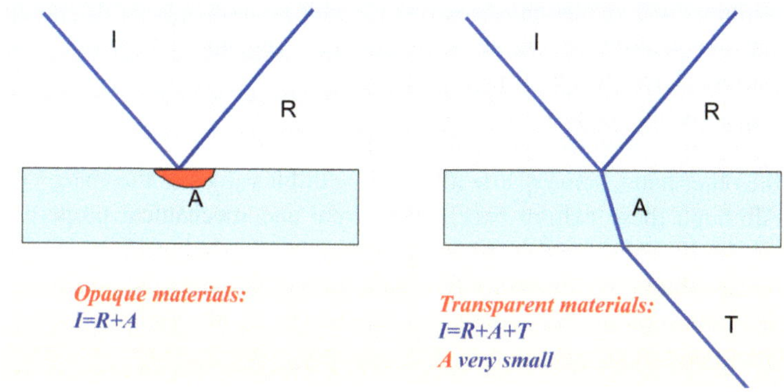

Fig. 6.63 Scheme of the different energy contributions that take place in the interaction of electromagnetic waves with materials

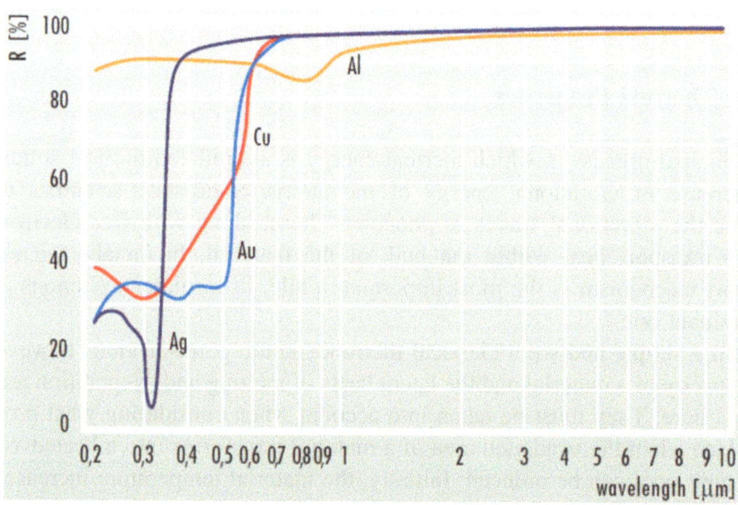

Fig. 6.64 Dependence of the reflectance with wavelength for different metals [82]

electromagnetic spectrum. It turns out the reason is that electronic transitions are possible because there is a continuous range of empty electronic states near the Fermi level.

It also can be used to understand why metal colour is grey in most cases, except in gold and copper alloys. The colour of a material is determined by the wavelength distribution of the radiation that is reflected. Figure 6.64 shows the reflectance of the some metals. Aluminium and silver reflects near 95 % of the light in all visible spectra, while the percentage of reflection in gold and copper in the high-energy range of the visible spectra, is much lower. This explains the colour of these two metals.

In insulators and semiconductors, it is possible to induce two different electron transitions: (i) electron excitations across the band gap, and (ii) electron transitions to impurity or defect levels. It is important to have in mind that the energy of the light in the visible range covers from 1.8 to 3.1 eV. For this reason, semiconductors can absorb visible light and Si and Ge are grey.

On the other hand, some quartz allotropes exhibit values of the energy gap near 9 eV. Although these exhibit excellent thermal and mechanical properties, they should never be considered to be used as protection windows for a visible or a Nd:YAG laser based system, since they are transparent to its fundamental emission wavelength of 1,064 nm (E = 1.16 eV). Obviously, an absorption spectrum is not the only variable in the selection of the protection window, other parameters, such as the laser intensity, are also relevant.

These arguments change, however, when the laser emission wavelength is different, as deduced from the fact that absorption changes as a function of wavelength for basically any material. For instance, silicon and germanium absorb near all the radiation in the visible but are nearly transparent in the IR.

6.7.2.4 Thermal Properties

The principal process in which thermal energy is assimilated in most solids is by the increase in vibrational energy of the atoms. Solid-state scientists usually describe the vibrational waves as phonons. These, along with free electrons, are able to transport heat within the bulk of the material. In metals, the electron transport mechanism is the most important, while, in insulators, phonons are the predominant one.

From a simple and strict classical thermodynamic point-of-view, however, the heat capacity of a material and the latent heats of melting and evaporation are quite relevant here. They must be taken into account when considering what processes take place when the irradiated area of a material transforms into a heated volume, and a melt pool can be induced. Initially, the material temperature increases until the melting temperature is reached. This process is controlled by the material heat capacity, and with sufficiently high intensity laser irradiation this process is quite fast. Then the material reaches a change of state, melting. Melting consumes lots of energy and requires time before the change of state of the entire material takes place. Latent heat controls this second stage. On the other hand, the latent heat of sublimation is relevant for direct solid–vapor transformations, which are very important in surface cleaning or desorption processes. In the latter case, very intense energy transfer to the solid is required, as the transformation of solid atoms into a gaseous state is energy demanding.

In any case, the fact that time affects the thermal transformations caused by lasers on materials, it is important to take into account also the thermal conductivity of the material being irradiated. This parameter determines the volume that is affected by laser radiation during the process. It is important to compare the time

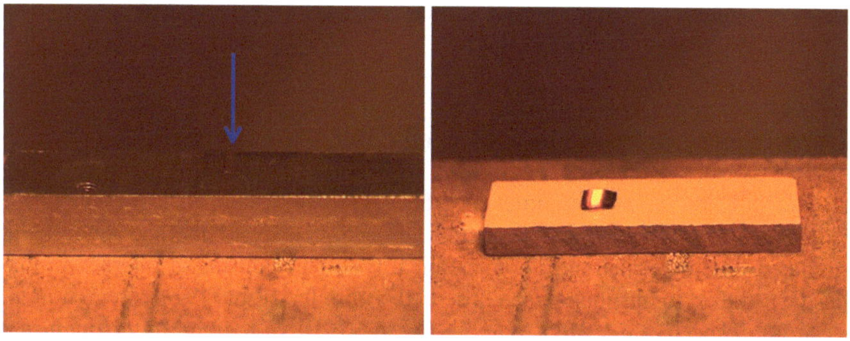

Fig. 6.65 Photographs showing the different thermal gradients induced in thermal conductor and thermal insulator materials

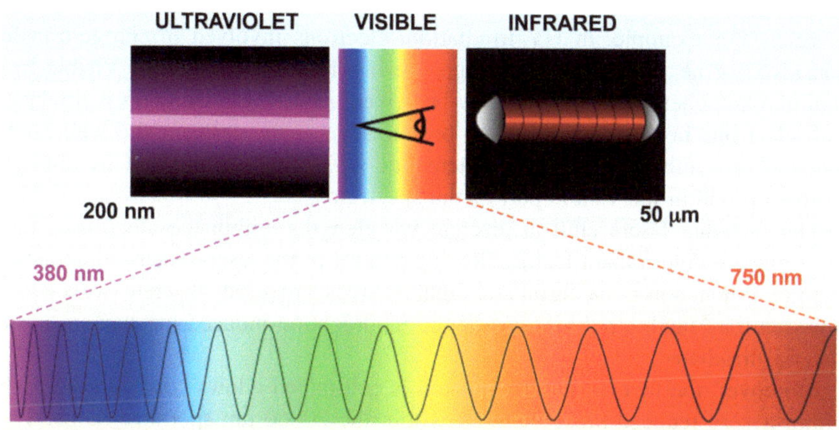

SPECTRAL EMISSION	WAVELENGTH, λ	PHOTON ENERGY
ULTRAVIOLET	200nm-380nm	6.2 eV – 3.2 eV
VISIBLE	380nm-750nm	3.2 eV – 1.7 eV
INFRARED	750nm-50μm	< 1.7 eV

Fig. 6.66 Range of energies associated with the different parts of electromagnetic spectra

in which heat is deposited on the sample and the time required to heat a given volume.

Figure 6.65 shows how these properties, particularly the thermal conductivity; affect laser irradiation of a metal and a ceramic substrate, respectively. In the case of the metal with a high thermal conductivity, heat deposited by laser radiation is spread over a large volume. By contrast, in a ceramic, a material with a low thermal conductivity, heat is concentrated in a small volume and a high local temperature increase can be obtained.

6.7.2.5 Induced Processes Depending on Laser Parameters

An additional interaction regime must be nevertheless defined as a function of wavelength and not only based on the absorption level of the substance being irradiated. Photothermal and Photochemical type mechanisms of interaction may thus be ascribed to the phenomena observed within the IR or UV–Vis region of the spectrum, respectively, as illustrated in Fig. 6.66. Three regions are distinguished in this figure: UV below 400 nm; Vis between 400 and 700 nm, and IR above 700 nm. Although all laser material interactions are based essentially on photon–electron interactions, a change in emission wavelength implies a change in photon energy, thus an interaction with electrons involved in transitions of different energies. For example, in UV irradiation, electrons involved in charge-transfer transitions or near the absorption edge, are involved. In the case of Visible irradiation, photochemical reactions may be driven, since the interaction is produced with electrons involved in chemical bonds. Such is the case for the well-established photosynthesis process, characteristic of plants and enabled by sunlight irradiation within the visible part of the spectrum.

The fact that lasers emit at discrete wavelengths within the electromagnetic spectrum is a determinant factor allowing control of the laser–material interaction regime. When non-monochromatic light sources are used, discrete wavelength emission may facilitate the choice of the modification-induced mechanism in the material irradiated.

Moreover, the fact that laser emission is collimated allows concentrating light within a very reduced area, typically within the 20–60 μm spot size range. This means that, for instance, a nominal 8 W (8 Js^{-1}) average-power laser emitting in pulsed mode (5 ns width) with a pulse repetition rate of 20 Hz and an irradiation spot of 300 μm^2 will reach Irradiance values, according to the following calculation:

$$I_{rr} = E_{SC}/\left(f_{rep} \cdot \tau \cdot A_{spot}\right)$$
$$I_{rr} = 8/\left(20 \cdot 5 \times 10^{-9} \cdot 300 \times 10^{-8}\right) = 2.66 \times 10^{13}\ \mathrm{Wcm}^{-2}$$

where I_{rr} is the irradiance or instant power density, E_{SC} is the energy contained in the laser pulse, f_{rep} is the pulse repetition rate, τ is the pulse width and A_{spot} is the focal spot area.

For comparison, the following analogous calculation for a 1 kW, broad emission wavelength (less than 400 to more than 2,500 nm) quartz-tungsten-halogen (QTH) lamp which, in the best of cases may be focused to a 3 cm^2 spot, the corresponding Irradiance (I_{rr}) value is 330 Wcm^{-2}. In this case, the I_{rr} or instant power density is the same as the average power density.

Conventional Laser processing is mainly performed with IR lasers and is based on what is known as "Photothermal processing". The emitted photons can excite

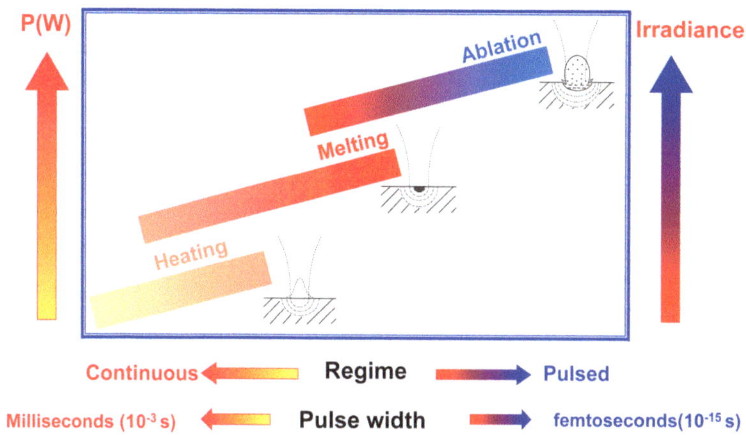

Fig. 6.67 Different processes that are induced depending on the irradiance of the laser

free electrons within a metal or vibrations within an insulator. In semiconductors, both types of excitations are possible within the bandgap. In general, the excitation energy is dissipated into heat within a very short time, above the ps regime, where electron–photon interactions start to become significant.

As illustrated in the virtual graph of Fig. 6.67, surface heating and melting may take place around the area of the laser focus, causing large thermal gradients within the material, solid-state diffusion, phase transformations, and other simple or complex phenomena. These provoked processes usually take place far away from equilibrium, specially due to the fast cooling rates intrinsic to this type of laser processing.

In addition, since the electron–phonon interaction takes place above the picosecond level, nanosecond pulses also induce significant local heating of the irradiated material, although these are far away from equilibrium, due to the extremely fast cooling rates associated to short pulse illumination. Although in Conservation the objective is to perform removal of undesired adsorbants via laser ablation, most of the lasers used nowadays are pulsed in the nanosecond regime. Care must be taken thus to avoid excessive heating of the substrate, which may cause irreversible damage to undesired layers of the material being decontaminated or restored. The restorer must therefore not forget that most lasers available for desorption of surface contaminants will cause large thermal gradients and, under many conditions, which are also material dependent, could generate irreversible damage on valuable surfaces.

This can be deduced from the illustration of the ablation process that appears in Fig. 6.68a. Such a complex process is a mixture of mechanisms which include direct sublimation of the substance being irradiated, heating, non-equilibrium melting and vapor plume formation, plasma formation, ejection of melt drops and removal of particles by combination of thermal stress with the intense acoustic shockwaves

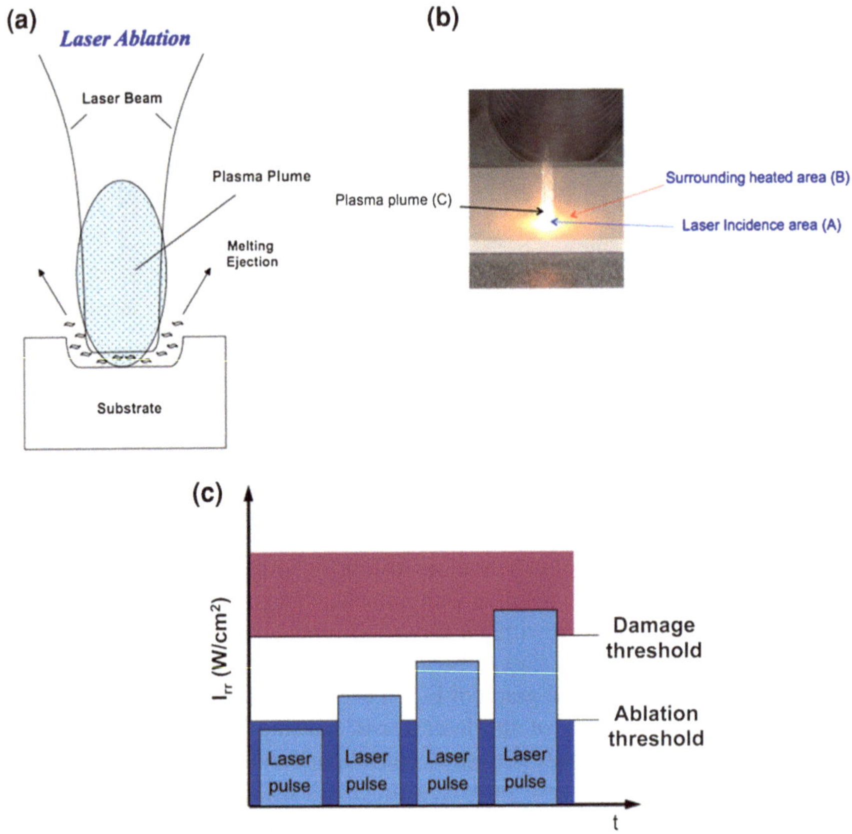

Fig. 6.68 **a** Scheme of the ablation process induced in the laser–matter interaction. **b** Example obtained in the processing of ceramic materials. **c** Scheme of the importance of the laser parameters selection to avoid damage threshold

associated to plasma formation. Part of the above effects may be visualised in Fig. 6.68b, where a heated area is observed on the surface of the substrate, with an intensity gradient between the area of laser incidence (*A*) and its outer limits (*B*). The plasma plume is easily observable above the laser focal spot (*C*). An infrared laser is incident from above and is not visible.

Ablation processes are the basics for cleaning in Cultural Heritage materials. As discussed from the previous figure ablation process includes complex phenomena, which may result in damage to the substrate. In order to avoid such negative effect, ablation and damage threshold has to be determined for both, the contaminants layers and the substrate (Fig. 6.68c) and the laser parameters have to be chosen before reaching the damage threshold.

Table 6.8 Industrial Laser systems available for cleaning a diversity of common substrate materials commonly found in relevant Cultural Heritage buildings and objects

PRODUCTS	MATERIALS										
	Stones	Biological pat.	Gilded surf.	Bronze	Silver	Iron	Stucco	Wall paint.	Wood	Parchment	Textiles
▲ EOS 1000	•		•			•		•			
▲ Smart Clean II	•		•			•		•			
◘ EOS 1000 LQS	•		•	•	•	•	•	•	•	•	
■ Thunder Art	•	•		•		•	•		•	•	•
■ Thunder Art with 532		•								•	•
■ Thunder Art with 355										•	•

▲ = Short Free Running: Pulse length of 50-120 µS
◘ = Long QS laser with pulse length of 100 ns
■ = QS laser with pulse length of 8 ns

Table 6.9 Emission parameters corresponding to the industrial Laser systems listed in Table 6.8

Laser system	Wavelength (nm)	Pulse width	Energy	Repetition rate (Hz)
Smart Clean II	1,064	50–110 µs	2 J	1–20
EOS 1000 SFR	1,064	60–120 µs	1 J	1–20
EOS 1000 LQS	1,064	100 ns	360 mJ	1–20
Compact Phoenix	1,064	10 ns	100 mJ	1–25
THUNDER ART	1,064	6 ns	1 J	1–20
THUNDER ART	532	6 ns	500 mJ	1–20
THUNDER ART	355	6 ns	200 mJ	1–20
Bramante	1,064	6 ns	1.6 J	1–20

6.7.3 Commercial Lasers Used in the Conservation of Cultural Heritage

The use of commercial laser systems for restoration work has been clearly limited by the necessity to perform such work either in situ at the historic sites of interest, or within a laboratory when the material can be reasonably transported. The former case is most common, although more and more applications, besides building restoration, are coming of age. It is thus clear that commercial laser system technology is the limit to laser desorption based restoration advances.

Initial systems used in the field were based on nanosecond pulsed, Q-switched Nd:YAG lasers with low repetition rates of 10–20 Hz and poor beam quality. During the past decade and a half, however, important advances have been achieved in laser technology, as gathered from the description found in Sect. 6.7.1 of this chapter. These have resulted in the availability of smaller, more portable lasers with much better beam quality, lower pulse widths and higher pulse repetition rates, thereby helping to improve significantly surface desorption efficiency and control of damage to the irradiated materials.

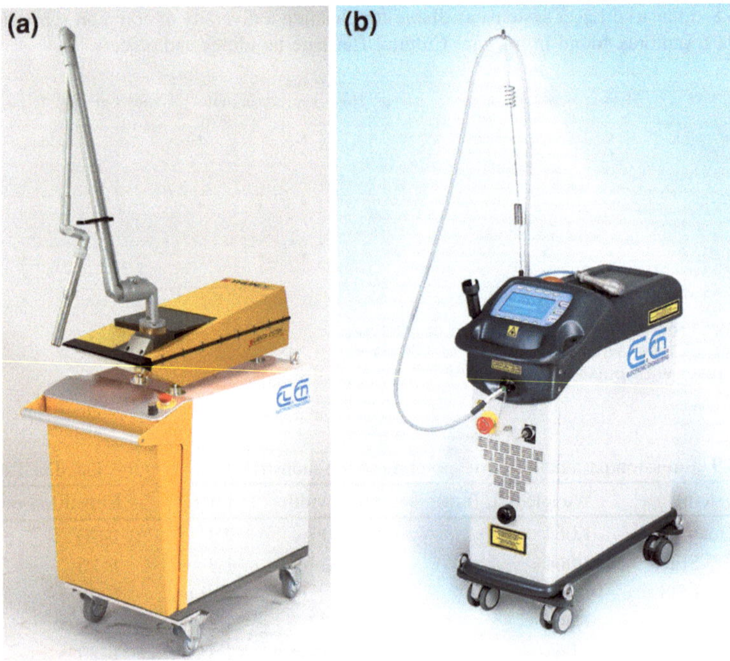

Fig. 6.69 **a** A Thunder Art (El.En Group) compact and portable industrial laserand, **b** the EOS 1000 LQS Laser.CourtesyEl.En Group http://www.elengroup.com

The most common industrial laser systems used in Cultural Heritage Restoration are gathered in Table 6.8, where they are classified with respect to the type of restoration work that they have been used for. The main difference between EOS 1000 to Smart Clean II is the maximum energy. EOS Long QS laser is the best solution for application on small metal objects, like coins.

In addition, specific emission characteristics are given in Table 6.9 for different laser models used in Cultural Heritage Restoration. In contrast to what may be concluded from Table 6.8, it is advisable to comment that within the microsecond pulse width regime, only cleaning of metal objects is usually recommended, if at all. With the advent of nanosecond solid state and fibre lasers, the older microsecond pulsed systems have been discarded as apt for most cleaning procedures in Cultural Heritage materials. It is well known that surface melting occurs even at the nanosecond regime, although an important threshold has been observed, for example, between >10 ns and <5 ns with regard to heat transfer into the substrate. Certainly below 5 ns, materials suffer much less thermal damage.

A recent nanosecond, portable laser system commercialised by El.En. S.p.A. (www.elengroup.com) is shown in Fig. 6.69 (left). The small size and compactness of this laser system, commercially sold as THUNDER ART and with the emission parameters shown in Table 6.9, may be observed in the photograph of the

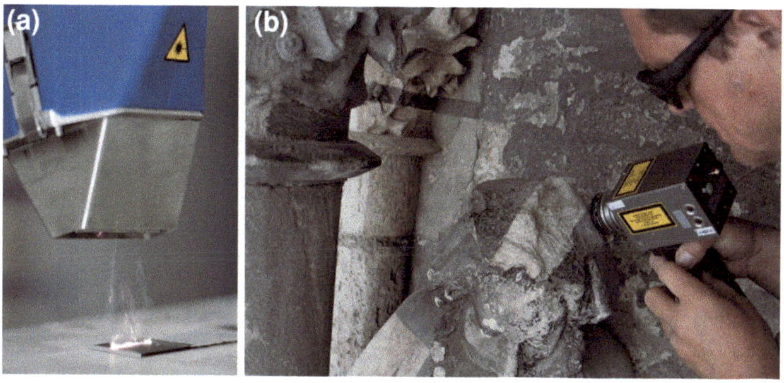

Fig. 6.70 **a** Novel wide-focus portable laser developed by CleanLaser systems, able to increase the cleaning efficiency substantially, compared to most lasers. **b** Example of the application of this laser to the cleaning of the surface of a stone statue (Courtesy of MPA, S.L. and Clean Lasersysteme GmbH)

figure. This is an extremely versatile system that it is adequate for fine, low efficiency cleaning work, but not for large areas, where it would not be economically feasible.

The EOS 1000 LQS Laser, shown in Fig. 6.69b, is, however, much more efficient than the THUNDER ART, and much more portable. It is approximate to the size of a personal computer with wheels and the laser beam is guided through a several metre long silica fibre. Nevertheless, this laser is still not sufficiently efficient to clean large areas, such as vertical walls, which are still being cleaned by conventional methods.

It is interesting to note at this time, however, that new developments have resulted in convincing advances in laser cleaning efficiency. As an example, the laser shown in Fig. 6.70a performs a several centimetre-wide scan over the substrate surface under irradiation, thereby enabling much higher cleaning rates than achieved with the previously shown lasers. Information about this laser may be found at http://cleanlaser.de/wEnglish/index.php, and it is recommended specially for metal substrates even though good results are also obtained in stone surfaces such as that shown on Fig. 6.70b. In this photograph, a wide cleaning track is observed following the plasma generated by the laser focused onto the stones statue's surface.

Fig. 6.71 Laser cleaning of a portal at Sant Denis Cathedral in Paris [83]

Fig. 6.72 Automated 1064 nm, 5 ns pulse width Laser cleaning of a brick pillar on the façade of the Cathedral in Tarazona (Spain) [84]

Fig. 6.73 Laser cleaning of roman coins. The photograph on the right reveals the inscripture after laser decontamination [85]

6.7.4 Laser Preservation: Practical Examples

6.7.4.1 Inorganic Surfaces (Sculptures, Building Materials, ...)

Beginnings of laser cleaning applied to conservation started with the application of this technique on stone. This material is unusually prone to be laser cleaned. Commonly, the contamination layers over this type of materials have lower ablation thresholds than the underlying rocks. This makes them highly suitable for selective laser cleaning procedures, along with elevated decontamination ablation rates.

Figure 6.71 shows the right column statues of one of the portals of Saint Denis Cathedral in Paris before and after cleaning. These cleaning procedures have been performed using laser cleaning alone, or combined with other techniques (micros and blasting, poultices, chemical washings, etc.). The brownish/reddish patina was preserved during the laser cleaning of the column statues despite its heterogeneity.

In this case, we are facing an example of selective laser cleaning, where the ablation threshold of the black crust is below the damage threshold of the white stone. These procedures have been performed for years with commercial solid-state lasers emitting with IR wavelengths, typically 1,064 nm, and with puse widths in the range of microseconds (μs).The technical advances of lasers in this last decade have allowed the reduction of pulse widths to the several nanoseconds (ns) range enabling a significantly lower thermal interaction with the underlying stones.

Cleaning of brick masonry is another example of laser cleaning of massive surfaces without causing any damage on the ceramic substrate material. Figure 6.72 shows the laser cleaning procedure of the north wall of the Tarazona Cathedral in Zaragoza (Spain) where urban pollution and encrustations have been successfully removed with IR laser wavelengths.

In this case, the image on the left shows a laser device combined with a scanning system in order to cover automatically the surface of the walls. The image on the right shows the results of the different cleaning tests performed at different distances and angles, with respect to the wall, at this site.

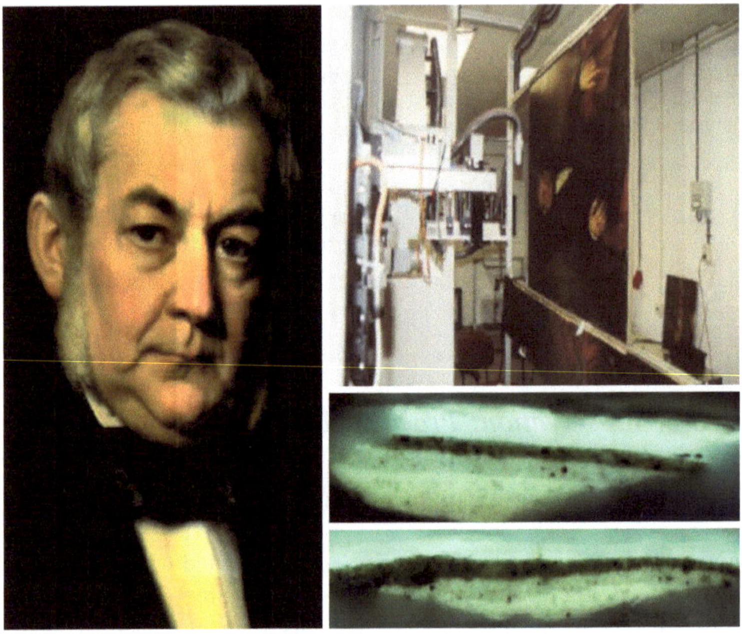

Fig. 6.74 Laser decontamination of a portrait by J.H. Neuman. A distinct contrast difference may be observed between the contaminated original painting (*left*) and the laser decontaminated surface (*right*) of the portrait's face. A cross-section micrograph shows the difference between the original (*top image*) and the laser treated surface (*bottom image*) [86]

6.7.4.2 Metals (Statues, Building Ornaments, Museum Artefacts, …)

Metals are the second best material for selective laser cleaning. Commonly, surface oxidation and even encrustations in archaeological metals have low ablation thresholds, as compared to other substrates.

Figure 6.73 shows a copper roman coin treated by laser ablation where the encrustation layer has been removed, enabling the study of the historic information contained therein. In this case, the use of commercial IR wavelength devices (as commented in the previous section) has been possible due to the nature and composition of the contamination layers.

However, in some other application on metals, visible wavelengths (typically 532 nm) apply better to this decontamination treatments since the undesired layers to remove absorb more efficiently these light sources.

6.7.4.3 Organic Surfaces (Paintings, …)

Organic materials usually face up to very difficult conservation treatments. Conventional treatments are commonly useless or cause some degree of surface

Fig. 6.75 Laser decontamination of a canvas containing a bright spring landscape by an unknown artist. The low levels of irradiation applied permit the easy ablation of the ash deposited during the fire without causing any damage either on the pigments or on the original surface of the painting [86]

damage. Laser cleaning may be a very interesting method to control the removal of these layers by not modifying the underlying materials.

For example, the laser ablation technique is a non-contact cleaning method very useful for the removal of darkened varnishes in paintings. However, thermal effects related to the common ablation interaction mechanisms could cause undesired structural modifications on these materials. This is the reason why different types of lasers are used for these applications. Lasers emitting in the UV range and with short pulse widths are more suitable for the selective and highly controlled removal of thin layers, without affecting the rest of the surrounding material.

Figure 6.74 shows a portrait of J.H. Neuman. The stratigraphic cross-section images of the canvas show that the system allows the removal of the varnish without causing any damage to the pigment layer. This is a very good example of non-selective laser cleaning. So, the picture is revealed again and the adequate conservation strategies can now be applied. Excimer lasers that emit within the UV range of wavelengths are usually employed to remove organic contaminants or undesired layers. The UV light is mainly absorbed by the varnish layer present on the outer surface, so that, even when the layer of varnished is partially removed the remaining layers protect the pigments from discolouration.

A final example of laser cleaning of paintings, in a selective laser cleaning mode, is presented in Fig. 6.75. This is a painting damaged during a fire and the ablation threshold of ashes are far below from the ablation threshold of the varnish layer, so the cleaning results are quite impressive. This procedure has been also performed with an excimer laser applying very low irradiance levels and, consequently, ablation rates.

6.7.5 Conclusions and Future Prospects

This chapter has been organised in order to present the following basic ideas. First of all, a minimum background on what lasers are and how they work is needed in order to be able to select the most adequate laser systems for any specific restoration work proposed. Second, minimum knowledge of laser–material interactions is necessary to help decide on the emission parameters of the chosen laser, as well as to define the laser cleaning conditions to avoid damage to the irradiated surface. For these reasons, photothermal processes were reviewed in order to transmit to the restorers how important it is to avoid certain lasers and irradiation conditions, knowing that laser melting processes, for example, are to be avoided at all costs during surface cleaning. Third, a minimum guidance towards the selection of appropriate, commercially available laser systems, is essential to plan and assess whether a laser cleaning job may be feasible, both from the economic and material damage points of view. Finally, representative examples help to understand what has been previously explained about lasers, laser–material interactions and how these interplay with the commercially available laser systems.

In addition, laser cleaning is expected to rely soon on the commercial availability of ultra-short pulse lasers. Probably, the difference between picosecond and femtosecond is not sufficient to justify a substantial investment in femtosecond lasers, and industrial efficiency, reliable and compact picosecond lasers are already available with emission in the nIR, Vis and UV parts of the electromagnetic spectrum. These will probably change how laser cleaning is viewed by professional restorers, and will make picosecond lasers, in the near future, the restorer's tool of choice. However, developments in auxiliary machinery (support and robotic mechanical devices) are necessary in order for the laser to become an essential tool for the Conservation of Cultural Heritage

References

1. Pye K, Schiavon N (1989) Cause of sulphate attack on concrete, render and stone indicated by sulphur isotope ratios. Nature 342(6250):663–664
2. Schiavon N, Zhou L (1996) Magnetic, chemical and microscopical characterization of urban soiling on historical monuments. Environ Sci Technol 30(12):3624–3629. doi:10.1021/es9604774
3. Zappia G, Sabbioni C, Riontino C, Gobbi G, Favoni O (1998) Exposure tests of building materials in urban atmosphere. Sci Total Environ 224:235–244
4. Ausset P, DelMonte M, Lefevre RA (1999) Embryonic sulphated black crusts on carbonate rocks in atmospheric simulation chamber and in the field: role of carbonaceous fly-ash. Atmos Environ 30:1525–1534
5. Schiavon N, Chiavari G, Fabbri D (2004) Soiling of limestone in an urban environment characterized by heavy vehicular exhaust emissions. Environ Geol 46:448–456. doi:10.1007/s00254-004-1046-8
6. Schiavon N (2007) Kaolinisation of granite in an urban environment. Environ Geol 52(2):399–407. doi:10.1007/s00254-006-0473-0
7. Bonazza A, Brimblecombe P, Grossi C, Sabbioni C (2007) Carbon in Black Crusts from the Tower of London. Environ Sci Technol 41:4199–4204

8. EU 2008 Directive 2008/50/EC of the European Parliament and the Council of 23 May 2008 on Ambient Air Quality and Cleaner Air for Europe. European Commission, Brussels, Belgium
9. Viles H, Gorbushina AA (2003) Soiling and microbial colonization on urban roadside limestone: a three year study in Oxford, England. Build Environ 38:1217–1224
10. Torok A (2008) Black crusts on travertine: factors controlling development and stability. Environ Geol 56:583–594
11. DelMonte M, Sabbioni C, Vittori (1984) Urban stone sulphation and oil-fired carbonaceous particles. Sci Total Environ 36:369–376
12. Schiavon N (1991) Gypsum crust growth and stratigraphy in building limestones: a SEM study of stone decay in the UK. In: Baer N, Sabbioni C, Sors AI (eds) Science, technology and the European cultural heritage. Butterworth-Heinemann Ltd, Oxford
13. Purcell D (1967) Cambridge stone. Faber & Faber Ltd, London
14. Pingitore NE (1976) Vadose and phreatic diagenesis processes, products and their recognition in corals. J Sediment Pet 46:985–1006
15. Tziafalias A (1994) Fifteen years of excavations in ancient Larissa. In Decourt J-C, Helly B, Gallis K (eds) La Thessalie. Quinze années de recherches archéologiques, 1975–1990. Bilans et perspectives. TAPA, Athens (in Greek)
16. Tziafalias A (2009) Ancient Theatre of Larisa? How a dream became reality. In: Proceedings of the 1st international conference on history and culture of Thessaly, Larisa, pp 207–223 (in Greek)
17. Caputo R, Helly B (2005) Archaeological evidences of past earthquakes: a contribution to the SHA of Thessaly, Central Greece. J Earthq Eng 9:199–222
18. Caputo R, Hinzen KG, Liberatore D, Schreiber S, Helly B, Tziafalias A (2011) Quantitative archaeoseismological investigation of the Great Theatre of Larissa, Greece. Bull Earthq Eng. doi:10.1007/s10518-010-9206-6
19. Coleman M, Walker S (1979) Stable isotope identification of Greek and Turkish marbles. Archaeometry 21:107–112
20. Herz N (1992) Provenance determination of Neolithic to classical Mediterranean marbles by stable isotopes. Archaeometry 34:185–194
21. Hermann JJ, Barbin V, Mentzos A, Reed R (2000) Architectural decoration and marble from Thasos: Macedonia, central Greece, Campania, and provenance. In: Lazzarini L (ed) Interdisciplinary studies on ancient stone. Bottega D'Erasmo Aldo Ausilio Editore, Padova
22. Capedri S, Venturelli G (2004) Accessory minerals as tracers in the provenancing of the archaeological marbles, used in combination with isotopic and petrographic data. Archaeometry 46:517–536
23. Maniatis Y, Papadopoulos S, Dotsika E, Kavoussanaki D, Tzavidopoulos E (2009) Provenance investigation of neolithic marble vases from Limeraria, Thassos: imported marble to Thassos? In: Maniatis Y (ed) ASMOSIA VII, proceedings of the 7th international conference of the association for the study of marble and other stones in antiquity. BCH Suppl, Thassos
24. Maniatis Y, Tambakopoulos D, Dotsika E, Tiveriou-Stephanidou Th (2010) Marble provenance investigation of Roman sarcophagi from Thessaloniki. Archaeometry 52:45–58
25. Al-Naddaf M, Al-Bashaireh K, Al-Waked F (2010) Characterization and provenance of marble Chancel Screens, northern Jordan. Mediterr Archaeol Archaeom 10:75–83
26. Germann K, Holzmann G, Winkler FJ (1980) Determination of marble provenance: limits of isotopic analyses. Archaeometry 22:99–106
27. Capedri S, Giampiero V, Photiades A (2004) Accessory minerals and δ18O and δ13C marbles from Mediterranean area. J Cult Heritage 5:27–47
28. Melfos V (2004) Mineralogical and stable isotopic study of ancient white marble quarries in Larissa, Thessaly, Greece. Bull Geol Soc Greece XXXVI/3:1164–1172
29. Melfos V, Voudouris P, Papadopoulou L, Sdrolia S, Helly B (2010) Mineralogical, petrographic and stable isotopic study of ancient white marble quarries in Thessaly, Greece—II. Chasanbali, Tempi, Atrax, Tisaion mountain. Bull Geol Soc Greece XLIII/2:845–855

30. Melfos V (2008) Green Thessalian Stone: the Byzantine quarries and the use of a unique architecture material from Larisa area, Greece. Petrographic and geochemical characterization. Oxf J Archaeol 27:387–405
31. Craig H, Craig V (1972) Greek marbles: determination of provenance by isotopic analyses. Science 176:401–403
32. Herz N (1987) Carbon and oxygen isotopic ratios: a data base for classical Greek and Roman marble. Archaeometry 29:35–43
33. Salzer R, Lunkwitz R (1998) Diagnose von Bauschäden mittels IR- und Ramanspektroskopie. GDCh-Monographie, Band 11, GDCh Frankfurt
34. Salzer R, Lunkwitz R et al (1998) Baustoffanalyse mittels Infrarotspektroskopie. Internat Zeitschr Bauinstandsetzen 4:209–232
35. Genestar C, Pons C (2003) Ancient covering plaster mortars from several convents and Islamic and Gothic palaces in Palma de Mallorca (Spain). Analytical characterisation. J Cult Heritage 4:291–298
36. Biscontin G, Birelli MP, Zendri E (2002) Characterization of binders employed in the manufacture of Venetian historical mortars. J Cult Heritage 3:31–37
37. Moropoulou A, Bakolas A, Bisbikou K (1995) Characterization of ancient, byzantine and later historic mortars by thermal and X-ray diffraction techniques. Thermochim Acta 269(270):779–795
38. Bakolas A, Biscontin G, Contardi V, Franceschi E, Moropoulou A, Palazzi D, Zendri E (1995) Thermoanalytical research on traditional mortars in Venice. Thermochim Acta 269(270):817–828
39. Bakolas A, Biscontin G, Moropoulou A, Zendri E (1998) Characterization of structural byzantine mortars by thermogravimetric analysis. Thermochim Acta 321:151–160
40. Vecchio S, La Ginestra A, Frezza A, Ferragina C (1993) The use of thermoanalytical techniques in the characterization of ancient mortars. Thermochim Acta 227:215–223
41. Xidas PI, Triantafyllidis KS (2010) Effect of the type of alkylammonium ion clay modifier on the structure and thermal/mechanical properties of glassy and rubbery epoxy-clay nanocomposites. Eur Polym J 46:404–417
42. Yeha J-M, Huanga H-Y, Chena C-L, Sua W-F, Yub Y-H (2006) Siloxane-modified epoxy resin–clay nanocomposite coatings with advanced anticorrosive properties prepared by a solution dispersion approach. Surf Coat Technol 200:2753–2763
43. Hang TTX, Truc TA, Nam TH, Oanh VK, Jorcin J-B, Pébère N (2007) Corrosion protection of carbon steel by an epoxy resin containing organically modified clay. Surf Coat Technol 201:7408–7415
44. Allie L, Thorn J, Aglan H (2008) Evaluation of nanosilicate filled poly (vinyl chloride-co-vinyl acetate) and epoxy coatings. Corros Sci 50:2189–2196
45. Truc TA, Hang TTX, Oanh VK, Dantras E, Lacabanne C, Oquab D, Pébère N (2008) Incorporation of an indole-3 butyric acid modified clay in epoxy resin for corrosion protection of carbon steel. Surf Coat Technol 202:4945–4951
46. Dai C-F, Li P-R, Yeh J-M (2008) Comparative studies for the effect of intercalating agent on the physical properties of epoxy resin-clay based nanocomposite materials. Eur Polymer J 44:2439–2447
47. Hosseinia MG, Raghibi-Boroujenia M, Ahadzadeha I, Najjarb R, Seyed Dorrajic MS (2009) Effect of polypyrrole–montmorillonite nanocomposites powder addition on corrosion performance of epoxy coatings on Al 5000. Prog Org Coat 66:321–327
48. Orazem M, Tribollet B (2008) Electrochemical impedance spectroscopy, the electrochemical society series. Wiley, New York. ISBN 978-0-470-04140-6
49. Winston Revie R (2000) Uhlig's corrosion handbook, 2nd edn, Wiley, New York, pp 949–1238. ISBN 0-471-15777-5
50. Baboian R (2005) Corrosion tests and standards: application and interpretation, 2nd edn. Wiley, New York, pp 107–130. ISBN 0-8031-2098-2
51. ASTM G 106. Standard practice for verification of algorithm and equipment for electrochemical impedance measurements

52. ASTM G 457. Standard test method for measurement of impedance of anodic coatings on aluminum
53. Cesareo R, Ridolfi S, Marabelli M, Castellano A, Buccolieri G, Donativi M, Gigante GE, Brunetti A, Rosales Medina MA (2008) Portable Systems for Energy-Dispersive X-Ray Fluorescence Analysis of Works of Art. In: Potts PJ, West M (eds) Portable X-ray fluorescence spectrometry: capabilities for in situ analysis. The royal society of chemistry, pp 206–243. ISBN-13: 9780854045525
54. Guida G, Artioli D, Ridolfi S, Gigante GE (2010) Study by mobile non destructive testing of the bronze statue of the "Satiro" of Marsala., science for cultural heritage; technological innovation and case studies in marine and land archaeology in the adriatic region and inland. World Scientific Publishing Co, Singapore, pp 23–30. ISBN-13 978-981-4307-06-2
55. Gigante GE et al (2008) Restoration of the funeral monument of Pope Sixtus IV by Antonio Pollaiolo (1493) in the Vatican Basilica: non invasive and microdestructive analysis—an operation protocol. ART2008—9th international conference, 2008. Jerusalem, Israel
56. Gabrielli N et al (2005) "Il restauro della sfera bronzea sulla cupola della Basilica di San Pietro", Materiali e Strutture, nuova serie anno III numeri 5–6, 38–87, editore Nuova Argos
57. Wikipedia (17 April 2010). https://upload.wikimedia.org/wikipedia/commons/8/8a/Electromagnetic-Spectrum.png
58. Demtröder, Experimentalphysik 2, Chapter 7 Elektromagnetische Wellen im Vakuum, Springer
59. Wikipedia (8 May2007). https://upload.wikimedia.org/wikipedia/commons/f/f5/Photoelectric_effect.svg
60. Wikipedia. https://upload.wikipedia.org/wikipedia/commons/b/bc/Doubleslitexperiment.svg
61. Wikipedia (6 Jan 2006). https://upload.wikimedia.org/wikipedia/commons/7/7e/Doubleslit_experiment_results_Tanamura_2.jpg
62. Spektrum der Wissenschaft
63. Wikipedia (4 Aug 2006). https://upload.wikimedia.org/wikipedia/commons/a/a2/Wiens_law.svg
64. Springer Handbook of Lasers and Optics, Chapter 11, Lasers and Coherent Light Sources
65. Gerthsen, Meschede: Physik, Chapter 15 Laserphysik, Springer, 2006
66. Widl G (2010) Am Anfang war das Licht. Ein kleines Laser Kompendium, Trumpf. ISBN 978-3-9813676-0-7
67. Limpert J: High power fiber lasers and amplifiers. http://www.optecbb.de/summerschool2006/lectures/15%20-%20Wed%2011.30%20-%20Jens%20Limpert.pdf, http://www.swisslaser.net/libraries.files/Vortrag_Limpert1.pdf
68. Diehl R (ed) (2000) High-power diode lasers. Topics Appl Phys 78:369–408
69. Trumpf TruDisk Scheibenlaser (2011). http://www.trumpf-laser.com/typo3temp/pics/edca33c6a6.jpg
70. Schubert F. http://www.ecse.rpi.edu/~schubert/Light-Emitting-Diodes-dot-org/http://www.ecse.rpi.edu/~schubert/Light-Emitting-Diodes-dot-org/chap07/F07-02%20Homo%20&%20heterojunction.jpg
71. Reithmaier JP (2004) Halbleiterlaser: Grundlagen und aktuelle Forschung, Uni Würzburg, SS 2004 formerly. http://www.physik.uni-wuerzburg.de/index.php?id=5297
72. Diehl R (ed) (2000) High-power diode lasers. Topics Appl Phys 78:289–301
73. Jenoptik. http://bilddatenbank.jenoptik.com/index.php/6222a68dd1764af9cf8e07bc5
74. Bäuerle D (1996) Laser processing and chemistry. Springer, Berlin
75. Ion JC (2005) Laser processing of engineering materials. Elsevier Butterworth-Heinemann, Oxford
76. Knowles MRH, Rutterford G, Karnakis D, Fergusin A (2005) Laser Micromachining of metals, Ceramics, Silicon and Polymers using Nanosecond Laserss, Oxford Lasers Ltd, Unit 8, Moorbrook Park, Didcot, OX11 7HP
77. Ready JF (editor in chief) (2001) LIA handbook of laser materials processing. Laser Institute of America, Magnolia Publishing, Inc, Magnolia
78. Rubahn HG (1999) Laser applications in surface science and technology. Wiley, New York

79. Steen WM (1998) Laser material processing, 3rd edn. Springer, Berlin
80. Jastrzebski ZD (1987) The nature and properties of engineering materials, 3rd edn. Wiley, New York
81. http://hyperphysics.phy-astr.gsu.edu/hbase/molecule/molec.html
82. http://www.layertec.de/en/capabilities/coatings/metallic
83. Bromblet P, Laboure M, Orial G (2003) Diversity of the cleaning procedures including laser for there storation of carved portals in France over the last 10 years. J Cult Heritage 4: 17s–26s
84. Lahoz R (2006) PhD Thesis, University of Zaragoza
85. Pini R et al (2000) J Cult Heritage 1:S129–S137
86. Teule R et al (2003) Controlled UV laser cleaning of painted artworks: a systematic effect study on egg tempera paint samples". J Cult Heritage 4:209s–215s

Index

E. A. Varella (ed.), *Conservation Science for the Cultural Heritage*,
Lecture Notes in Chemistry 79, DOI: 10.1007/978-3-642-30985-4,

MIX
Papier aus verantwortungsvollen Quellen
Paper from responsible sources
FSC® C105338

If you have any concerns about our products,
you can contact us on
ProductSafety@springernature.com

In case Publisher is established outside the EU,
the EU authorized representative is:
Springer Nature Customer Service Center GmbH
Europaplatz 3, 69115 Heidelberg, Germany

Printed by Libri Plureos GmbH
in Hamburg, Germany